Dirofilariasis

Editors

Peter F. L. Boreham, Ph.D.
Principal Research Fellow
Experimental Parasitology Unit
Queensland Institute of Medical Research
Brisbane, Australia

Richard B. Atwell, Ph.D.
Senior Lecturer
Department of Companion Animal Medicine and Surgery
University of Queensland
Brisbane, Australia

CRC Press
Taylor & Francis Group
Boca Raton London New York

CRC Press is an imprint of the
Taylor & Francis Group, an **informa** business

First published 1988 by CRC Press
Taylor & Francis Group
6000 Broken Sound Parkway NW, Suite 300
Boca Raton, FL 33487-2742

Reissued 2018 by CRC Press

© 1988 by CRC Press, Inc.
CRC Press is an imprint of Taylor & Francis Group, an Informa business

No claim to original U.S. Government works

Library of Congress Cataloging-in-Publication Data

Dirofilariasis.

   Includes bibliographies and index.
   1. Canine heartworm disease.   2. Dirofilaria immitis.
I. Boreham, Peter F. L.   II. Atwell, Richard B.   [DNLM:
1. Dirofilariasis—veterinary.   2. Dog Diseases.
3. Filarioidea.   SF 992.H4 D599]
SF992.H4D57   1988   636.7'08969652   87-10348
ISBN 0-8493-6488-4

A Library of Congress record exists under LC control number: 87010348

ISBN 13: 978-1-315-89232-0 (hbk)
ISBN 13: 978-1-351-07142-0 (ebk)

Visit the Taylor & Francis Web site at http://www.taylorandfrancis.com and the
CRC Press Web site at http://www.crcpress.com

PREFACE

Heartworm disease of dogs has been recognized as an important clinical entity for well over 100 years. Since the original description of the parasite by a Philadelphia physician, Joseph Leidy (1823-1891), in 1850, many hundreds of papers have been written about the parasite, the disease it causes, and its transmission by mosquitoes. Information on *Dirofilaria* has grown exponentially with an explosion of data in the last 20 years.

Dirofilaria immitis has been used as a model for major fundamental discoveries in medicine and biology. For example, in 1901 in Brisbane, Australia, Thomas Bancroft (1860-1933) was the first person to complete the life cycle of a filarial worm when he demonstrated that the infective larvae of *D. immitis* were transferred to the host during the act of mosquito feeding.[1] Previous to this, it was considered that they were transferred via water.

Heartworm disease is probably the most well-known disease of dogs, and its epidemiology is being recorded in those countries in which the disease is routinely treated in the domestic dog. Because the disease is complex and because research development over the last few years has been very rapid, it was felt that a bench text drawing together all aspects (not just the clinical aspects) would be of significant benefit to undergraduate and postgraduate students at teaching institutions, to researchers (both to veterinary colleagues and to other biologists generally), and to the practicing veterinarians who have been exposed to a large number of journal publications and to periodic reviews in major veterinary text books.

This dual aspect of basic and applied research is the central theme of this book. It has been designed to review current information on all aspects of the parasite and the disease and so provide a reference source for both the clinical veterinarian and the research scientist.

Research on dirofilariasis is at a crossroads with detailed descriptions of most aspects of the clinical disease, pathology, epidemiology, and the parasite being available. The new era on which we are embarking involves the exciting research tools of molecular biology and immunology. When fully applied to the problem of dirofilariasis they will lead to more specific and sensitive methods of diagnosis and even, in time, to the development of vaccines. A better understanding of the mode of action of currently available drugs and the development of new adulticide compounds will enhance the armament of treatment available to the clinician.

In the preparation of this text it was obvious to us that many people have produced a wealth of material on this particular topic. Dr. R. Jackson and Dr. G. Otto and their colleagues were responsible for much of the earlier work, as were Drs. Kume, Fuji, Isihara, and Osihi, and other colleagues in Japan. In fact, in Japan a lot of work has been done over the years, some of which has, unfortunately, been inaccessible due to non-English publications. Dr. D. Knight performed introductory work with regard to the pulmonary hypertension associated with the disease, and, since then, the work at Georgia, U.S. by Drs. Rawlings, McCall, and colleagues has added greatly to our knowledge about the vascular aspects of the disease. In Australia, Drs. Carlisle and Winter did introductory work on arsenical use and pathophysiology, and, since then, workers in Queensland have continued to investigate the DEC reaction, caval syndrome, and other aspects of pathophysiology and treatment. The establishment of the American Heartworm Society, and, recently, a similar society in Australia, has allowed practicing veterinarians access to the latest research findings which has enabled research and clinical findings to be more widely and more rapidly dispersed.

This text, finalized after the latest Heartworm Symposium (1986), will allow the reader access to both clinical and research data and will further advance the understanding of this fascinating disease.

A clinically based book has been written by Dr. Rawlings.[2] However, when our text was conceived it was designed to provide information to research workers (who may be without existing knowledge of the disease), medical researchers (who may mistakenly work with the possibly infected dog), and, of course, veterinarians in practice (who want access to a complete coverage of all aspects of the disease — diagnosis, prognosis, pathology, treatment, etc.).

It is hoped that the text will be of benefit to us all and that our understanding of this disease will continue to grow and so enable us to resolve some of the existing problems associated with the treatment of this disease. To this end, we plan to project some hypotheses in different areas that, while unproven, seem to be the most probable, based on our current knowledge.

Richard Atwell
Peter Boreham

REFERENCES

1. **Boreham, P. F. L. and Marks, E. N.,** Human filariasis in Australia: introduction, investigation and elimination, *Proc. R. Soc. Queensl.,* 97, 23, 1986.
2. **Rawlings, C. A.,** *Heartworm Disease in Dogs and Cats,* W. B. Saunders, Philadelphia, 1986.

ACKNOWLEDGMENTS

This text was written in the interests of furthering the understanding of this complex disease. Its complexity is emphasized in that it is probably the only veterinary disease for which there exists a society of veterinarians with a basic aim to keep its members informed of latest developments in this disease.

This book is dedicated to those researchers and veterinary clinicians who, over the last 40 to 50 years, have enabled us to know what we now know. In particular, we personally draw attention to the work at the University of Queensland and the Queensland Institute of Medical Research over the last ten years and to the late Dr. E. Moodie who, as Head of Department and as a Ph.D. supervisor (R.A.), gave much encouragement and enabled much of this work to be performed.

Richard Atwell
Peter Boreham

THE EDITORS

Peter F. L. Boreham, Ph.D., is Head of the Experimental Parasitology Unit at the Queensland Institute of Medical Research, Brisbane, Australia. Dr. Boreham graduated with a B.Pharm. (Honours) degree from the School of Pharmacy, University of London and obtained a Ph.D. also from London University working at the Nuffield Institute of Comparative Medicine. He subsequently became a Research Fellow and later Lecturer in Immunology at Imperial College of Science and Technology, London. While at Imperial College, Dr. Boreham was responsible for running a research unit concerned with the identification of bloodmeals of hematophagous arthropods and undertaking research on the pathogenesis of African trypanosomiasis. Since moving to Queensland in 1979, Dr. Boreham has carried out research on dirofiliariasis and giardiasis with particular emphasis on the chemotherapy and, more recently, the molecular biology of the parasites. In 1974, Dr. Boreham was designated a Fellow of the Pharmaceutical Society of Great Britain for research excellence and in 1976 was awarded their Science Award. He has undertaken several consultantships for the World Health Organization and has published 150 papers and reviews on parasites and their vectors. Dr. Boreham is a member of several professional organizations including the Australian Society for Parasitology, the Royal Society of Tropical Medicine and Hygiene and the British Pharmacological Society.

Richard Atwell, Ph.D., has been Senior Lecturer in the Department of Companion Animal Medicine and Surgery, formerly the Department of Veterinary Medicine, at the University of Queensland, Brisbane, Australia, since 1983. He graduated in 1973 with a B.V.Sc., and worked in general practice for three years after which he was appointed Lecturer at the University of Queensland. He was Director of the Hospital and Clinics (1977 to 1979) and obtained a Membership (1980) and Fellowship (1983) of the Australian College of Veterinary Scientists. In 1983 he obtained his Ph.D. in Experimental Canine Dirofilariasis and has produced many research and clinical papers on different aspects of dirofilariasis. He is also Scientific Editor of the *Australian Veterinary Practitioner* journal, Director of the Australian Heartworm Society, and conducts referral clinics in cardiopulmonary medicine within the School of Veterinary Science at the University of Queensland.

CONTRIBUTORS

David Abraham, Ph.D.
Assistant Professor
Department of Microbiology
Jefferson Medical College
Thomas Jefferson University
Philadelphia, Pennsylvania

Richard B. Atwell, Ph.D., F.A.C.V.Sc.
Senior Lecturer
Department of Companion Animal
  Medicine and Surgery
University of Queensland
St. Lucia, Queensland, Australia

Peter F. L. Boreham, Ph.D., F.P.S.
Principal Research Fellow
Experimental Parasitology Unit
Queensland Institute of Medical Research
Brisbane, Queensland, Australia

**Christopher Bryant, Ph.D., F.I.Biol.,
  F.A.I.Biol.**
Professor
Department of Zoology
Australian National University
Canberra, ACT, Australia

Ibrahim B. J. Buoro, B.V.Sc.
Department of Clinical Studies
Faculty of Veterinary Science
University of Nairobi
Kabete, Kenya

**Carol H. Carlisle, M.V.Sc., D.V.R.,
  F.A.C.V.Sc., M.R.C.V.S.**
Associate Professor
Department of Companion Animal
  Medicine and Surgery
University of Queensland
St. Lucia, Queensland, Australia

Charles H. Courtney, D.V.M., Ph.D.
Associate Professor
Department of Infectious Diseases
College of Veterinary Medicine
University of Florida
Gainesville, Florida

Ray Dillon, D.V.M., M.S.
Professor and Head
Section of Medicine
Scott-Ritchey Research Program
Department of Small Animal Surgery and
  Medicine
College of Veterinary Medicine
Auburn University
Auburn, Alabama

Robert B. Grieve, Ph.D.
Associate Professor
Department of Pathology
College of Veterinary Medicine and
  Biomedical Sciences
Colorado State University
Fort Collins, Colorado

James B. Lok, Ph.D.
Assistant Professor of Parasitology
Department of Pathobiology
School of Veterinary Medicine
University of Pennsylvania
Philadelphia, Pennsylvania

**Richard H. Sutton, Ph.D.,
  F.A.C.V.Sc., M.R.C.V.S.**
Senior Lecturer
Department of Veterinary Pathology and
  Public Health
University of Queensland
Brisbane, Queensland, Australia

TABLE OF CONTENTS

Chapter 1

DIROFILARIA SP.: TAXONOMY AND DISTRIBUTION

James B. Lok

TABLE OF CONTENTS

I. SYSTEMATICS

Dirofilaria immitis, as the agent of canine cardiovascular dirofilariasis, is a parasite of paramount veterinary importance. It is increasingly recognized as the cause of zoonotic infections, most frequently as the agent of human pulmonary dirofilariasis.[1] The parasite, now known to occur throughout the world in many species of host, was first described in 1850 in Philadelphia, by physician J. Leidy who assigned the name *Filaria canis cordis*.[2] The worm was renamed *Filaria immitis* by Leidy in 1856.[3] and the genus *Dirofilaria* was erected in 1911 by Railliet and Henry[4] with *Filaria immitis* as its type species.[2] Synonyms in addition to those listed above include *Filaria spirocauda*, *F. haematica*, *F. sanguinis*, *F. hebetata*, *Dirofilaria nasuae*, *D. pongoi*, *D. fausti*, *D. louisianensis*, *D. indica*,[5] and *D. magalhaesi*.[1] The taxonomy of the genus *Dirofilaria* has been reviewed by Anderson.[6]

Like other members of the superfamily Filarioidea the genus *Dirofilaria* consists of elongate, thin nematodes which live outside the alimentary tract and which are viviparous giving birth to motile first-stage larvae called microfilariae.[1] As is the case with all filariae, members of this genus require an arthropod intermediate host, generally a mosquito or black fly, to complete their life cycles.[5-7] The species generally recognized as being of primary veterinary importance belong to two subgenera. The subgenus *Dirofilaria* contains those worms such as *D. immitis* which have smooth cuticles and which, as adults, are usually found in the cardiovascular system. Members of the subgenus *Nochtiella* such as *D. ursi*, *D. repens*, *D. striata*, and *D. tenuis* generally dwell in subcutaneous tissues of the definitive host and have prominent longitudinal ridges and fine striations on their cuticles.[1] These differential subgeneric characters are discussed more fully below.

II. MORPHOLOGY

A. Adults

1. Gross Morphology

The genus *Dirofilaria* consists of elongate, thin filarioid nematodes, whitish in coloration, with bluntly rounded anterior extremities, rudimentary buccal capsules without lips, small cephalic papillae and an abbreviated esophagus indistinctly differentiated into muscular and glandular regions (Figure 1A, B). The caudal extremities of female worms are also bluntly rounded (Figure 1E), and the vulvar openings are located just posterior to the junction of the esophagus and intestine. The caudal ends of males are coiled spirally and are more conical in shape then those of females. In all species of *Dirofilaria* spicules are unequal in length (Figure 1C). The gubernaculum is absent.[5,6]

Dirofilaria (Dirofilaria) immitis is the longest of the *Dirofilaria* spp. of medical and veterinary importance. Females measure 250 to 310 mm in length and 1.0 to 1.3 mm in width. They have an obtuse caudal end with the vulvar opening some 2.7 mm from the anterior end. Males of *D. immitis* are 120 to 200 mm long and 0.7 to 0.9 mm wide. The spirally coiled caudal end bears five pairs of preanal and six pairs of postanal papillae (Figure 1C). The left spicule is 300 to 375 μm long, and the right 175 to 299 μm. In both sexes the stoma is surrounded externally by six inconspicuous papillae (Figure 1B).[5]

Other species of *Dirofilaria* which occur in dogs and cats and which may occasionally affect man are readily distinguished from *D. immitis* on the basis of size. Females of *Dirofilaria repens*, a parasite of the subcutaneous tissues of dogs and cats throughout Africa, Southern Europe, India, and Southeast Asia, measure 100 to 170 mm in length and 0.46 to 0.65 mm in width. Males are 50 to 70 mm long and 0.37 to 0.45 mm wide with two to six preanal papillae to the right of the anus and four to five to the left. The spicules are highly dissimilar, the left being 460 to 590 μm long and the right being 180 to 210 μm long. Microfilariae of *D. repens* occur in the subcutaneous lymph spaces and in the peripheral

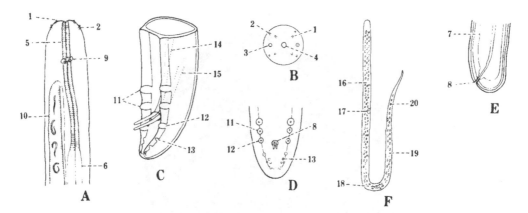

FIGURE 1. Gross morphology of *D. immitis*. (A) Anterior end of adult female showing (1) inner circle of cephalic papillae, (2) outer circle of cephalic papillae, (5) muscular esophagus, (6) glandular esophagus, (9) nerve ring and (10) uterus. (B) *En face* view of anterior end of adult worm showing (1) inner and (2) outer circles of cephalic papillae, (3) amphids, and (4) stoma. (C) Caudal end of adult male showing (11) preanal papillae, (12) adanal papillae, (13) postanal papillae, (14) long spicule, (15) and short spicule. (D) Ventral view of caudal end of male showing (11) preanal, (12) adanal, and (13) postanal papillae and (8) anus. (E) Caudal view of adult female showing (7) intestine and (8) anus. (F) Microfilaria from peripheral blood showing (16) nerve ring, (17) excretory cell, (18) G_1 cell, (19) anal space, (20) and tail cells. (From Olsen, O. W., *Animal Parasites, Their Life Cycles and Ecology*, 3rd ed., University Park Press, Baltimore, 1974, 491. With permission.)

blood. On the basis of size alone it would appear difficult to distinguish the microfilariae of *D. repens* from those of *D. immitis*. The range of lengths, 268 to 360 μm, reported for these parasites overlaps that reported for *D. immitis* microfilariae. The relative positions of somatic structures such as the nerve ring, the excretory cell, the G_1 cell, and the anal space (Figure 1F) along the length of the microfilaria provide some means of diagnosis.

2. Morphology Apparent in Tissue Sections

Human pulmonary infections with *Dirofilaria immitis* and subcutaneous infections with *D. tenuis*, *D. repens*, and *D. ursi* are being reported with increasing frequency[8] (see also Boreham, chapter 13). Moreover, *D. immitis* has been reported from numerous ectopic locations in both dogs and alternative, definitive, and aberrant hosts.[9,10] For these reasons the clinician may be required to recognize the characteristics of the parasite in tissue cross sections. Gutierrez[8] describes the characteristics of adult *Dirofilaria* spp. as seen in histological sections. Adults of the genus *Dirofilaria* have relatively thick multilaminate cuticles with individual layers most apparent in proximity to the lateral chords (Figures 2 and 3). The external surfaces of the cuticles of *D. ursi*, *D. tenuis*, and *D. repens* bear longitudinal ridges which are prominent in cross sections. The cuticle of *D. immitis*, on the other hand, is relatively smooth with longitudinal ridges occurring only in the caudal area on the ventral aspect. Two of the cuticular layers are fibrous in structure with fibers of one layer running perpendicular to those of the other and the whole array running oblique to the axis of the body. These cuticular fibers appear as a criss-cross network when viewed in tangential sections of adult worms. The lateral cords are prominent in cross sections with well-developed musculature and numerous cells (Figure 3).

Identification of *Dirofilaria* spp. in cross section is done primarily on the basis of structure and arrangement of the longitudinal cuticular ridges. Longitudinal ridges of adult *D. repens* are conspicuous and sharp, separated by a space three to four times the width of the ridge itself. *Dirofilaria tenuis* has a distinctive pattern of low, rounded ridges in a branching network with spaces appearing narrower than the ridges themselves. As mentioned above, the cuticle of *D. immitis* is smooth with ridges occurring only on the ventral aspect of the caudal extremity of the male worms.

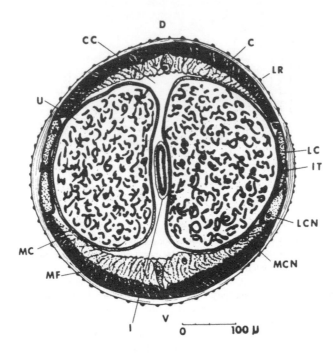

FIGURE 2.   Typical cross-sectional anatomy of a filarial parasite illustrated with drawing of *D. ursi*. Section is through mid portion of worm. Note the two uteri and intestine. Anatomical landmarks indicated are (C) the cuticle, (D) dorsal aspect of body, (I) intestine, (IT) internal thickening of the cuticle, (LC) lateral chord, (LCN) lateral chord nucleus, (LR) longitudinal ridge, (MC) muscle cell (MCN) muscle cell nucleus, (MF) muscle fibers, (P) pseudocoelom (body cavity), (U) uterus, and ventral aspect of body. (From Gutierrez, Y., *Hum. Pathol.*, 15, 514, 1984. With permission.)

The sexes of adult *Dirofilaria* seen in histological sections may also be determined. The female reproductive tract is generally seen as two tubular structures (Figures 1A and 2) except at the anterior and posterior extremities where the vagina and ovaries, respectively, loop several times. The male tract appears as a single tube in cross sections except at the anterior end where the testes are looped a number of times.[8]

B. Microfilariae

Microfilariae of *D. immitis* measure on the average 308 μm in length and 7 μm in width. They are generally found in the peripheral circulation at concentrations ranging from 10^3 to 10^5/mℓ of blood (Table 1). In direct smears of fresh blood with anticoagulant, microfilariae of *D. immitis* exhibit nonprogressive motility. This characteristic has been proposed to aid in differentiation of *D. immitis* from other species of microfilariae which may occur in dogs. Microfilariae of *D. immitis* seen in routine Knott's preparations exhibit a relatively straight, fully extended body with a tapering cephalic end and straight caudal end (Figures 4, 5, and 6). Hematoxylin staining, although not a routine part of the diagnostic procedure, reveals a number of other characters which may be used to distinguish microfilariae of *D. immitis* from those of nonpathogenic filariae (Figures 7 and 8). The cephalic and caudal spaces of *D. immitis* measure 11.5 μm and 28.0 μm on average, respectively. Moreover, hematoxylin staining reveals a distinctive pattern of annular striations along the entire length of the microfilaria being most conspicuous in the cephalic and caudal spaces (Figure 7).[11]

Histochemical techniques have been used to study the distribution of acid phosphatase activity in microfilariae of *D. immitis*. In three studies[12-14] microfilariae displayed a consistent

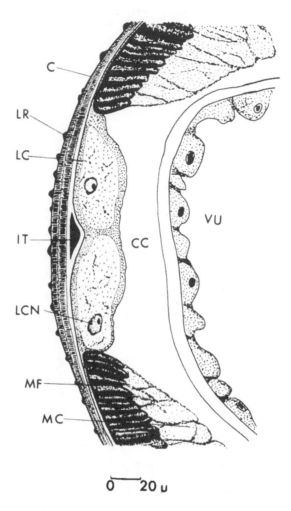

FIGURE 3. Cross section of lateral chord are of female *D. ursi*. Anatomical landmarks are as indicated in Figure 2 with the exception of (VU) the vagina uterina. (From Gutierrez, Y., *Hum. Pathol.*, 15, 514, 1984. With permission.)

pattern of acid phosphatase staining with activity limited to two well-defined areas at the levels of the excretory and anal pores (Figure 1F). Staining was most intense at the level of the excretory pore.

III. GEOGRAPHIC DISTRIBUTION

In compiling information on the geographic distribution of *Dirofilaria immitis* and other agents of canine filariasis, three types of records were examined. First, there are published reports of autochthonous cases of *D. immitis* or *Dipetalonema* sp.* infection. These cases are indicated by darkly shaded areas of distribution maps (Figures 9 to 14). Second, there are published records of *D. immitis* infections in introduced animals. Origins of this type of report are indicated by lightly shaded areas of the distribution maps. Finally, where there is a paucity of published information on the occurrence of *D. immitis*, the results of a survey for animal parasitism conducted in 1983 by agencies of the World Health Organization

* When abbreviated, *D.* will refer to *Dirofilaria* and *Di.* to *Dipetalonema*.

Table 1

DISTINGUISHING CHARACTERISTICS AMONG MICROFILARIAE OF *D. IMMITIS, DI. RECONDITUM*, IRISH *DIPETALONEMA* SP., AND FLORIDA *DIROFILARIA* SP.

Characteristic	*D. immitis* (μm)	*Di. reconditum* (μm)	Irish *Dipetalonema* sp. (μm)	Florida *Dirofilaria* sp. (μm)
Number in blood	Few to numerous	Usually few	Few	Few
Body movement	Stationary	Progressive	Progressive	Stationary
Body shape (K, H)[a]	Usually straight	Usually curved	Usually straight	Usually curved
Body length				
(K)	308 (295—325)	263 (250—288)	244 (232—256)	372 (360—285)
(H)	299 (285—315)	266 (260—274)	240 (237—259)	366 (353—384)
Body width				
(K)	5—7.5	4.5—5.5	4.5—5.5	5—6
(H)	4—5	4—5	3—5	5—6
Cephalic end, shape (K)	Tapered	Blunt	Blunt	Tapered
Caudal end, shape (K, H)	Straight	Usually curved or buttonhooked	Straight	Curved (variable)
Annular striations (H)	Present	Absent	Absent	Present
Cephalic space, length (H)	11.5 (7.5—14)	9.5 (8.5—10)	11 (10—13)	13 (10—15)
Caudal space, length (H)	28 (22.5—32.5)	23 (21—26)	19.5 (15—23)	48 (38—56)

[a] K = routine Knott material, H = hematoxylin-stained Knott material.

From Redington, B. C., Jackson, R. F., Seymour, W. G., and Otto, G. F., Proceedings of the Heartworm Symposium '77, Otto, G. F., Ed., Veterinary Medicine Publ., Bonner Springs, Kan., 1975, 17. With permission.

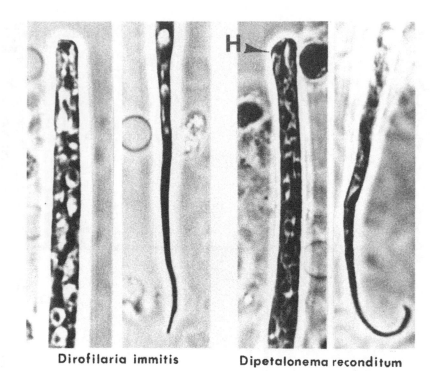

Dirofilaria immitis   Dipetalonema reconditum

FIGURE 4. Microfilariae of *D. immitis* and *Di. reconditum*. Note (H) the cephalic hook of *Di. reconditum*. (From Georgi, J. R., *Parasitology for Veterinarians*, 4th ed., W. B. Saunders, Philadelphia, 1985, 262. With permission.)

(WHO)[15] have been relied upon to construct the approximate global distribution of canine filariasis. The usefulness of this information is limited in that the identity of the filarial agents involved is not specified. In addition, records are presented by country so that exact geographic distributions cannot be inferred. The WHO/FAO/OIE survey[15] is valuable, however, in that it indicates areas where anecdotal or unpublished accounts of the occurrence of *D. immitis* or other filariae infecting dogs could be substantiated by systematic, published field studies.

A. North America

The results of the WHO/FAO/OIE survey indicate a low, sporadic occurrence for *Dirofilaria* sp. among dogs in Canada.[15] Until recently this level of infection was attributed not to ongoing transmission but to movement of infected dogs and cats into Canada from enzootic areas to the south. However, in 1973 the detection of a *D. immitis* infection in a coyote in Ontario was taken as evidence that the parasite is enzootic at least in the southern portion of that province.[16] Subsequently, annual questionnaire surveys identified *D. immitis* in dogs in every province except Prince Edward Island.[17-22] Locally acquired cases have been confirmed from Montreal, Quebec; Calgary, Alberta; Winnepeg, Manitoba; and from the southern Ontario peninsula (Figure 9). However, only the latter two areas are thought to constitute significant enzootic foci. It is worth noting that these areas are adjacent to enzootic foci in the U.S.A.

Overall prevalence rates for *D. immitis* infection among Canadian dogs ranged from 1.31% in 1978 to 1.79% in 1981. Locally higher prevalence rates were recorded in the southeastern Ontario focus, but none of these exceeded 2.94%.[22] There is no evidence of a significant increase in the prevalence of canine dirofilariasis as has been documented in Australia and the U.S. (see discussion below). In discussing the latest survey results Slocombe and

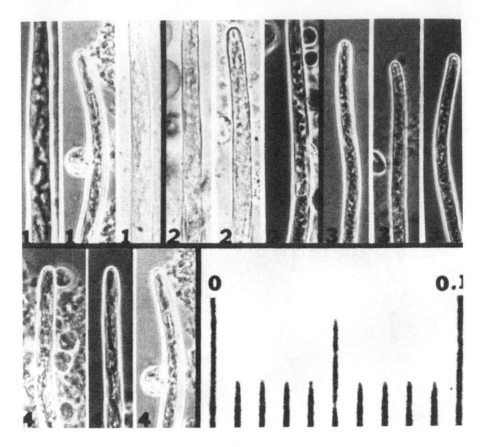

FIGURE 5. Cephalic ends of microfilariae from the routine Knott preparation; oil immersion. Three specimens of each species, more clearly illustrating the tapering cephalic ends of (1) *D. immitis*, (4) the long slender Florida *Dirofilaria* sp., (2) the blunt cephalic end of *Di. reconditum*, and (3) the Irish *Dipetelonema* sp. (Magnification × 1000.) (From Redington, B. C., Jackson, R. F., Seymour, W. G., and Otto, G. F., *Proceedings of the Heartworm Symposium '77*, Otto, G. F., Ed., Veterinary Medicine Publ., Bonner Springs, Kan., 1978, 17. With permission.)

McMillan[22] debate the appropriateness of regular blood testing and preventive chemotherapy for dogs in the face of such a low prevalence of infection. They conclude that regular blood testing of Canadian dogs is to be encouraged in order to monitor fluctuations in the prevalence of heartworm infection. On the other hand, they recommend preventive medication only for dogs in the most highly enzootic areas (namely a strip of the southeast Ontario peninsula comprising the cities of Forest, Sarnia, Chatham, and Windsor, and their immediate surroundings as well as localities along the Lake Erie shoreline). Cats and foxes have also been reported infected with *D. immitis* in Canada.[22] *Di. reconditum* infections were also reported in dogs although at a much lower frequency than *D. immitis* infections. This finding points to the necessity of differential diagnosis in interpreting the presence of microfilariae in canine blood samples from Canada.

The WHO/FAO/OIE report[15] notes the occurrence of canine dirofilariasis in the U.S. and characterizes its distribution as "limited to certain regions". *D. immitis* is enzootic in an Atlantic and Gulf Coastal zone stretching from Massachusetts through eastern Texas[23-40] (Figure 9). The enzootic focus extends throughout the Mississippi River valley from Tennessee and Arkansas northward through the states of Kentucky, Illinois, Indiana, Ohio, Missouri, Iowa, Minnesota, Wisconsin, and Michigan.[41-51] The parasite is being recognized with increasing frequency in areas to the west of this long standing enzootic focus. Locally

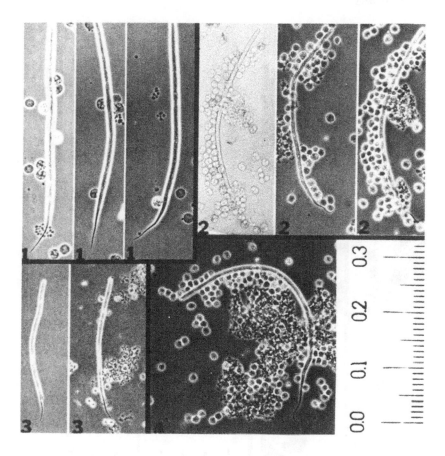

FIGURE 6. Specimens of four species of microfilariae from the routine Knott sediment seen at low power. (1) Three specimens of *D. immitis* showing the essentially straight body, straight caudal end and tapering cephalic end. (2) Three specimens of *Di. reconditum*, illustrating the smaller curved body, buttonhooked caudal end, and blunt cephalic end (like the end of a broom handle). (3) Two specimens of the even smaller Irish *Dipetalonema* sp. with essentially straight body, straight caudal end, and blunt cephalic end. (4) One specimen of the Florida *Dirofilaria* sp. showing the long slender curved body, curved caudal end, and tapering cephalic end. (Magnification × 250) (From Redington, B. C., Jackson, R. F., Seymour, W. G., and Otto, G. F., *Proceedings of the Heartworm Symposium '77*, Otto, G. F., Ed., Veterinary Medicine Publ., Bonner Springs, Kan., 1978, 14.)

acquired infections have now been documented in Kansas, Nebraska, and West Central Colorado,[52-53] and there is good documentation of a relatively new focus of infection emerging in northern California.[54-60] *D. immitis* is also enzootic in Hawaii.[64-66] *Di. reconditum* has been encountered virtually throughout the range of *D. immitis* thus necessitating differential diagnosis of the two species of microfilariae in examinations of peripheral blood.

The coastal region of the Southeastern U.S. constituted the first recognized focus of *D. immitis* in North America, and remains the zone of highest prevalence of the parasite.[67] In general, the rate of infection is highest among dogs in the southernmost portion of this zone. The overall infection rate among dogs sampled in a New Orleans, La. study was 44%.[38] *Di. reconditum* microfilariae were observed in 2.4% of the dogs surveyed. Fifty-five percent of dogs sampled from a Mississippi population were positive at necropsy[36] and a field study in the vicinity of Clarksdale, Miss. revealed a microfilarial infection rate among dogs of 33%.[37] In a survey conducted in 1958-59 in Alabama 42% of dogs examined in the Mobile area were positive for *D. immitis* microfilariae while 23% were positive for *Di. reconditum*.[35] The prevalences of these microfilarial infections among dogs in Auburn, some 100 miles

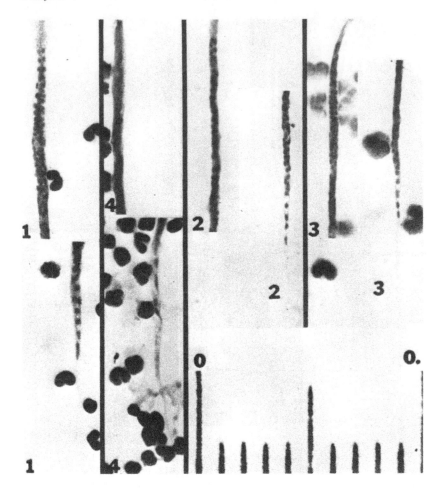

FIGURE 7.   Cephalic and caudal ends of microfilariae from the Knott sediment subsequently reacted with ethanol, ether, and Delafield's hematoxylin; oil immersion. These illustrate both the relative lengths of the cephalic and caudal spaces of the four species (1) *D. immitis*, (2) *Di. reconditum*, (3) Irish *Dipetalonema* sp., (4) the Florida *Dirofilaria* sp., and the annular striations of (1) *D. immitis* and the (4) Florida *Dirofilaria* sp. (Magnification × 1000.) (From Redington, B. C., Jackson, R. F., Seymour, W. G., and Otto, G. F., *Proceedings of the Heartworm Symposium '77*, Otto, G. F., Ed., Veterinary Medicine Publ., Bonner Springs, Kan., 1978, 14.)

inland and to the north were 3 and 32%, respectively. In Jacksonville, Fla. 23% of dogs examined had circulating microfilariae of *D. immitis* and 37% had microfilariae of *Di. reconditum*.[35] Thrasher and Clanton[33] examined dogs from 15 communities in Georgia and found 19.6% harboring *D. immitis* microfilariae. *Di. reconditum* was also present in the majority of the communities with infection rates ranging from 1.4% to 15.7%. The prevalence of microfilarial infections among dogs in Georgia varied by locality, ranging from 0% in Quitman and Winder to 40.9% in Thomasville. In general, the highest microfilarial infection rates were encountered in coastal or swamp areas. A study of dogs in the Atlanta area[34] revealed a microfilarial infection rate of 5.4% among privately owned dogs and 12.5% among animals at the municipal pound. *Di. reconditum* was also found in these populations at a rate of 14.6% and 37.5%, respectively. The overall rate of infection with *D. immitis* microfilariae among dogs from five North Carolina counties was 26.8% with the highest local prevalence (43.7%) occurring in coastal Edgecomb County.[31] An observation of 37.7% microfilaria positives from three additional counties in North Carolina[32] includes dogs har-

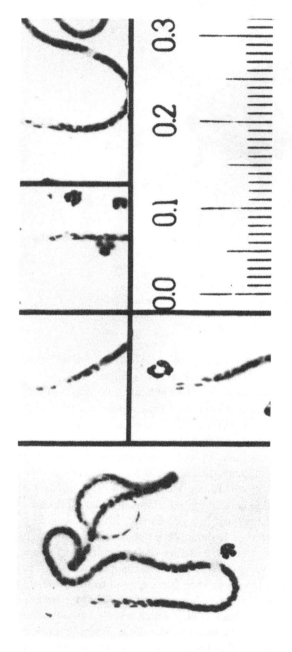

FIGURE 8. Microfilariae of *D. striata*, oil immersion. Note specifically the two anterior nuclei separated from the column of nuclei. (Hematoxylin stain; magnification × 1000.) (From Redington, B. C., Jackson, R. F., Seymour, W. G., and Otto, G. F., *Proceedings of the Heartworm Symposium '77*, Otto, G. F., Ed., Veterinary Medicine Publ., Bonner Springs, Kan., 1978, 14.)

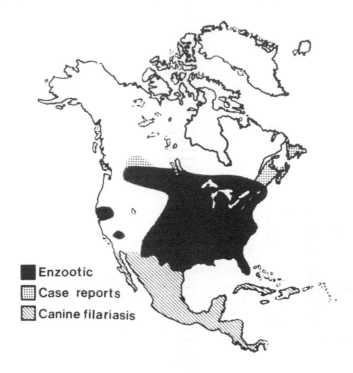

FIGURE 9.   Distribution of *D. immitis* and canine filariasis in North
America. Dark shading indicates confirmed enzootic zone for *D. immitis*.
Light shading indicates areas reporting cases of *D. immitis* infection in
introduced dogs. Cross-hatched areas represent countries reporting canine
filariasis (agent not specified) in WHO/FAO/OIE survey.[15]

boring *Di. reconditum* microfilariae. When corrected for the 6% of *Di. reconditum* infection
observed previously,[31] the two estimates for the state are remarkably similar.

Survey results from the mid-Atlantic and New England states indicate a decreasing trend
in the prevalence of *D. immitis* infections. A rate of 13.6% microfilaria positives from four
Virginia counties was significantly lower than the infection rate in a similar study population
in North Carolina.[31] On the other hand, infection rates with *Di. reconditum* remained constant
at 6%. Estimates from other northern coastal states based upon examinations for microfilariae
are 8% in Maryland,[29] 28% in Delaware,[28] 2% in Pennsylvania,[25] 1 to 2% in New York,[62,114]
9.3% in New Jersey,[27] 8% in Connecticut,[26] 2% in Rhode Island,[25] and 10% in Massachusetts.[23,24] A questionnaire survey conducted in New York State indicated that most cases
of *D. immitis* infection were located in Long Island and in a geographic belt corresponding
to a major roadway, the New York State Thruway.[63] It should be emphasized that although
a declining prevalence of infection in the cooler northern extremes of *D. immitis'* range may
be assumed, there is great potential for local variation due in part to characteristics of the
dog population under study, possible strain differences in the parasite, and to the presence
of isolated landscapes which favor high populations of vector mosquitoes. The importance
of the former factor is illustrated by the study of Mallack et al.[30] which revealed an infection
rate with *D. immitis* microfilariae of 44% among foxhounds in Maryland. When compared
to a previous estimate of 8% for dogs in the state[29] this result underscores the potential for
extremely high prevalence among dogs which spend the majority of time outdoors where
they are exposed to the bites of infective mosquitoes.

Otto[67] discusses the increasingly frequent recognition of *D. immitis* infections in dogs in
the American midwest. From highly enzootic foci in Louisiana and Mississipi, the range of
D. immitis extends northward throughout virtually the entire Mississippi River valley. Eyles

et al.[41] found 5% of the dogs examined in Memphis, Tenn. to be infected with *D. immitis* microfilariae. *D. immitis* has been detected in dogs throughout the state of Illinois with overall infection rates ranging from 35% in southern portions of the state to 10% in northern counties. *Di. reconditum* microfilariae were detected in 7.2% of dogs examined.[42] Questionnaire surveys conducted in Illinois from 1971 to 1976 by Noyes[43] indicate a decrease in the incidence of infection among dogs statewide from 20.4% in 1971 to 3.9% in 1976. A similar trend is observed in Lake County, Ill., where the rate of microfilarial infections declined from 7.2% in 1971 to 1.6% in 1976. The author attributes this decreasing trend in the prevalence to increased use of preventive medication. Kazacos[44] reports that 15.2% of pound dogs from Tippecanoe County, Ind. exhibited infections with adults of *D. immitis* in the heart at necropsy. In Ohio, 4.8% of 500 dogs necropsied at a Columbus humane shelter harbored adults in the heart.[45] Only 54% of these dogs had detectable circulating microfilariae. Similar results were obtained in a survey spanning five counties in northwestern Ohio.[46] In this case 2.5% of 239 animals exhibited circulating *D. immitis* microfilariae. Prevalence estimates for *Di. reconditum* in Ohio range from 0.2% in the Columbus area[45] to 3.8% in the northeastern portion of the state.[46] *D. immitis* is also found in Michigan. Prevalence estimates in the southeastern portion of the state range from 2% in the Detroit area[47] to 22% in the town of Bellview.[48] Circulating microfilariae were found in 4.2% of dogs examined in central Michigan.[68] *Di. reconditum* also occurred in 2.8% of the animals sampled in the Detroit area.[48] Records from the University of Minnesota Veterinary Clinic presented by Schlotthauer and Griffiths[49] documented an increase in the reported cases of heartworm infection in Minnesota from 1950 to 1962. *D. immitis* is considered endemic in Hennipen County which includes metropolitan Minneapolis/St. Paul, and isolated cases have been found in native dogs throughout the southern portion of the state. *Di. reconditum* is also present in Minnesota but little published information on its prevalence is available.

There are relatively few records of *D. immitis* infection in dogs from areas west of the Mississippi River. A questionnaire survey conducted in Missouri in 1981[50] revealed 3.8% of 22,414 dogs examined were positive for microfilariae and/or adults of *D. immitis*. *Di. reconditum* infections occurred in 0.2% of these dogs. Results of prior surveys were also presented and indicate a decrease in the incidence of infection from 10.8% in 1975-76 to 3.8% in 1979. A similar finding in 1981 was taken as evidence that the incidence of heartworm in Missouri is gradually stabilizing. In the same study, results of a questionnaire survey in Nebraska revealed microfilariae and/or adults of *D. immitis* in 3.3% and *Di. reconditum* in 0.4% of 2598 dogs examined. The authors attribute the similar prevalences of *D. immitis* infection in Missouri and Nebraska to similar distribution of bodies of water and, presumably, similar population dynamics of vector mosquitoes in the two states. The Iowa State University veterinary school reported a very high proportional morbidity due to heartworm (ratio of *Dirofilaria* infections to all clinical diagnoses) compared to other northern localities.[51] In Levenworth County, Kan. 16.7% of 288 dogs examined harbored microfilariae of *D. immitis* and 7.6% harbored *Di. reconditum*.[52] In the Grand Junction area of west central Colorado 27 of 801 dogs (3.3%) were infected with *D. immitis*. Sixteen of the positive animals had never left the study county. The authors point out the arid nature of west central Colorado and speculate that irrigation projects may serve to increase vector populations to a level which will support transmission of *D. immitis*.[53]

Although filariasis in dogs native to California was first recognized in 1946,[54] the parasite has only been well documented in that state since 1970 when McGreevy et al.[55] found microfilariae of *D. immitis* in only 0.13% of pound dogs from northern California. At that time a search of records in the University of California School of Veterinary Medicine revealed only 12 cases of *D. immitis* in dogs from 1957 to 1968. Since that original report, there have been numerous findings of canine dirofilariasis in northern California with evidence of an increase in incidence and geographic range of the parasite.[56] To date, *D. immitis* infections have been reported in dogs from at least 11 counties in northern California.[56-58] Point prevalence estimates vary widely by locality ranging from 0.3 to 4.0% in Marin County to 40.0% in Sonoma county. The mean prevalence rate based on a questionnaire survey of veterinarians in nine northern counties was 7.0% (n = 3476).[57] The mean rate of infection with *D. immitis* microfilariae among dogs from four northern counties was 5.7%. These same surveys[55-58] reveal *Di. reconditum* throughout the range of *D. immitis* in northern California. In contrast to *D. immitis*, estimates of this parasite's prevalence are rather constant from locality to locality ranging from 4.9 to 6.8%. The presence of a large population of introduced dogs in northern California raises some question as to the occurrence of natural transmission in the region. However, the presence of *D. immitis* infection in native dogs[54-56,58] and the isolation of advanced-stage filarial larvae presumed to be *D. immitis* from field-captured vector mosquitoes[59] suggests that the parasite is enzootic. As in the Canadian studies cited above,[16] the observation of adult *D. immitis* infections in coyotes from five northern California counties[60] also indicates that the parasite is enzootic and, moreover, that a sylvatic reservoir has become established.[58,60]

There is little information on the prevalence of *D. immitis* in southern California. The parasite appears to be enzootic in at least one county, however, since Corselli and Platzer[61] report that 4% of 193 dogs examined in heartworm clinics in central and eastern Riverside County had circulating microfilariae of *D. immitis* and that 6 of the 7 infected animals were native to the region. Knott's testing of 366 pound dogs from western Riverside County revealed only *Di. reconditum* microfilariae at a prevalence of 2.2%. As part of the same study, a questionnaire survey covering approximately 70% of the practicing veterinarians in Riverside revealed that 22 cases of *D. immitis* infection had been diagnosed in native dogs in 1978-79.

Both *D. immitis* and *Di. reconditum* are enzootic in Hawaii. Results of necropsy surveys indicated prevalence rates for *D. immitis* of 19% in 1962,[64] 30.2% in 1965,[65] and 41% in 1974.[66] Prevalence estimates for *Di. reconditum* obtained in these same studies were 16, 8.3 and 1%, respectively. The prevalence estimate in the latter study[66] was based on necropsy data and demonstrated the potential for underestimation in studies based exclusively on examination of peripheral blood. In this case only 25 of 41 infected animals had detectable circulating microfilariae.

D. immitis has been recorded in wild canids from many parts of the U.S. The parasite was found in 71% of coyotes (*Canis latrans*), 100% of red wolves (*Canis rufus*), and in 83% of coyote/red wolf hybrids examined in southeastern Texas and southwestern Louisiana.[40] Circulating microfilariae were observed in 46% of the wild canids sampled. In another Louisiana study[39] a high incidence of infection (58%) with *D. immitis* was found in coyotes. In addition, both microfilariae and adults of *D. immitis* were found in red foxes (*Vulpes vulpes*) and gray foxes (*Urocyon cinereoargenteus*).

B. South America

Results of the WHO/FAO/OIE survey[15] indicate that canine filariasis is present in every country of South America except Chile (Figure 10). These records also indicate that the presence of canine filariasis in Ecuador is suspected but not confirmed. No data are given for French Guiana, and the Falkland (Malvinas) Islands are listed as negative. In the countries of Brazil, Argentina, Uruguay, Paraguay, Bolivia, Peru, and Venezuela the occurrence of canine filariasis is described as low and sporadic. Moderate occurrence is reported for Colombia, Guyana, and Surinam. This latter finding is supported by the report of Panday et al.[69] that 26% of dogs screened by blood examination at a veterinary clinic in Paramaribo, Surinam harbored *D. immitis* microfilariae. None of the 521 dogs examined had microfilariae of *Di. reconditum*. While the WHO/FAO/OIE report does not indicate the species of filariae involved in South America, the report from Surinam[69] implies that *D. immitis* is present at least in tropical and subtropical areas of the continent.

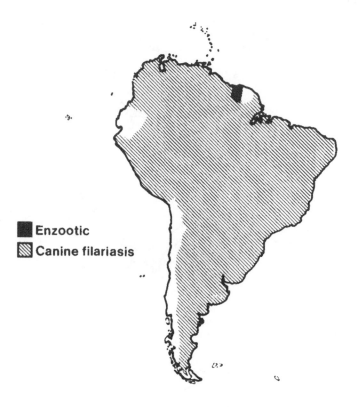

Enzootic

Canine filariasis

FIGURE 10.   Distribution of *D. immitis* and canine filariasis in South America. Dark shaded and cross-hatched areas are as described for Figure 9.

C. Europe

Figure 11 presents the distribution of canine filariasis in Europe based on published accounts. The WHO/FAO/OIE survey[15] records canine filariasis only in the U.K., Portugal, Italy, Hungary, and Greece. The peculiar discontinuous distribution thus depicted and the known disinclination of these parasites to respect political boundaries would seem to warrant the assumption that canine filariasis exists throughout southern Europe. The designation of the Republic of Ireland as free of canine filariasis by WHO/FAO/OIE is erroneous since Jackson et al.[70] report microfilariae of the genus *Dipetalonema* in greyhounds from County Cork and County Limerick. *D. immitis* infection in the U.K. appears to be limited to dogs imported from enzootic areas overseas. Thomas[71] reports the parasite in a dog released from quarantine after importation from the Seychelles. Effective quarantine measures have, in fact, been cited as a reason for the failure of *D. immitis* to become established in the U.K.[72] However, the presence of introduced cases and the ability of the parasite to adapt to temperate climates (see discussion below) argues for continued surveillance.

D. Africa

Figure 12 presents the distribution of countries reporting canine filariasis to the 1983 WHO/FAO/OIE survey.[15] The information suggests that filariasis is enzootic in dogs in a zone of western Africa extending from Morocco, Algeria, and Tunisia on the Mediterranean to Ghana, Burkina Faso, and Nigeria on the Atlantic Coast. The Republic of Congo and Angola are also listed as positive. In eastern Africa, canine filariasis has been reported from countries forming a continuous band extending from the Republic of South Africa northward through to the Republic of Sudan. With the exception of Sierra Leone, Liberia, and Burkina Faso which report a moderate occurrence, all African countries report a low, sporadic

FIGURE 11.   Distribution of *D. immitis* and canine filariasis in Europe. Light-shaded and cross-hatched areas are as described for Figure 9.

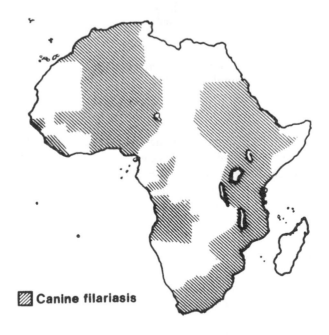

FIGURE 12.   Distribution of African countries reporting canine filariasis (agents not specified) in 1983 WHO/FAO/OIE survey.[15]

occurrence of canine filariasis. Levine[1] remarks that *D. immitis* is rare in Africa. However, the widespread occurrence of canine filariasis indicated by the WHO/FAO/OIE survey and the paucity of other published information argue for a strong effort to investigate and document the geographic distribution of this parasite in Africa.

E. Asia and the Pacific

Figure 13 shows the distribution of *D. immitis* and other canine filariasis in Asia. Countries reporting canine filariasis in the 1983 WHO/FAO/OIE survey form a band extending from

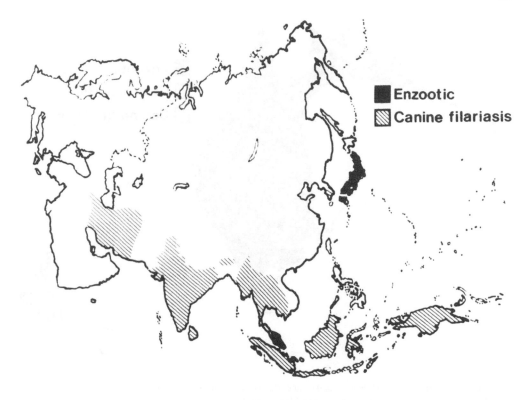

FIGURE 13. Distribution of *D. immitis* and canine filariasis in Asia. Darkly shaded and cross-hatched areas are as described for Figure 9.

Iran on through the Indian subcontinent, Burma, Malaysia, Indonesia, and Papua New Guinea. *D. immitis* is enzootic throughout Malaysia. Prevalence estimates range from 9.6% in an urban area surrounding Kuala Lumpur[73] to 32.4% in Seremban[74] and 70% in Sabah, eastern Malaysia.[75] *D. immitis* is also prevalent in French Polynesia. A mean microfilarial infection rate of 47.5% was determined in a survey conducted on the Society, Tuamotu, and Marquesas Islands.[76] The occurrence of *D. immitis* in Japan is also well documented. Mizuno et al.[77] report that 25.2% of dogs examined in suburban Tokyo exhibited circulating microfilariae. These authors also provide a valuable summary of previous reports of canine dirofilariasis in Japan. This compilation of records suggests that *D. immitis* is enzootic throughout the islands. Point prevalence estimates range from 1.2% in Okinawa Prefecture to 74.0% in Tochigi Prefecture. The mean infection rate among dogs from the 13 reports cited by Mizuno et al.[77] is 23%. Canine filariasis also occurs at a low, sporadic rate in South Korea.[15]

F. Australia

Published accounts indicate that *D. immitis* is enzootic along the northern coasts of western Australia and the Northern Territory[78] and in an eastern zone comprising the states of Queensland,[78-82] New South Wales,[83-87] and Victoria[88,89] (Figure 14). Within this region the prevalence of heartworm infection generally increases along a northerly axis with maximum infection rates approaching 90% in tropical localities of northern Queensland.[78] Even in such areas of high overall prevalence local geography can bring about drastic variation in infection rates among dogs. For instance, tropical Queensland sites at high elevation (Hope Vale and Atherton) and an island (Mornington Island) were free of heartworm in a survey conducted in 1969 to 1972 and 1981 to 1982.[81] In a similar manner, proximity to bodies of standing

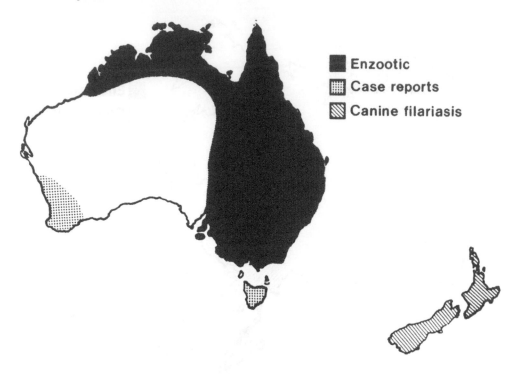

FIGURE 14.   Distribution of *D. immitis* and canine filariasis in Australia and New Zealand. Darkly shaded, lightly shaded, and cross-hatched zones are as described for Figure 9.

water generally seems to increase the risk of heartworm infection in dogs especially in more temperate regions of the south; a case in point being the significant impoundment of surface water associated with the Murrumbidgee Irrigation Area.[83] Canine filariasis exists in New Zealand, but its distribution and prevalence have not been documented.[15]

The range of *D. immitis* in eastern Australia appears to be expanding from its long-standing tropical and subtropical focus into more temperate localities in the south. Infection rates among dogs in tropical Queensland were high and generally stable over the period from 1969 to 1976.[78] In contrast, significant increases in the infection rate have occurred in southern Queensland, New South Wales, and northeastern Victoria. For example, in 1973, Watson et al.[85] reported heartworm infections in 5% of dogs examined in Sydney, N.S.W. A survey of dogs in the same area conducted from 1982 to 1983 indicated an infection rate of at least 12.6% as indicated by the Knott's test. Similarly a study of pound dogs in the Brisbane area in 1969 revealed an infection rate of 24% at necropsy and a rate of 12% microfilaria positives.[79] In the same locality 35.8% of dogs examined in 1979 had circulating microfilariae of *D. immitis*.[82] This increase would be even more marked if one took into account the potentially high incidence of occult infections. In 1946, Pullar[91] found no evidence of *D. immitis* infection among pound dogs in Melbourne, Victoria. More recently, however, an infection rate of 2.7% among pound dogs in that area was observed.[89] Several authors have theorized that unrestricted transport of infected dogs is the most likely mode of dispersal by *D. immitis* in Australia.[89,90] Interstate transport of greyhounds in particular is cited as a possible cause of the parasite's recent establishment in Victoria.[89]

There is little published information on the distribution of heartworm outside the eastern states of Australia. In fact, it was only recently concluded that the enzootic zone extends to the west of the Great Dividing Range in these states.[90] An infection rate of 13% was recorded in Alice Springs, central Australia in a 1972 to 1976 survey.[78] However, the dogs involved were thought to have been introduced, and it was concluded that *D. immitis* was not yet enzootic in that locality. Similarly, respondents to a 1981/1982 survey[90] report cases of *D. immitis* infection in dogs from localities surrounding the City of Perth. However, practitioners in that area maintain that the parasite is not enzootic. In the same questionnaire survey respondents from Tasmania reported heartworm in local dogs but, again, considered their area to be nonenzootic. In view of the parasite's capacity to move into previously nonenzootic areas, these sites appear to be at some risk of becoming new foci of transmission. All dogs examined in a number of remote aboriginal settlements of central Australia were found to be free of heartworm.[78]

D. immitis also infects wild canids in Australia. Mulley and Starr[86] found *D. immitis* infections in 8.8% (n = 68) of red foxes (*Vulpes vulpes*) surveyed in 1982/1983 in the vicinity of Sydney, N.S.W. Two of the six infected foxes were microfilaremic. It is possible, therefore, that this animal could serve as a reservoir of infection. Infections with *D. immitis* have also been observed in dingoes (*Canis dingo*).[79,82,90]

The WHO/FAO/OIE survey[15] indicates that canine filariasis in Australia is caused by *D. immitis* only. This report would appear to be in error, however, since there are several reports of *Di. reconditum* infection in Australian dogs.[82,87,92] Rates of infection appear low compared to *D. immitis* and the parasite may occur either alone or in mixed infections with *D. immitis*.[82,92] In the Brisbane area where the prevalence of *D. immitis* infection is on the increase there appears to be no corresponding increase in infection rates with *Di. reconditum*.[92]

G. The Changing World Distribution of *Dirofilaria immitis*

Heartworm in dogs due to *D. immitis* has long been considered a disease limited to the warm coastal areas of the world. For example, as late as 1972 Otto[93] recorded in the U.S. significant levels of disease to occur in dogs in an Atlantic and Gulf coastal band extending from southern New Jersey southward through Texas and Mexico. An apparently unique and isolated enzootic focus of *D. immitis* was also described in central Minnesota. Similarly, in Australia the incidence of infection with *D. immitis* has classically been in the highest in the tropical and subtropical regions of the northern provinces.[80,81] In a more recent study, however, Otto[67] discusses an apparent increase in the incidence of infection in dogs in more temperate localities such as the mid-Atlantic states of Pennsylvania, Maryland, Connecticut, and Rhode Island and the northcentral states of Illinois, Ohio, and Michigan. A similar trend is apparently occurring in Australia with an increasing incidence of *D. immitis* in the more temperate east coast zone and apparently new foci appearing in the southernmost provinces of Victoria and New South Wales.[72,78,83-85,87] Pacheco[94] cites evidence from Kume for an expanding range of *D. immitis* in Japan.

Several authors have suggested possible explanations for the upsurge of heartworm infections in relatively temperate climates. The unrestricted movement of infected dogs is cited as a primary mechanism by several authors.[66,72,95] Moreover, Murdoch[72] theorizes that the failure to date of *D. immitis* to become enzootic in the U.K. may be attributed to effective quarantine measures restricting the influx and dispersal of infected animals. A second possible explanation for the increasing prevalence of *D. immitis* throughout the world is the decreased use or efficacy of chemical insecticides due to environmental considerations and to the increase of insecticide resistance in mosquito populations.[66,95] Otto[67] has formulated another theory which takes into account the temperature requirements for development in the mosquito vector. He observes that in temperate foci of *D. immitis*, such as the Minnesota focus, prevailing temperatures are generally below those originally deemed necessary for development in the mosquito vector (21°C for 10 to 14 days). He further postulates that isolated temperate foci of this type may arise due to the appearance of genotypes of the parasite which are capable of completing development at temperatures below the normal threshold for development. Such isolated geographic populations might then act as sources for the further expansion of the parasite throughout the temperate zone. More recent work in this area[96,97] shows that *D. immitis* originating in temperate climates actually will complete development at temperatures as low as 18°C and that larvae maintained in mosquitoes below this threshold will resume development when environmental temperatures increase.

IV. FILARIOIDS OTHER THAN *DIROFILARIA IMMITIS* IN DOGS AND CATS

In addition to *D. immitis*, dogs and cats throughout the world have been found infected with filariae from the genera *Dirofilaria*, *Dipetalonema*, and *Brugia*. While no significant pathological conditions have been attributed to these various species they are of concern to the clinician because their circulating microfilariae tend to complicate the diagnosis of heartworm infection. It is not known to what extent these filariae may interfere with the various immunodiagnostic methods. Systematic studies by Redington et al.[11] and Price[14] detected five apparently distinct species of circulating microfilariae in dogs in Florida. The following are summaries of the morphologies and bionomics of these nonpathogenic species and the differential morphologic characteristics which may be used to distinguish them from *D. immitis*.

A. *Dipetalonema reconditum*

This parasite, originally described by Grassi and Calandruccio[98] from the subcutaneous connective tissues of dogs, is perhaps the best studied of the nonpathogenic filariae in dogs. Adults of *Di. reconditum* are generally located in subcutaneous connective tissues of the definitive host.[99] The adult females and males, measuring 32 and 17 mm in length, respectively, are readily distinguishable from the far larger adults of *D. immitis*. Unlike *D. immitis*, microfilariae of *Di. reconditum* in smears of fresh blood with anticoagulant undergo intermittent periods of rapid progressive motility.[11] Routine Knott's preparations from dogs infected with *Di. reconditum* reveal microfilariae which are significantly shorter and somewhat narrower than those of *D. immitis* (Table 1). Microfilariae of *Di. reconditum* Knott's preparations exhibit a cephalic end which is blunt compared to the distinctly tapered anterior extremity of *D. immitis* (Figures 4, 5[1,2], and 6[1,2]). The microfilaria of *Di. reconditum* also bears a prominent cephalic hook (Figure 4). The caudal ends of *Di. reconditum* from Knott's samples have been described as curved or buttonhooked in the majority of cases compared to the generally straight tails of *D. immitis* (Figures 4 and 15[1,2]). In general the entire body of the *Di. reconditum* microfilaria is curved in a Knott's preparation while that of *D. immitis* is essentially straight (Figure 6[1,2]). Finally, the number of circulating microfilariae of *Di. reconditum* is generally lower than that of *D. immitis* (Table 1). Hematoxylin-stained microfilariae of *Di. reconditum* reveal caudal and cephalic spaces which are generally shorter than those of *D. immitis*, and the cuticles of *Di. reconditum* lack the annular striations evident in hematoxylin-stained *D. immitis* (Figure 7[1,2]). The distribution of acid phosphatase activity in microfilariae of *Di. reconditum* has been described variously as uniform over the entire body,[12] localized in three to four areas,[13] and concentrated around the excretory and anal pores.[14] In the latter study, although a similar staining pattern to *D. immitis* was obtained, staining in the area of the anal pore of *Di. reconditum* was described as more intense than that observed in *D. immitis*. Despite this potential for variation in staining patterns, acid phosphatase staining combined with relative length measurements remains the most reliable method for distinguishing between microfilariae of *D. immitis* and *Di. reconditum*.[100]

B. The "Irish" *Dipetalonema*

Jackson et al.[70] reported the discovery of a microfilaria from the blood of dogs in southern Ireland. The initial observation was made on greyhounds imported into Florida and was subsequently confirmed by examination of dogs on the breeder's premises in County Lim-

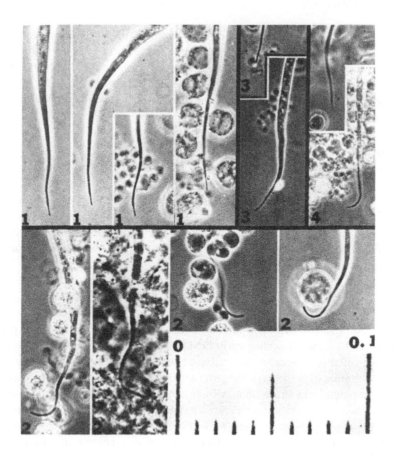

FIGURE 15. Caudal ends of microfilariae from the routine Knott sediment; oil immersion. These illustrate the straight caudal end of (1) *D. immitis*, (2) the buttonhooked or curved caudal end of *Di. reconditum*, (3) the essentially straight caudal end of the Irish *Dipetalonema* sp., and (4) the curved caudal end of the Florida *Dirofilaria* sp. (Magnification × 1000.) (From Redington, B. C., Jackson, R. F., Seymour, W. G., and Otto, G. F., *Proceedings of the Heartworm Symposium '77*, Otto, G. F., Ed., Veterinary Medicine Publ., Bonner Springs, Kan., 1978, 14.)

erick, Ireland. These newly discovered microfilariae resembled those of *Di. reconditum* in a number of respects. They were present in relatively low numbers in the peripheral circulation and exhibited progressive motility in fresh blood smears (Table 1). In Knott's preparations these microfilariae exhibited the blunt cephalic end common to *Di. reconditum* (Figures 5[3] and 6[3]) but lacked the characteristic "buttonhooked" tail (Figures 15[3] and 6[3]). The Irish microfilaria was significantly shorter on average than *Di. reconditum* and some 20% shorter than *D. immitis* (Table 1). The ranges of measurements of cephalic and caudal spaces in hematoxylin-stained specimens overlapped those of *Di. reconditum*, and the Irish microfilaria, like *Di. reconditum*, lacked the annular striations characteristic of *D. immitis* (Table 1, Figure 7[3]). Redington et al.[11] state that, although the parasite has been reported from imported greyhounds in Florida, there is no evidence that it has become established there. However, in a more recent paper, Price[14] speculates that due to the importation of greyhounds over a number of years the Irish *Dipetalonema* may possibly have become enzootic.

C. *Dirofilaria striata*

This parasite, originally described from bobcats (*Lynx rufus*) in the U.S.[101] was reported as an accidental parasite of the domestic dog in Florida.[102] Adults were recovered from deep intramuscular fascia. Formalin-fixed microfilariae of *D. striata* stained with hematoxylin appear significantly longer and more slender than those of similarly prepared *D. immitis* (348 μm × 4 to 5 μm vs. 299 μm × 5 to 6.5 μm).[11] Microfilariae of *D. striata* exhibit within the cephalic space two prominent nuclei separate from the main body of the nuclear column[101,102] (Figure 8).

D. The "Florida" *Dirofilaria*

It is noteworthy that Redington et al.[11] described a microfilaria from dogs in Florida apparently in the genus *Dirofilaria*, which shares some characteristics with *D. striata*. The microfilariae are of similar dimensions both being significantly longer than *D. immitis* (348 μm × 4 to 5 μm for *D. striata* and 366 μm × 5 to 6 μm for the Florida *Dirofilaria*, see Table 1). However, the Florida *Dirofilaria* lacks the prominent pair of cephalic nuclei common to *D. striata* (Figure 8), and it is believed that the two microfilariae represent distinct species.

V. VECTORS OF *DIROFILARIA IMMITIS*

The first, second, and early third larval stages of *D. immitis* are obligate parasites of mosquitoes. In addition to supporting early larval development, mosquitoes serve as vectors actively moving infective stages to susceptible vertebrates and, to the extent that their flight range allows, disseminating the parasite geographically. Transmission of *D. immitis* is then restricted geographically to those areas which can support adequate populations of adult mosquitoes and seasonally within temperate climates to those periods when temperature is sufficiently high to support development in the cold-blooded vector. Vector relationships of *D. immitis* are discussed at length in Chapter two of this volume and in several recent reviews.[5,95,103]

Although its association with a mosquito vector limits its transmission both seasonally and geographically, *D. immitis* seems unique among the filariae in its ability to exploit a wide range of mosquito species indigenous to a variety of habitat types. Well in excess of 60 species of mosquito throughout the world have been found to be susceptible.[104] Although innate susceptibility to infection is only one component of vector competence, field isolation of infective larval *D. immitis* from at least 13 species of mosquito in the U.S. alone underscores the ability of this parasite to exploit a diversity of vector species under natural conditions. Known vectors of *D. immitis* and their predilection for particular habitat types are summarized in Table 2. The table includes only those species in which presumed *D. immitis* infections have been observed in field-captured mosquitoes. As was noted earlier,[95] many of these species, particularly *Aedes vexans*, *Ae. togoi*, and *Culex quinquefasciatus* are serious pests of humans, a fact which creates a significant zoonotic potential of *D. immitis*.

Ae. vexans figures most prominently among the mosquito species incriminated as vectors in domestic and peridomestic habitats in North America.[103,105] Its range spans the continent, and it is one of the primary pests of humans in much of North America.[102,105] Females of this species feed avidly on dogs under natural conditions, and they are highly susceptible to infection.[106-108] Infective larvae presumed to be *D. immitis* have been isolated from field-captured *Ae. vexans* in central Michigan,[109] Alabama,[110,111] Connecticut,[112] New Jersey,[113] New York,[114] and California.[59] In domestic habitats throughout the southeastern U.S. *Cx. quinquefasciatus* has been incriminated as the primary vector species. It feeds aggressively on dogs and has yielded presumed field isolations of the infective-stage larvae presumed to be *D. immitis* in southeastern Louisiana,[115] Alabama,[110] and Florida.[116] *Cx. nigripalpus* has also yielded natural isolates of larval *D. immitis* in Florida.[116] *Anopheles quadrimaculatus* and *An. punctipennis* are serious pests of both humans and animals in domestic and peri-

Table 2
VECTORS OF *D. IMMITIS*

Region	Locality	Habitat type	Species	Ref.
North America	U.S., Canada	Domestic, peridomestic	*Ae. vexans*	59, 109—114
			Cx. quinquefasciatus	110, 115, 116
			Cx. nigripalpus	116
			An. punctipennis	110, 111, 117
	Eastern U.S., Canada	Woodland, savannah	*Ae. trivittatus*	117, 118
			Ae. canadensis	24, 112, 113,
			Ae. sticticus	24, 110
			Ae. stimulans	112
			Ae. excrucians	112, 113
			Ps. ferox	112
	Eastern U.S.	Coastal salt marsh	*Ae. sollicitans*	112, 113
			Ae. cantatos	113
			Cx. salinarius	113
			Ae. taeniorhynchus	116
	West coast, U.S.	Woodland, savannah	*Ae. sierrensis*	59
Australia	Eastern Australia	Domestic, peridomestic	*Cx. annulirostris*	120, 121
			Cx. quinquefasciatus	120, 121
			Ae. vigilax	120
			Ae. notoscriptus	120, 121
			An. annulipes	121
Japan	Coastal	Domestic, peridomestic	*Cx. quinquefasciatus*	123
			Ae. togoi	125
South Pacific	Fiji, French Polynesia	Domestic, peridomestic	Ae. pseudoscutellaris	76, 122
			Ae. fijiensis	122
			Cx. quinquefasciatus	122
			Cx. annulirostris	122
				76

domestic situations in the eastern U.S. Both of these species feed readily on dogs under natural conditions[108] and are susceptible to infection. Natural isolations of larval *D. immitis* from field-captured *An. punctipennis* have been made in Alabama[110,111] and Iowa.[117]

Numerous reports (see above section) of *D. immitis* in wild canids in the U.S. and Canada suggest an increasingly well-entrenched sylvatic reservoir of infection. It is interesting in this light that there are numerous records of natural *D. immitis* infections in species of mosquito, particularly *Aedes* spp., which inhabit woodlands and are seldom encountered in domestic habitats. *Ae. trivittatus* probably best characterizes such a potential sylvatic vector. This mosquito is encountered throughout much of the eastern U.S. and southern Canada. It is attracted to dogs, and in some localities, is the most abundant mosquito species attacking these animals.[108,117,118] It is susceptible to infection,[117,118] and larval *D. immitis* have been recovered from field-captured females in Iowa[117] and Tennessee.[118] There is evidence that a number of other *Aedes* spp. common to woodlands and savannahs of North America are involved in transmission of *D. immitis*. Natural infections have been observed in *Ae. sticticus* in Alabama[110] and Massachusetts.[24] Natural isolations of larval *D. immitis* suggest that *Ae. canadensis*,[24,112,113] *Ae. excrucians*,[112,113] *Ae. stimulans*,[112] and *Psorophora ferox*[112] are also involved in transmission in sylvan habitats of eastern North America. There is evidence that *Ae. sierrensis* acts as the sylvatic counterpart to *Ae. vexans* in maintaining transmission of *D. immitis* in the emerging focus in northern California.[59,119] Brackish marshlands of the eastern coastal U.S. also constitute foci of *D. immitis* transmission, and isolations of infective larvae have been made from several mosquitoes which exploit this ecotype. These include *Ae. sollicitans* in Connecticut[112] and New Jersey,[111] *Ae. cantator* and *Cx. salinarius* from New Jersey,[113] and *Ae. taeniorhynchus* from Florida.[116]

Globally there are few published findings to indicate the vectors of *D. immitis*. In Australia *Cx. annulirostris, Cx. quinquefasciatus, Ae. vigilax, Ae. notoscriptus,* and *Anopheles annulipes* are considered potential vectors,[120,121] while *Ae. alboannulatus, Ae. rubrithorax* and *Cx. australius* may play a minor role as vectors.[121] In Fiji, natural infections with third-stage larval *D. immitis* were observed in *Ae. fijiensis, Ae. pseudoscutellaris, Cx. quinquefasciatus,* and *Cx. annulirostris.*[122] *Cx. annulirostris* and *Ae. polynesiensis* have also been incriminated as vectors in French Oceania,[76] and there is evidence linking *Cx. quinquefasciatus* to transmission in Japan[123] and the Phillipines.[124] In coastal areas of Japan, *Ae. togoi* has been shown to feed avidly on dogs and natural infections have been observed in field-collected females.[125]

ACKNOWLEDGMENTS

This work was supported in part by NIH grants AI-19995 and BRSG S07 RR 05464 and by the World Health Organization Onchocerciasis Chemotherapy Project of the Onchocerciasis Control Programme in the Volta River area (RP 85017). The author is grateful to Mr. René Morris and Ms. Winnie Mapp for assistance in preparing the manuscript. Finally the author is indebted to Drs. O. W. Olsen, Y. Gutierrez, J. R. Georgi, and B. C. Redington, for providing the photographs and drawings used to prepare some of the figures in this chapter.

REFERENCES

1. **Levine, N.,** *Nematode Parasites of Domestic Animals and Man,* Burgess, Minneapolis, Minn., 1968, 356.
2. **Schlotthauer, J. C.,** Host-Parasite Relationships of *Dirofilaria immitis* in the Dog, Ph.D. thesis, University of Minnesota, Minneapolis, 1965, 1.
3. **Leidy, J.,** A synopsis of entozoa and some of their ectocongeners observed by the author, *Proc. Acad. Natl. Sci. Philadelphia,* 8, 42, 1856.
4. **Railliet, A. and Henry, A.,** Recherches sur les ascarides des carnivores, *C. R. Seances Soc. Biol. Paris,* 70, 12, 1911.
5. **Barriga, O.,** Dirofilariasis, in *Handbook Series in Zoonoses,* Steele, J. H., Ed.; *Section C: Parasitic Zoonoses,* Vol. 2, Schulz, M. B., Ed., CRC Press, Boca Raton, Fla., 1982, 93.
6. **Anderson, R. C.,** Description and relationships of *Dirofilaria ursi* Yamaguti, 1941, and a review of the genus *Dirofilaria* Railliet and Henry, 1911, *Trans. R. Can. Inst.,* 29, 35, 1952.
7. **Addison, E. M.,** Transmission of *Dirofilaria ursi* Yamaguti, 1941 (Nematoda: Onchocercidae) of black bears *(Ursus americanus)* by blackflies (Simuliidae), *Can. J. Zool.,* 58, 1913, 1980.
8. **Gutierrez, Y.,** Diagnostic features of zoonotic filariae in tissue sections, *Hum. Pathol.,* 15, 514, 1984.
9. **Otto, G. F.,** Occurrence of the heartworm in unusual locations and in unusual hosts, in *Proceedings of the Heartworm Symposium '74,* Morgan, H. C., Ed., Veterinary Medicine Publ., Bonner Springs, Kan., 1975, 6.
10. **Carlisle, M. S., Webb, S. M., Sutton, R. H., Hampson, E. C., and Blum, A. J.,** Case report — adult *Dirofilaria immitis* in the brain of a dog, *Aust. Vet. Practit.,* 14, 10, 1984.
11. **Redington, B. C., Jackson, R. F., Seymour, W. G., and Otto, G. F.,** The various microfilariae found in dogs in the United States, in *Proceedings of the Heartworm Symposium '77,* Otto, G. F., Ed., Veterinary Medicine Publ., Bonner Springs, Kan., 1978, 14.
12. **Chalifoux, L. and Hunt, R. D.,** Histochemical differentiation of *Dirofilaria immitis* and *Dipetalonema reconditum, J. Am. Vet Med. Assoc.,* 158, 601, 1971.
13. **Kuecks, R. W. and Slocombe, J. O. D.,** Rapid histochemical differentiation of *Dirofilaria immitis* and *Dipelatonema reconditum,* in *Proceedings of the Heartworm Symposium '80,* Otto, G. F., Ed., Veterinary Medicine Publ., Edwardsville, Kan., 1981, 48.
14. **Price, D. L.,** Microfilariae other than those of *Dirofilaria immitis* in dogs in Florida, in *Proceedings of the Heartworm Symposium '83,* Otto, G. F., Ed., Veterinary Medicine Publ., Edwardsville, Kan., 1983, 8.

15. WHO/FAO/OIE, *Animal Health Yearbook 1983*, Animal Health Service, Animal Production and Health Division, World Health Organization, Food and Agriculture Organization, Organization International des Epizooties, Rome, 1984.

16. **Slocombe, J. O. D. and McMillan, I.,** Heartworm in dogs in Canada in 1977, *Can. Vet. J.,* 19, 244, 1978.

17. **Slocombe, J. O. D. and McMillan, I.,** Heartworm in dogs in Canada in 1978, *Can. Vet. J.,* 20, 284, 1979.

18. **Slocombe, J. O. D. and McMillan, I.,** Heartworm in dogs in Canada in 1979, *Can. Vet. J.,* 21, 159, 1980.

19. **Slocombe, J. O. D. and McMillan, I.,** Heartworm in dogs in Canada in 1980, *Can. Vet. J.,* 22, 201, 1981.

20. **Slocombe, J. O. D. and McMillan, I.,** Heartworm in dogs in Canada in 1981, *Can. Vet. J.,* 23, 219, 1982.

21. **Slocombe, J. O. D. and McMillan, I.,** Heartworm in dogs in Canada in 1982, *Can. Vet. J.,* 24, 227, 1983.

22. **Slocombe, J. O. D. and McMillan, I.,** Heartworm in dogs in Canada in 1983, *Can. Vet. J.,* 25, 347, 1984.

23. **Augustine, D. L.,** Observations on the occurrence of heartworms, *Dirofilaria immitis,* in New England dogs, *Am. J. Hyg.,* 28, 390, 1938.

24. **Arnott, J. J. and Edman, J. D.,** Mosquito vectors of dog heartworm, *Dirofilaria immitis,* in western Massachusetts, *Mosq. News,* 38, 222, 1978.

25. **Rothstein, N., Kinnanion, K. E., Brown, M. L., and Carrithers, R. W.,** Canine microfilariasis in eastern United States, *J. Parasitol.,* 47, 661, 1961.

26. **Hirth, R. S., Huizinga, H. W., and Nielson, S. W.,** Dirofilariasis in Connecticut dogs, *J. Am. Vet. Med. Assoc.,* 148, 1508, 1966.

27. **Lilis, G.,** *Dirofilaria immitis* in dogs and cats from south central New Jersey, *J. Parasitol.,* 50, 802, 1964.

28. **Maddux, T. C. and Cohen, R. B.,** A survey of canine heartworm in one county in Delaware, *Vet. Med. Small Anim. Clin.,* 67, 545, 1972.

29. **Wallenstein, W. L. and Tibola, B. J.,** Survey of canine filariasis in a Maryland area — incidence of *Dirofilaria immitis* and *Dipetalonema, J. Am. Vet. Med. Assoc.,* 137, 712, 1960.

30. **Mallack, J., Sass, B., and Ludlam, K. W.,** *Dirofilaria immitis* in hunting dogs from an area in Maryland, *J. Am. Vet. Med. Assoc.,* 159, 177, 1971.

31. **Falls, R. K. and Platt, T. R.,** Survey for heartworm, *Dirofilaria immitis* and *Dipetalonema reconditum* (Nematoda: Filarioidea) in dogs from Virginia and North Carolina, *Am. J. Vet. Res.,* 43, 738, 1982.

32. **Rowley, J.,** The prevalence of heartworm infection in three counties in North Carolina, *Canine Pract.,* 8, 45, 1981.

33. **Thrasher, J. P. and Clanton, J. R., Jr.,** Epizootiologic observations of canine filariasis in Georgia, *J. Am. Vet. Med. Assoc.,* 152, 1517, 1968.

34. **Thrasher, J. P., Gould, K. G., Lynch, M. J., and Harris, C. C.,** Filarial infections of dogs in Atlanta, Georgia, *J. Am. Vet. Med. Assoc.,* 153, 1059, 1968.

35. **Lindsey, J. R.,** Diagnosis of filarial infections in dogs. I. Microfilarial surveys, *J. Parasitol.,* 47, 695, 1961.

36. **Godfrey, W. D., Neely, W. A., Elliott, R. L., and Grogan, J. B.,** Canine heartworms in experimental cardiac and pulmonary surgery, *J. Surg. Res.,* 6, 331, 1966.

37. **Ward, J. W. and Franklin, M. A.,** Further studies on the occurrence of the dog heartworm, *Dirofilaria immitis,* in dogs in Mississippi, *J. Parasitol.,* 39, 570, 1953.

38. **Thrasher, J. P., Ash, L. R., and Little, M. D.,** Filarial infection of dogs in New Orleans, *J. Am. Vet. Med Assoc.,* 143, 605, 1963.

39. **Crowell, W. A., Klei, T. R., Hall, D. I., Smith, N. K., and Newsom, J. D.,** Occurrence of *Dirofilaria immitis* and associated pathology in coyotes and foxes from Louisiana, in *Proceedings of the Heartworm Symposium '77,* Otto, G. F., Ed., Veterinary Medicine Publ., Bonner Springs, Kan., 1978, 10.

40. **Custer, J. W. and Pence, D. B.,** Ecological analyses of helminth populations of wild canids from the gulf coastal prairies of Texas and Louisiana, *J. Parasitol.,* 67, 289, 1981.

41. **Eyles, D. E., Gibson, C. L., Jones, F. E., and Cunningham, M. E. G.,** Prevalence of *Dirofilaria immitis* in Memphis, Tennessee, *J. Parasitol.,* 40, 216, 1954.

42. **Marquardt, W. C. and Fabian, W. E.,** Distribution in Illinois of filariids of dogs, *J. Parasitol.,* 52, 318, 1966.

43. **Noyes, J. D.,** Illinois state veterinary medical association six-year heartworm survey, in *Proceedings of the Heartworm Symposium '77,* Otto, G. F., Ed., Veterinary Medicine Publ., Bonner Springs, Kan., 1978, 1.

44. **Kazacos, K. R.,** The prevalence of heartworms *(Dirofilaria immitis)* in dogs from Indiana, *J. Parasitol.,* 64, 959, 1978.

45. **Streitel, R. H., Stromberg, P. C., and Dubey, J. P.,** Prevalence of *Dirofilaria immitis* infection in dogs from a humane shelter in Ohio, *J. Am. Vet. Med. Assoc.,* 170, 720, 1977.
46. **Rabalais, F. C. and Votava, C. L.,** Canine filariasis in northwestern Ohio, *J. Am. Vet. Med. Assoc.,* 160, 202, 1972.
47. **Zydeck, F. A., Chodkowski, I., and Bennett, R. R.,** Incidence of microfilariasis in dogs in Detroit, Michigan, *J. Am. Vet. Med. Assoc.,* 156, 890, 1970.
48. **Prouty, D. L.,** Canine heartworm disease in southeastern Michigan, *J. Am. Vet. Med. Assoc.,* 161, 1675, 1972.
49. **Schlotthauer, J. C. and Griffiths, H. J.,** Canine filariasis in Minnesota, *J. Am. Vet. Med. Assoc.,* 144, 991, 1964.
50. **Pratt, S. E. and Corwin, R. M.,** *Dirofilaria immitis* and *Dipetalonema reconditum* in dogs in Nebraska and Missouri (1981), *Vet. Med. Small Anim. Clin.,* 79, 180, 1984.
51. **Nzabanita, E., Priester, W., and Farver, T.,** Distribution of canine filariasis, *Calif. Vet.,* 36, 24, 1982.
52. **Graham, J. M.,** Canine filariasis in Northeastern Kansas, *J. Parasitol.,* 60, 322, 1974.
53. **Sears, B. W., McCallister, G. L., and Heideman, J. C.,** *Dirofilaria immitis* in west central Colorado, *J. Parasitol.,* 66, 1070, 1980.
54. **Roberts, I. M. and Roberts, S. R.,** Canine filariasis — a report, *J. Am. Vet. Med. Assoc.,* 109, 490, 1946.
55. **McGreevy, P. B., Conrad, R. D., Bulgin, M. S., and Stitzel, K. A.,** Canine filariasis in northern California, *Am. J. Vet. Res.,* 31, 1325, 1970.
56. **Walters, L. L., Lavoipierre, M. M. J., and Kimsey, R. B.,** Endemicity of *Dirofilaria immitis* and *Dipetalonema reconditum* in dogs of Pleasants Valley, northern California, *Am. J. Vet. Res.,* 42, 151, 1981.
57. **Acevedo, R. and Theis, J. H.,** Report of a preliminary survey in 9 counties of California to assess the prevalence of heartworm in dogs, *Calif. Vet.,* 34, 15, 1980.
58. **Walters, L. L. and Lavoipierre, M. M. J.,** Landscape epidemiology of mosquito-borne canine heartworm *(Dirofilaria immitis)* in northern California, USA. I. Community based survey of domestic dogs in three landscapes, *J. Med. Entomol.,* 21, 1, 1984.
59. **Walters, L. L. and Lavoipierre, M. M. J.,** *Aedes vexans* and *Aedes sierrensis* (Diptera: Culicidae): potential vectors of *Dirofilaria immitis* in Tehama County, northern California, USA, *J. Med. Entomol.,* 19, 15, 1982.
60. **Acevedo, R. A. and Theis, J. H.,** Prevalence of heartworm *(Dirofilaria immitis)* in coyotes from five northern California counties, *Am. J. Trop. Med. Hyg.,* 31, 968, 1982.
61. **Corselli, N. J. and Platzer, E. G.,** Canine heartworm infection in southern California, *J. Am. Vet. Med. Assoc.,* 178, 1278, 1981.
62. **Sengbusch, H., Sartori, P., and Wade, S.,** Prevalence of *Dirofilaria immitis* in a stray dog population in western New York, *Am. J. Vet Res.,* 36, 1035, 1975.
63. **Georgi, J. R. and Cupp, E. W.,** *Dirofilaria immitis* survey in New York State, *Cornell Vet.,* 65, 286, 1975.
64. **Ash, L. R.,** Helminth parasites of dogs and cats in Hawaii, *J. Parasitol.,* 48, 63, 1962.
65. **Gubler, D. J.,** A comparative study on the distribution, incidence and periodicity of the canine filarial worms *Dirofilaria immitis* Leidy and *Dipetalonema reconditum* Grassi in Hawaii, *J. Med. Entomol.,* 3, 159, 1966.
66. **Myahara, A., Chung, N. Y., and Chung, G.,** Increasing incidence of canine heartworm disease on Oahu, *Vet. Med. Small Anim. Clin.,* 71, 1429, 1976.
67. **Otto, G. F.,** Changing geographic distribution of heartworm disease in the United States, in *Proceedings of the Heartworm Symposium '74,* Morgan, H. C., Ed., Veterinary Medicine Publ., Bonner Springs, Kan., 1975, 1.
68. **Williams, J. F., Williams, C. S. F., Signs, M., and Hokama, L.,** Evaluation of a polycarbonate filter for detection of microfilaremia in dogs in central Michigan, *J. Am. Vet. Med. Assoc.,* 170, 714, 1977.
69. **Panday, R. S., Lieuw, A Joe, R. G. H. M., Moll, K. F. G., and Oemrawsing, I.,** *Dirofilaria* in dogs of Surinam, *Vet. Q.,* 3, 25, 1981.
70. **Jackson, R. F., Redington, B. C., Seymour, W. G., and Otto, G. F.,** A filarial parasite in greyhounds in southern Ireland, *Vet. Rec.,* 97, 476, 1975.
71. **Thomas, R. E.,** A case of canine heartworm disease *(Dirofilaria immitis)* in the U.K., *Vet. Rec.,* 117, 114, 1985.
72. **Murdoch, D. B.,** Heartworm in the United Kingdom, *J. Small Anim. Pract.,* 25, 299, 1984.
73. **Rajamanickam, C., Wiesenhutter, E., Zin, F. Md., and Hamid, J.,** The incidence of canine haematozoa in peninsular Malaysia, *Vet. Parasitol.,* 17, 151, 1985.
74. **Kan, S. P., Rajah, K. V., and Dissanaike, A. S.,** Survey of dirofilariasis among dogs in Seremban, Malaysia, *Vet. Parasitol.,* 3, 177, 1977.

75. **Macadam, I., Gudan, D., Timbs, D. V., Urquhart, H. R., and Sewell, M. M. H.,** Metazoan parasites of dogs in Sabah, Malaysia, *Trop. Anim. Health Prod.,* 16, 34, 1984.

76. **Rosen, L.,** Observations on *Dirofilaria immitis* in French Oceania, *Ann. Trop. Med. Parasitol.,* 48, 318, 1954.

77. **Mizuno, F., Higashio, T., and Matsumura, T.,** A survey of canine *Dirofilaria* invasion in a suburb of Tokyo (Nerima Ward), *Kobe J. Med. Sci.,* 27, 113, 1981.

78. **Welch, J. S., Dobson, C., and Freeman, C.,** Distribution and diagnosis of dirofilariasis and toxocariasis in Australia, *Aust. Vet. J.,* 55, 265, 1979.

79. **Carlisle, C. H.,** The incidence of *Dirofilaria immitis* (heartworm) in dogs in Queensland, *Aust. Vet. J.,* 45, 535, 1969.

80. **Aubry, J. N. and Copeman, D. B.,** Canine dirofilariasis — an evaluation of bimonthly diethylcarbamazine therapy in prophylaxis, *Aust Vet. J.,* 48, 310, 1972.

81. **Welch, J. S. and Dobson, C.,** The prevalence of antibodies to *Dirofilaria immitis* in aboriginal and caucasian Australians, *Trans. R. Soc. Trop. Med. Hyg.,* 68, 466, 1974.

82. **Atwell, R. B. and Carlisle, C.,** Canine dirofilariasis in the metropolitan area of Brisbane, *Aust. Vet. J.,* 55, 399, 1979.

83. **Rees, D., Sharrock, A. G., and Lillecrapp, J. A.,** Canine dirofilariasis in the Murrumbidgee irrigation area, *Aust. Vet. J.,* 42, 61, 1966.

84. **Whitlock, L. E.,** The incidence of microfilaria in canine blood in Sydney, *Aust. Vet. J.,* 45, 136, 1969.

85. **Watson, A. D. J., Porges, W. L., and Testoni, F. J.,** A survey of canine filariasis in Sydney, *Aust. Vet. J.,* 49, 31, 1973.

86. **Mulley, R. C. and Starr, T. W.,** *Dirofilaria immitis* in red foxes *(Vulpes vulpes)* in an endemic area near Sydney, Australia, *J. Wildl. Dis.,* 20, 152, 1984.

87. **Holmes, P. R. and Kelly, J. D.,** The incidence of *Dirofilaria immitis* and *Dipetalonema reconditum* in dogs and cats in Sydney, *Aust. Vet. J.,* 49, 55, 1973.

88. **Gee, R. W. and Auty, J. H.,** The heartworm *Dirofilaria immitis* in Victoria. An unusual cause of death in a dog, *Aust. Vet. J.,* 33, 152, 1957.

89. **Blake, R. T. and Overend, D. J.,** The prevalence of *Dirofilaria immitis* and other parasites in urban pound dogs in north-eastern Victoria, *Aust. Vet. J.,* 58, 111, 1982.

90. **Carlisle, C. H. and Atwell, R. B.,** A survey of heartworm in dogs in Australia, *Aust. Vet. J.,* 61, 356, 1984.

91. **Pullar, E. M.,** A survey of Victorian canine and vulpine parasites. IV. Nematoda, *Aust. Vet. J.,* 22, 85, 1946.

92. **Boreham, P. F. L. and Atwell, R. B.,** *Dipetalonema reconditum* in dogs with microfilaraemia, *Aust. Vet. J.,* 62, 27, 1985.

93. **Otto, G. F.,** Epizootiology of canine heartworm disease, in *Canine Heartworm Disease: The Current Knowledge,* Bradley, R. E., Ed., University of Florida, Gainesville, Fla., 1972, 1.

94. **Pacheco, G.,** Synopsis of Dr. Seiji Kume's reports at the first international symposium on canine heartworm disease, in *Canine Heartworm Disease: The Current Knowledge,* Bradley, R. E., Ed., University of Florida, Gainesville, Fla., 1972, 137.

95. **Grieve, R. B., Lok, J. B., and Glickman, L. T.,** Epidemiology of canine heartworm infection, *Epidemiol. Rev.,* 5, 220, 1983.

96. **Hendrix, C. M., Schlotthauer, J. C., Bemrick, W. J., and Averback, G. A.,** Temperature effects on development of *Dirofilaria immitis* in *Aedes vexans,* in *Proceedings of the Heartworm Symposium '80,* Otto, G. F., Ed., Veterinary Medicine Publ., Edwardsville, Kan., 1981, 9.

97. **Fortin, J. F. and Slocombe, J. O. D.,** Survival of *Dirofilaria immitis* in *Aedes triseriatus* exposed to low temperatures, in *Proceedings of the Heartworm Symposium '80,* Otto, G. F., Ed., Veterinary Medicine Publ., Edwardsville, Kan., 1981, 13.

98. **Grassi, B. and Calandruccio, S.,** Uber Hematozoon Lewis: Entwicklungscyklus einer Filaria (Filaria recondita Grassi) des Hundes, *Cent. Bakteriol. Parasitol.,* 3, 18, 1890.

99. **Newton, W. L. and Wright, W. H.,** The occurrence of a dog filariid other than *Dirofilaria immitis* in the United States, *J. Parasitol.,* 42, 246, 1956.

100. **Acevedo, R. A., Theis, J. H., Kraus, J. F., and Longhurst, W. M.,** Combination of filtration and histochemical stain for detection and differentiation of *Dirofilaria immitis* and *Dipetalonema reconditum* in the dog, *Am. J. Vet. Res.,* 42, 537, 1981.

101. **Orihel, T. C. and Ash, L. R.,** Occurrence of *Dirofilaria striata* in the bobcat *(Lynx rufus)* in Louisiana with observations on its larval development, *J. Parasitol.,* 50, 590, 1964.

102. **Pacheco, G. and Tulloch, G. S.,** Microfilariae of *Dirofilaria striata* in a dog, *J. Parasitol.,* 56, 248, 1970.

103. **Otto, G. F. and Jachowski, L. A., Jr.,** Mosquitoes and canine heartworm disease, in *Proceedings of the Heartworm Symposium '83,* Otto, G. F., Ed., Veterinary Medicine Publ., Edwardsville, Kan., 1983, 17.

104. **Ludlam, K. W., Jachowski, L. A., Jr., and Otto, G. F.**, Potential vectors of *Dirofilaria immitis, J. Am. Vet. Med. Assoc.*, 157, 1354, 1970.
105. **Horsfall, W. R.**, *Mosquitoes: Their Bionomics and Relation to Disease*, Hafner, New York, 1972.
106. **Bemrick, W. J. and Sandholm, H. A.**, *Aedes vexans* and other potential mosquito vectors of *Dirofilaria immitis* in Minnesota, *J. Parasitol.*, 52, 762, 1966.
107. **Ernst, J. and Slocombe, J. O. D.**, Mosquito vectors of *Dirofilaria immitis* in southwestern Ontario, *Can. J. Zool.*, 62, 212, 1984.
108. **Pinger, R. R.**, Species composition and feeding success of mosquitoes attracted to caged dogs in Indiana, *J. Am. Mosq. Control Assoc.*, 1, 181, 1985.
109. **Lewandowski, H. B., Hooper, G. R., and Newson, H. D.**, Determination of some important natural potential vectors of dog heartworm in central Michigan, *Mosq. News*, 40, 73, 1980.
110. **Buxton, B. A. and Mullen, G. R.**, Field isolations of *Dirofilaria* from mosquitoes in Alabama, *J. Parasitol.*, 66, 140, 1980.
111. **Tolbert, R. H. and Johnson, W. E.**, Potential vectors of *Dirofilaria immitis* in Macon County, Alabama, *Am. J. Vet. Res.*, 43, 2054, 1982.
112. **Magnarelli, L. A.**, Presumed *Dirofilaria immitis* infections in natural mosquito populations of Connecticut, *J. Med. Entomol.*, 15, 84, 1978.
113. **Crans, W. J. and Feldlaufer, M. F.**, The likely vectors of dog heartworm in a coastal area of southern New Jersey, *Proc. Calif. Mosq. Control Assoc.*, 42, 168, 1974.
114. **Todaro, W. S., Morris, L. D., and Haecock, N. A.**, *Dirofilaria immitis* and its potential mosquito vectors in central New York State, *Am. J. Vet. Res.*, 38, 1197, 1977.
115. **Villavaso, E. J. and Steelman, C. D.**, Laboratory and field studies of the southern house mosquito, *Culex pipiens quinquefasciatus* Say, infected with the dog heartworm, *Dirofilaria immitis*, in Louisiana, *J. Med. Entomol.*, 4, 411, 1970.
116. **Sauerman, D. M., Jr. and Nayar, J. K.**, A survey for natural potential vectors of *Dirofilaria immitis* in Vero Beach, Florida, *Mosq. News*, 43, 222, 1983.
117. **Christensen, B. M. and Andrews, W. N.**, Natural infections of *Aedes trivittatus* (Coq.) with *Dirofilaria immitis* in central Iowa, *J. Parasitol.*, 62, 276, 1976.
118. **Hribar, L. J. and Gerhardt, R. R.**, Wild-caught *Aedes trivittatus* naturally infected with filarial worms in Knox County, Tennessee, *J. Am. Mosq. Control Assoc.*, 1, 250, 1985.
119. **Weinman, C. J. and Garcia, R.**, Canine heartworm in California, with observations on *Aedes sierrensis* as a potential vector, *Calif. Vector Views*, 21, 45, 1974.
120. **Bemrick, W. J. and Moorhouse, D. E.**, Potential vectors of *Dirofilaria immitis* in Brisbane area of Queensland, Australia, *J. Med. Entomol.*, 5, 269, 1968.
121. **Russell, R. C.**, Report of a field study on mosquito (Diptera: Culicidae) vectors of dog heartworm, *Dirofilaria immitis* Leidy (Spirurida:Onchocercidae) near Sydney, N.S.W., and the implications for veterinary and public health concern, *Aust. J. Zool.*, 33, 461, 1985.
122. **Symes, C. B.**, A note on *Dirofilaria immitis* and its vectors in Fiji, *J. Helminthol.*, 34, 39, 1960.
123. **Keegan, H. L., Betchley, W. W., Haberkorn, T. B., Nakasone, A. Y., Sugiyama, H., and Warne, R. J.**, Laboratory and field studies of some entomological aspects of the canine dirofilariasis problem in Japan, *Jpn. J. Sanit. Zool.*, 18, 6, 1967.
124. **Del Rosario, F.**, *Dirofilaria immitis* Leidy and its culicine intermediate hosts in Manila, *Philipp. J. Sci.*, 60, 45, 1936.
125. **Intermill, R. W. and Frederick, R. M.**, A study of potential mosquito vectors of *Dirofilaria immitis* Leidy on Okinawa, Ryukyu Islands, *J. Med. Entomol.*, 7, 455, 1970.

Chapter 2

BIOLOGY OF *DIROFILARIA IMMITIS*

David Abraham

TABLE OF CONTENTS

I. INTRODUCTION

The biology of *Dirofilaria immitis* must be studied on two levels to understand the requirements for the parasite's survival. The first approach is to study the biology of individual organisms, the second is to observe the dynamics of the total population. The general outline for this discussion on the biology of *D. immitis* will be to first describe the development and migrations of *D. immitis* worms in the dog and mosquito. Then, biological adaptions developed to promote parasite transmission and survival will be reviewed. Finally, experimental systems developed to study the organismal and population biology of *D. immitis* will be discussed.

II. LIFE CYCLE AND DEVELOPMENTAL STAGES

A. Adult Worms

Adult heartworms, first noted by Leidy[1,2] in 1850, were described as "the cruel thread worms".[3] The predilection site of the adult worm is the right ventricle and adjacent blood vessels including pulmonary artery, right atrium, and vena cavae, though there have been numerous ectopic findings.[4] Male worms have a mean length of 16 cm and females have a mean of 25 cm.[5] The anterior end of *D. immitis* is unadorned but does contain amphids and cephalic papillae. The male tail is coiled (Figure 1), has small lateral alae, and has several pairs of papillae located pre- and postcloacally. Two spicules of dissimilar length and shape are also present.[6,9] The somatic musculature of *D. immitis* has morphological features of an obliquely striated muscle.[10] The digestive tract consists of a single tube divided into the esophagus anteriorly and the intestine posteriorly. The intestinal cells are simple, with only a few rudimentary mitochondria, as may be expected in a blood feeding anaerobe.[11] The male genital system consists of a looped testes in the anterior body connected to a single tube which proceeds posteriorly as the seminal vesicle and terminates as the vas deferens and ejaculatory duct in the cloaca.[5,6,12] The female genital system consists of two sets of sex organs. The dual ovaries are located posteriorly. They connect through oviducts with the seminal receptacles which then connect to the uteri. The uteri join to form a single vagina which turns into the vulva, at the region of the esophagus.[5,6,12]

Spermatozoa are amoeboid and contain four or five chromosomes, the former presumably giving rise to males and the latter to females. Adult males therefore have nine chromosomes and adult females have ten.[13] Sperm is stored in the seminal receptacle of the female worm. There is experimental evidence that sperm remain viable in the seminal receptacle for several weeks before there is a need for reinsemination.[14] Fertilization of eggs occurs in the seminal receptacle and involves the penetration of the egg by the entire spermatozoon. The nuclear materials of *D. immitis* sperm, as well as other sperm contents, then dissolve into the ooplasm. As no pronuclei are seen, it is probable that the male genome enters the female nucleus in the form of ribosome-like granules, thus achieving genetic recombination without fusion of pronuclei.[15] Development is then ovoviviparous, leading to the production of microfilariae which are released into the bloodstream of the host.

B. Microfilariae

Microfilariae measure approximately 300 μm long and reside in the peripheral blood, as well as in various organs, especially the lungs.[16] They are capable of extravascular migration as evidenced by the fact that they are capable of passing through the placenta and infecting the pups.[17-19] Ultrastructural examination of *D. immitis* microfilariae has revealed a well-developed cuticle, hypodermis, and muscle cells, comparable to that found in the adult.[20-23] Additionally, a carbohydrate-rich surface-coat matrix has been seen on the external surface of the cuticle of *D. immitis* microfilariae.[24] Structures specific to microfilariae have

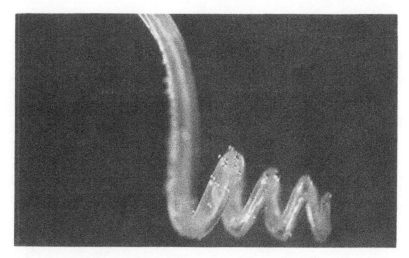

FIGURE 1. Posterior end of adult male *D. immitis.*

also been identified. A wedge-shaped hook and lip-like processes are present at the anterior end. At the narrow part of the hook a sharp stylet is found. The composition of the hook is mostly muscular and tubular attachments, whereas the stylet appears to consist of cuticular material. Posterior, there is a nerve ring consisting of two cells from which originate two or more masses of axons which extend to the anterior and posterior regions of the microfilaria. The excretory complex is found posterior to the nerve ring and consists of a cell and a vesicle with a connecting cytoplasmic bridge. The pore of the excretory vesicle, which is on the surface of the worm, is covered by one or more lip-like cuticular modifications. The anal vesicle is a cavity which occupies the ventral two thirds of the body. This vesicle is filled with microvilli and has an opening to the outside which is closed by two small cuticular lips.[21-23] Neurosecretory cells have been seen associated with the nerve ring and excretory and anal vesicles, as well as the dorsal and ventral nerve trunks.[25] In the cephalic space and in the region posterior to the anal vesicle, microfilariae have paired, cuticularized channels. Each of the cephalic channels contain eight or nine cilia, whereas only a single cilia is found in the caudal channels. These cilia lack basal bodies and have microtubular patterns ranging from $1 + 11 + 4$ to $9 + 2$. They are apparently associated with elements of the nervous system and may therefore be sensory structures.[26] It should be noted that microfilariae lack a digestive system, i.e., there is no esophagus or intestine. In addition no genital primordium has been identified in microfilariae.

C. Development in Mosquito

Microfilariae must be ingested by a mosquito for development to continue. Female mosquitoes, in the process of taking a blood meal, ingest microfilariae. The microfilariae remain in the mid-gut of the mosquito for approximately 24 hr. Larvae at this stage appear almost identical to microfilariae found in the blood of a dog. During the next 24 hr the larvae migrate into the malpighian tubules where they reside in the cytoplasm of the primary cells[27] (Figure 2). The larvae then become shorter and stouter as they develop into the "sausage" stage. By the fifth day, a gut may be differentiated in the larvae, consisting of an esophagus, intestine, and rectum. On day six or seven, larvae leave the cells of the malpighian tubules and enter the lumen of the tubules. The first molt occurs on approximately day ten, with the sausage-stage larvae becoming second-stage larvae. The second molt in the mosquito occurs an approximately day thirteen. The larvae have thus developed into infective third-stage larvae. They now measure about 1300 μm in length, having increased their size by fourfold while in the mosquito. Infective larvae then migrate through the mosquito where most come to lie in the cephalic spaces of the head or in the proboscis[28] (Figure 3).

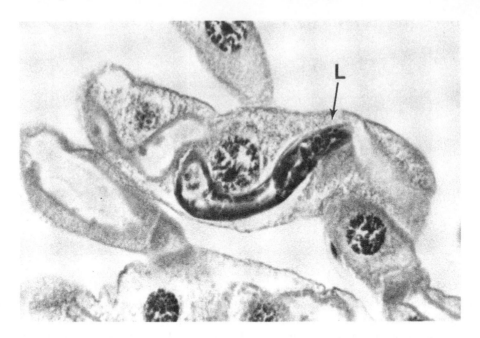

FIGURE 2.   Larval (L) *D. immitis* within primary cell of mosquito malpighian tubule, 4 days postinfection.

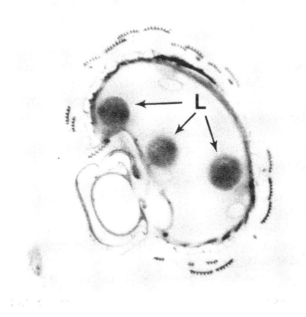

FIGURE 3.   Cross section of mosquito proboscis showing third-stage infective larvae (L) of *D. immitis* within.

The rates at which larvae survive, develop, and migrate within the mosquito are dependent on several factors. Genetic differences are seen in the capacity of mosquitoes to support *D. immitis*. Different species and strains of mosquitoes have different rates of larval *D. immitis* development.[29,30] The temperature at which mosquitoes are maintained is another important factor with larvae developing faster at warmer temperatures. The reverse is also true, to the extent that at low temperatures in the range of 15°C, larval development in the mosquito is arrested. However, when these mosquitoes are removed from the cold environment and are placed at a warm temperature (26.5°C) the larvae complete their development. This finding suggests that larvae within mosquitoes have the capability of overwintering in cold climates.[31,32] Another important factor is the presence and condition of the blood meal. Mosquitoes can be infected with microfilariae that have been removed from the blood. These microfilariae develop into infective larvae but in low numbers. This is due to a discharge of microfilariae from the mid-gut prior to their penetration into the malpighian tubules. It was hypothesized that the blood in which microfilariae are normally found hinders the discharge of microfilariae from the mosquito.[33] Mosquitoes lacking in anticoagulants are poor hosts for *D. immitis* because the clotted blood prevents the microfilariae from moving freely out of the mid-gut and into the malpighian tubules.[34] The viscosity of the blood in which microfilariae are bathed in the mid-gut of the mosquito must therefore be dense enough to prevent microfilarial discharge, but fluid enough to allow microfilarial migration.

In susceptible mosquito hosts and at optimal environmental conditions, developing larvae are still pathogenic to the mosquito. The pathogenicity is dependent upon the presence of live microfilariae, as dead worms do not increase the mosquitoes' mortality.[35] The times of increased mortality in infected mosquitoes coincides with specific aspects of the parasite's life cycle. The first period when the increase occurs coincides with the time microfilariae are migrating from the mid-gut to the malpighian tubule; a 32 × increase in the rate of infected mosquito deaths was seen at this time. While the larvae are in the malpighian tubules there is no difference between infected and uninfected mosquitoes mortality rates. However, as the third-stage larvae migrate into the head and mouthparts the mortality rates in these mosquitoes increased again.[36]

D. Development in Dog

Third-stage larvae leave the mosquito while the mosquito is ingesting a blood meal. The infective larvae emerge from the tip of the labellum or in few instances from the mid-portion of the labium (Figure 4). The stimulus for larvae to escape from the mosquito appears to be the bending of the labium during feeding or probing. A drop of hemolymph is released from the mosquito with the larvae. Larvae are thereby maintained in a moist environment while they search for a means of entering the dog's skin. After the mosquito has finished its blood meal and the fasicle is removed from the puncture wound, the larvae can then use this wound site as their portal of entry into the dog.[37,38]

Following successful penetration into the dog's skin the larvae must find their way to the adult worms' predilection site. Three days after leaving the mosquito, almost all larvae are still found in the subcutaneous tissues near the site of their entry into the skin. By day 21 most larvae have migrated to the abdomen of the dog, and by day 41 larvae can be recovered from either the abdomen or thorax. Worms first reach the dog's heart on day 70 to 85 and all worms will have entered the heart by day 90 to 120. Third- and fourth-stage larvae appear to travel between muscle fibers during their migration, whereas fifth-stage immature adults appear to penetrate muscle. The immature adults may use this enhanced migratory ability to travel to the upper abdomen, thorax, head, neck, and forelimbs, where they penetrate the jugular or other veins, thus leading them directly to the heart. There has been some experimental evidence demonstrating that worms enter the heart via veins, and not by direct tissue penetration of the heart, nor by migration through lymphatic tissue.[5,39,40]

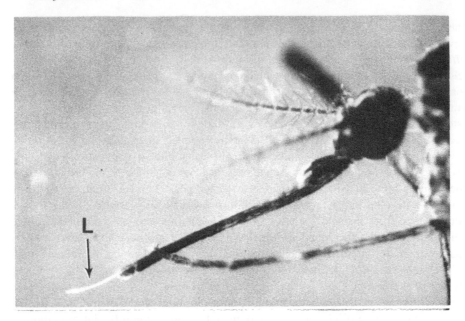

FIGURE 4.   Third-stage infective larva (L) of *D. immitis* leaving mosquito proboscis.

D. immitis third-stage larvae measuring approximately 1.2 mm enter the dog. They must then molt twice and increase in size more than 100-fold to develop into mature adults. The anterior end of the third-stage larva has a slit-like mouth surrounded by a 4-2-4 configuration of papillae. The cuticle has numerous transverse striations, with the exception of the terminal 2.5 μm of the anterior end.[41] Sex may be differentiated at this stage by locating the genital primordium. In the male the genital primordium is located ventrally just anterior to the mid-body. The female genital primordium is found just posterior to the junction of the muscular and glandular esophagus.[5,42] The molt from third- to fourth-stage larvae begins, according to one study, prior to day three[40,42,43] and, according to another, between day 9 to 12.[5] The most evident developmental changes seen in fourth-stage larvae, aside form general growth of the worm, are related to the development of the reproductive systems. By day 70 all fourth-stage larvae have molted to the fifth stage and have attained a length of approximately 18 mm. Thus, the first worms entering the dog's heart, on day 70 to 85 are 2 to 4 cm in length. Worms reaching the heart will then increase in length by almost tenfold by the time sexual maturity is reached.[5,40,42,43] Ova are seen in the distal portions of the ovaries of female worms at 100 days postinfections. Inseminated females are found by day 120 postinfection. Fully developed microfilariae are seen in the uteri and vagina of female worms during the 6th month postinfection,[5] and dogs develop patent microfilaremias by 7 to 9 months post-infection.[5,40,44-46] The life cycle is summarized in Table 1.

II. PARASITE POPULATION CONTROL

A. Adult Worm Burden

A dog receiving a single *D. immitis* infection will retain a patent microfilaremia for up to $7^{1}/_{2}$ years.[47] The life expectancy of a microfilaria in a dog is as long as $2^{1}/_{2}$ years.[48] Combining the results from these two studies shows that adults and microfilariae of *D. immitis* can be very long lived. The fact that *D. immitis* infections persist for extended time periods and that dogs in endemic areas are subject to repeated infection should lead to a situation where dogs in endemic areas suffer from massive worm burdens. This, however, is not the case. Mean recoveries of 7,[49] 12.2,[50] and 15.2[51] adult worms per infected dog

Table 1
DEVELOPMENT OF MICROFILARIA TO ADULT FEMALE WORM[28,40]

Day	Stage	Length (cm)	Host	Location
0	Microfilaria	0.03	Dog	Blood
1	Microfilaria	0.03	Mosquito	Mid-gut
5	Sausage stage	0.015	Mosquito	Malpighian tubule cells
10	Second-stage larva	0.05	Mosquito	Malpighian tubule lumen
15	Third-stage larva	0.12	Mosquito	Proboscis
15(0)[a]	Third-stage larva	0.12	Dog	Subcutaneous tissue
18(3)[a]	Fourth-stage larva	0.12	Dog	Subcutaneous tissue
85(70)[a]	Fifth-stage adult	2.4	Dog	Head, neck forelimbs, thorax, abdomen
115(100)[a]	Fifth-stage adult	5.9	Dog	Heart
211(196)[a]	Fifth-stage gravid adult	26.8	Dog	Heart

[a]   Number in parentheses is age in dog.

found in endemic areas have been reported. Experimental infections of dogs with *D. immitis* larvae yield only approximately 40% of injected larvae as adults.[5,40,52-54] One may conclude from the results obtained from natural and experimental infections of dogs with *D. immitis* that there are mechanisms controlling and thus limiting the development of larvae into adults. The mechanisms involved, be they immunological, physiological, or behavioral, remain unclear.

B. Control of Microfilaremia

Another area where parasite control is essential is the production and circulation of microfilariae. As mentioned above, microfilariae may circulate in the dog for up to $2^1/_2$ years.[48] If microfilariae are produced at a constant rate and survive for several years, increasing numbers of microfilariae should be found in the peripheral circulation of the dog. Mosquitoes feeding on highly parasitized blood would die as has been shown experimentally[55,56] and dogs would suffer from having their blood vessels plugged by the microfilariae. Microfilaremias found in naturally or experimentally infected dogs do not, however, increase throughout the infection nor is there even a relationship between numbers of circulating microfilariae and numbers of adults worms found in a dog.[57-59]

Investigations have been conducted to determine what mechanisms might control the level of peripheral microfilaremia found in dogs infected with *D. immitis*. Microfilaremic blood was transfused out of a dog infected with *D. immitis* and an equal volume of uninfected blood was used to replace it. There was no resultant decrease in the peripheral microfilaremia of the infected dog. When the same procedure was done on a dog which had no adult worms but did have microfilariae in the circulation from a transfusion, again no significant change in peripheral microfilaremia was observed after the exchange transfusion. Furthermore, it was found that only 7% of microfilariae transfused into a dog were found in the peripheral circulation. It can therefore be concluded that adult worms play a minor role in control of microfilaremias. A reservoir of microfilariae exists in the dog that is capable of replenishing microfilariae lost from the blood. A homeostatic mechanism exists that maintains the peripheral levels of microfilaremia, not necessarily by affecting the female worms parturition rate, but by controlling the release of microfilariae from organ reservoirs.[60,61]

An important qualification must be made to the concept of stable peripheral microfilaremias in *D. immitis* infections. Microfilaremias do not expand uncontrollably, nor do they vary greatly day to day, but there is a great deal of variation in microfilaremia levels in individual dogs during the day. There is little agreement in the literature on when peak microfilaremias occur. It may be generalized that the maximum levels of microfilaremia occur from late afternoon through late evening. This type of periodicity is classified as subperiodic, because 5 to 20% of microfilariae still circulate when minimum microfilaremia levels are reached.[16,45,62-70]

Efforts have been made to determine the mechanisms controlling microfilarial periodicity. Initially, microfilariae were thought to be released synchronously daily and that they only lived for a few hours. The microfilaremia peaks thus correlated with worm parturition and the troughs correlated with microfilarial death. This theory was disproved by transfusing microfilariae into a dog and showing that worms survived for extended time periods and were capable of showing a periodicity.[62] Further studies demonstrated that the periodicity was related to the behavior of the dog and to the emotional or environmental stresses to which the dog was subjected. The periodic cycle could be reversed by reversing the light-dark cycle in the dog's environment.[64,69] Microfilariae have been shown to be sequestered in the small vessels of the lungs during off-peak time periods. Day-night change in the oxygen tension found in lungs may serve as the stimulus for microfilarial sequestration and release.[16]

In addition to the daily microfilarial periodicity, a seasonal periodicity has been observed for the microfilariae of *D. immitis*. Microfilariae are more abundant in the circulation during the spring and summer than during the fall and winter.[47,71,72] Seasonal periodicity is not dependent on adult worms as shown by the presence of seasonal periodicity in dogs whose microfilariae were derived from a transfusion. Furthermore, seasonal periodicity does not appear to be dependent on mosquito bites or the inoculation of third-stage larvae as stimuli. This was demonstrated by transplanting adult worms into puppies protected from contact with mosquitoes. The microfilaremias that developed in these puppies displayed seasonal periodicities. The source of control therefore appears to be the microfilariae and/or the host. There is some evidence that temperature changes may stimulate alterations in microfilaremias. When dogs are placed in heated rooms during the winter, an increase in microfilaremia levels was noted. If, however, dogs were placed in a temperature-controlled environment during the summer, microfilaremias in these dogs decreased even though the temperature at which the dogs were kept did not. Ambient temperature therefore seems to play a role in maintaining seasonal periodicities, but is not solely responsible.[71]

A coincidence exists between the time microfilariae are most abundant in the peripheral blood and the time the mosquito vector bites. Periodicity, whether daily or seasonal, may be an evolutionary adaption placing microfilariae in a location accessible to biting mosquitoes only during the time mosquitoes are feeding. The optimal location for microfilariae to be may be in the visceral blood, and microfilariae only leave these preferred sites and enter the peripheral circulation at times when there is the possibility of transmission by mosquitoes.[16]

An extreme example of control over peripheral microfilaremias is found in dogs with occult infections. These are infections where adult worms are found in a dog but no microfilariae circulate in the blood. Four sources of occult infections have been described. These include prepatent infections, unisexual infections, drug-induced sterility of adult heartworms, and immune-mediated infections. Occult infections caused by the immune response may be viewed from the parasite's perspective as failure in the host-parasite relationship. The host has controlled the peripheral microfilaremia to the degree that the infection can no longer be transmitted.[73-75]

Mosquitoes are not totally dependent on the host and parasite to produce microfilaremia levels which are nonpathogenic to the mosquito. Feeding to repletion, mosquitoes ingest a little more than half the number of microfilariae as would be expected, based on microfilaremia levels and volume of blood ingested.[55] Mosquitoes thus have a means of limiting the number of microfilariae ingested, thereby diminishing the pathology associated with developing a filarial infection.

Table 2
ANIMALS REPORTED INFECTED WITH *D. IMMITIS*[4]

Animal	Species	Ref.[a]
Canines		
Timber or Gray wolf	*Canis lupus* and subspecies	
Red wolf	*Canis rufus*	
Maned wolf	*Chrysocyon brachyurus*	
Dingo	*Canis dingo*	
Coyote	*Canis latrans* and spp.	76—82
Red fox	*Vulpes vulpes* and spp.	80, 81, 86, 88, 89
Gray fox	*Urocyon cinereoargenteus*	80, 81, 86, 90, 91
Bengal fox	*Vulpes bengalensis*	
Pale fox	*Vulpes pallida*	
Racoon dog	*Nyctereutes procyonoides*	87
Felines		
Domestic cat	*Felis catus*	
Jaguar	*Panthera onca*	
Tiger	*Panthera tigris*	
Jaguarondi	*Felix yagouaroundi*	
Clouded leopard	*Neofelis neburosa*	
Wild cat	Listed as *Felis bangsi costaricensis*	
Other mammals		
California sea lion	*Zalophus californianus*	
Common seal	*Phoca vitulina*	94
Wolverine	*Gulo gulo*	95
Muskrat	*Ondatra zibethica*	
Racoon	*Procyon lotor*	
Beaver	*Castor canadensis*	96
Coati	*Nasua nasua*	
Ferret	*Mustela putorius*	97, 98
Otter	*Lutra* sp. listed as *L. brasiliense*	
Giant Brasilian otter	*Pteronura brasiliensis*	
Rabbit	*Oryctolagus cuniculus*	99
Red panda	*Ailurus fulgens*	100, 101
Black bear	*Urus americanus*	102, 103
Bear	Listed as *Ursus torguatus japonicum*	
Sika deer	*Cervus nippon*	
Horse	*Equus caballus*	104, 105
Rhesus monkey	*Macaca mulatta*	106
Orangutan	*Pongo pygmaeus*	
Gibbon	*Hylobates lar*	
Man	*Homo sapiens*	

Note: Where the nomenclature is uncertain the original name is retained.

[a] References for reports after 1974. For reports prior to 1974 see Reference 4.

C. Host Range

A large number of mammals have been found infected with adult *D. immitis* (Table 2). The majority of these reports was based on incidental post-mortem findings. Microfilariae were rarely seen and usually only a single adult or a few infertile worms were recovered in locations outside of the heart. There were a few animals, aside from dogs, which were observed with circulating *D. immitis* microfilariae. These include the red wolf,[76] coyote,[76-86] gray fox,[80] red fox,[80] cat,[4] California sea lion,[4,93] wolverine,[95] and ferret.[97] Of these animals, only the coyote appears to have any potential as a reservoir host. This is based on the percentage of infected individuals found within a population and the proximity of coyote habitats to that of domestic dogs. Whether coyotes actually play a significant role in maintaining heartworm infections in dogs remains a question.[76,80,84]

IV. EXPERIMENTAL MODELS

A. Experimental Infections

Since 1900, when Grassi and Noè showed that dogs could be infected with *D. immitis* by an infected mosquitoe's bite,[37] dogs and cats[107,108] have been experimentally infected to study heartworm infections. These studies have elucidated details on the life cycle, development, and parasite population dynamics of heartworm infections that could not be learned from naturally acquired infections. Other animals, aside from dogs and cats have also been experimentally infected with *D. immitis*.

Ferrets can be experimentally infected with *D. immitis*, but due to their small size, they can only tolerate small numbers of adult worms. The prepatent period for *D. immitis* in the ferret is 8 months,[109] as it is in the dog. The advantage of using ferrets for experimental studies is that they cost less to purchase and maintain than dogs, but still can maintain the complete life cycle. Additionally, they are susceptible to other filarial infections, thus allowing the study of effects of concomitant infections.[110]

Gibbons are also capable of supporting the development of *D. immitis* to the adult stage. Following infection with third-stage larvae, adult worms are found in the heart. The female worms contained microfilaria in their uteri, but no patent microfilaremia developed.[111] Monkeys (*Macaca fascicularis*, *M. arctoides*, and *M. mulatta*) are capable of supporting *D. immitis* for up to 2 months. However, if monkeys were immunosuppressed, adult worms were recovered from the heart.[112] Study of *D. immitis* infections in gibbons and monkeys, may give insight into the requirements for development of human infections with *D. immitis*.

Jirds (*Meriones unguiculatus*) have been used extensively in the study of a variety of filarial infections. When third-stage larvae of *D. immitis* were injected intraperitoneally into jirds, only 3% of the larvae could be recovered live after 16 to 44 days. However, in a single instance, a male and female worm were recovered alive from the right heart and pulmonary artery of a jird 132 days postinfection.[113] This finding suggests that *D. immitis* is capable of developing to adults in jirds; the question remains why is does not routinely occur.

Mice have also been used as experimental hosts for *D. immitis*. C57BL and ddy mice were infected with third-stage larvae of *D. immitis*. One week postinfection 17% of the larvae were recovered live. After 3 weeks only approximately 2% could be recovered. The growth rates of larvae recovered from mice were equivalent to the rates reported for worms recovered from dogs. Mice have also been used to study immunity to *D. immitis* and to test drug efficacies against *D. immitis*.[114] Using a mouse model for study of *D. immitis* infections has the advantage of permitting specific immunological and genetic studies to be done, which would presently be impossible to do in the previously described hosts. The major disadvantage, is that only a small segment of the life cycle can be maintained in mice.

An alternative approach used in the study of experimental *D. immitis* infections is the use of micropore chambers. Larvae are sequestered within a chamber which is implanted in an animal. Host cells and serum can enter the chamber, but the larvae cannot escape due to the pore size of the chamber's membranes. Larvae within chambers have been implanted into BALB/c mice, jirds, cotton rats, and ferrets. Three days after implantation into any of these hosts, 94% of implanted larvae were recovered live. At this time 74 to 87% of the third-stage larvae had molted to the fourth stage. Larvae implanted in the aforementioned hosts grew at equivalent rates, and when larvae recovered from mice after 8 days were compared to larvae recovered from dogs after 7 days no difference in lengths were noted.[115] Micropore chambers thus provide a useful tool for the study of early larval stages of *D. immitis* in a variety of hosts.

B. Transplanted Infections

Adult and microfilarial stages of *D. immitis* have been recovered from infected dogs and transplanted into uninfected animals. These experiments were performed to assess factors such as longevity of specific stages, fecundity of individual worms, and the role of different parasitic stages in controlling peripheral microfilaremias. Adult worms collected from a dog's heart were transplanted subcutaneously into a normal dog. These worms were capable of migrating out of the subcutaneous tissues and back to the heart. However, if worms were transplanted into the abdominal or thoracic cavities of a normal dog, the worms remained in the implantation site.[39] Adult worms have also been implanted into the jugular vein of normal dogs. Implantation of as little as a single female worm, yielded a stable and long-lasting microfilaremia.[116-118] These experiments show that worms recovered from a heart and transplanted subcutaneously could still migrate back to the heart. Worms implanted in the abdominal or thoracic cavities or in the jugular vein remain at the implantation site.

Adult *D. immitis* have also been transplanted into the peritoneal cavity of several different rodents.[119] Worms implanted intraperitoneally into Lewis rats lived for as long as 11 weeks. Microfilariae, however, were only detected in the circulation or in the peritoneal cavity for 3 weeks. Female worms, remaining in rats for longer than 3 weeks, were devoid of microfilariae.[14] These results suggest that the Lewis rat is a poor host for both adult and microfilarial *D. immitis*. Furthermore, these results show that reinsemination of female *D. immitis* worms is required approximately every 3 weeks and that this does not occur in the peritoneal cavity of the Lewis rat.

Microfilariae, recovered from the blood of infected dogs, have been transfused into normal dogs. The transfused microfilariae circulate in the recipient dog for as long as $2^1/_2$years.[48] The majority of transfused microfilariae do not circulate in the peripheral blood, but are located within organ reservoirs.[48,60-62] Microfilariae recovered from in vitro culture have also been transfused into normal dogs. The recipient dogs maintain these microfilariae in their peripheral blood for several weeks, but in low numbers.[120]

Mice and rats have also served as hosts for transfused microfilariae of *D. immitis*. Microfilariae will persist in the blood of the rodent host for up to 3 weeks.[114,121-125] If mice are sublethally irradiated, microfilariae will circulate for up to 5 weeks.[125] Transfused microfilaremia in mice and multimammate rats initially had periodicities which were identical to that of the donor dog, however, a few days after the transfusion the periodicities diverged. The donor dog had maximal microfilaremia during the evening and the recipient mice or rats had maximal microfilaremia during the morning.[124] Studies were performed to determine if the microfilarial periodicity seen in mice could be altered by changing the behavior of the mice. In this study, microfilariae in BALB/c mice had nocturnal periodicities and in the donor dog they had a diurnal periodicity. When microfilaremic mice were placed in an environment where their light-dark cycles were reversed the time of maximal microfilaremia reversed to coincide with the time of darkness.[125] Use of transfused microfilariae thus allows experiments to be conducted on the microfilarial role in pathology and population dynamics of heartworm infections.

C. In Vitro Culture

Efforts have been made to cultivate the various stages of *D. immitis* in vitro for a number of reasons. These include: (1) to provide a source of parasite stages for study. Fourth-stage larvae, for example, are difficult to obtain from a dog. They may, however, be very important as a source of antigens for immunizations. (2) Screening for parasite death induced by drugs or immunized-host factors. It is much more efficient to screen for parasiticidal activity in vitro than in vivo. Furthermore, direct responses can be observed. (3) To study the metabolism of different stages. This is done by viewing what is required by the parasite to survive in vitro, as well as seeing what nutrients are consumed. (4) To study stimuli responsible for molting of larval parasites. This type of study can only be performed in vitro where all environmental conditions can be controlled. Prevention of larva moltings may be a very effective means of preventing the development of disease. (5) The collection of excretory-secretory products. These products may prove useful as diagnostic antigens or immunogens.

1. Adult Worms

There are two objectives in culturing adult heartworms. The first is to keep the worms alive, and the second is to recover microfilariae produced in vitro. Adult worms will survive in vitro for 65 days in Eagle's HeLa medium with 10% heat-inactivated serum. Microfilariae, however, are released for only 4 to 7 days.[126] Adult worms will produce microfilariae for $2^{1}/_{2}$ months if maintained in RPMI-1640. No serum is required under these culture conditions, thus this culture system may be a good source for adult excretory-secretory products. Microfilariae recovered from this culture system were inoculated into dogs and were found to circulate. Mosquitoes were allowed to feed on dogs with in vitro derived microfilaremias. Third-stage infective larvae developed in these mosquitoes, thus demonstrating that the microfilariae produced in vitro were viable.[120]

2. Microfilariae

Four factors have been identified which play a role in stimulating microfilariae to develop into the sausage stage. These are osmotic pressure, temperature, serum, and cells. The most favorable osmotic pressure for development of microfilariae into the sausage stage is between 330 and 390 mOsm/kg. This is considerably higher than that found in dog plasma. Osmotic pressures in the mosquito mid-gut may be higher than that found in the dog's blood, and this difference may stimulate microfilarial development.[127] Microfilariae will develop into the sausage stage in vitro if kept at 27°C, but will remain indefinitely as microfilariae at 36°C.[128,129] This finding was confirmed in vivo by using a surrogate vector host. Microfilariae developed to the sausage stage in the wax moth caterpillar, if the caterpillar was maintained at 26 to 30°C. No development occurred in the microfilariae if the caterpillar was kept at 37°C.[130] The change in temperature experienced by microfilariae when they leave their invertebrate host and enter the vertebrate host may stimulate them to resume their development.

Sera from a variety of sources have been added to in vitro culture systems for the culture of microfilariae. Dog, horse, pony, cow, and fetal calf sera have all shown some efficacy at either enhancing microfilarial development[129,131,132] or survival.[126,133] The results obtained for any serum type is dependent on the batch of serum used, the concentration it is used at, and the culture media with which it is used. The specific stimulatory factors found in serum have not been identified.

Erythrocytes from the dog[128,134,135] and mosquito cells, used alone or in combination,[136] have been shown to stimulate microfilarial development. The mechanism by which cells stimulate microfilariae to develop is unknown. Cells may contribute a stimulatory factor or they may remove an inhibitory factor from the culture medium.

When culture conditions are optimum for microfilarial survival, worms will survive approximately 2 months. If culture conditions are optimized for microfilarial development, they will develop to the sausage stage but not any further. A culture technique has been developed to produce second-stage larvae in vitro. Malpighian tubules, removed from mosquitoes 24 hr after ingestion of microfilaremic blood, were placed in culture. Fifteen percent of the larvae found in the tubules developed into second-stage larvae. There was no further development of the second-stage larvae into third-stage larvae.[137]

3. Infective Third-Stage Larvae

Third-stage larvae have been maintained in vitro under a number of culture conditions. Under optimal conditions larvae will molt to the fourth stage, and survive and grow for up to 30 days.[138] Human, horse, dog, and fetal calf sera[138-142] have proven beneficial for the culture of this stage as have dog erythrocytes[138] and sarcoma cells.[143] Approximately 24 hr after cultures are initiated larval movements become lethargic. Motility returns to normal during the next 24 to 48 hr. The molt for third to fourth stage occurs at 48 to 96 hr after the initiation of the culture.[138-142] The timing of the third molt, at 2 to 3 days after the larva leaves the mosquito, has been confirmed by in vivo studies.[40,115,144] Ultrastructural studies were performed on the in vitro molting process to determine if it stimulates the process seen in vivo. It was demonstrated that the fourth-stage cuticle was synthesized in vitro, and that the molting process did not differ from that seen in vivo.[143] Development and utilization of in vitro systems thus provide an excellent tool for the study of the biology of *D. immitis*.

REFERENCES

1. **Leidy, J.,** Descriptions of three filariae, *Proc. Acad. Natl. Sci. Philadelphia*, 5, 117, 1850.
2. **Leidy, J.,** A synopsis of entozoa and some of their ectocongeners observed by the author, *Proc. Acad. Natl. Sci. Philadelphia*, 8, 42, 1856.
3. **Leidy, J.,** Notice of the cruel thread worm, *Filaria immitis* of the dog, *Proc. Acad. Natl. Sci. Philadelphia*, 32, 10, 1880.
4. **Otto, G. F.,** Occurrence of the heartworm in unusual locations and in unusual hosts, in *Proceedings of the Heartworm Symposium '74*, Morgan, H. C., Ed., Veterinary Medicine Publ., Bonner Springs, Kan., 1975, 6.
5. **Orihel, T. C.,** Morphology of the larval stages of *Dirofilaria immitis* in the dog, *J. Parasitol.*, 47, 251, 1961.
6. **Fülleborn, F.,** Zur Morphologie der *Dirofilaria immitis* Leidy 1856, *Cent. Bakt. Parasiten. Infekt.*, 65, 341, 1912.
7. **Tulloch, G. S, Pacheco, G., Anderson, R. A., and Miller, F. H., Jr.,** Scanning electron microscopy of adult male *Dirofilaria immitis*, in *Canine Heartworm Disease: The Current Knowledge*, Bradley, R. E., Ed., University of Florida, Gainesville, Fla., 1972, 113.
8. **Uni, S.,** Scanning electron microscopic study of *Dirofilaria* species (Filarioidea, Nematoda) of Japan and a review of the genus *Dirofilaria*, *J. Osaka City Med. Cent.*, 27, 439, 1978.
9. **Wong, M. M. and Brummer, M. E. G.,** Cuticular morphology of five species of *Dirofilaria*: a scanning electron microscopy study, *J. Parasitol.*, 64, 108, 1978.
10. **Lee, C. C. and Miller, J. H.,** Fine structure of *Dirofilaria immitis* body-wall musculature, *Exp. Parasitol.*, 20, 334, 1967.
11. **Lee, C. C. and Miller, J. H.,** Fine structure of the intestinal epithelium of *Dirofilaria immitis* and changes occurring after vermicidal treatment with caparsolate sodium, *J. Parasitol.*, 55, 1035, 1969.
12. **Gutierrez, Y.,** Diagnostic features of zoonotic filariae in tissue sections, *Hum. Pathol.*, 15, 514, 1984.
13. **Taylor, A. E. R.,** The spermatogenesis and embryology of *Litomosoides carinii* and *Dirofilaria immitis*, *J. Helminthol.*, 34, 3, 1960.
14. **Grieve, R. B., Griffing, S. A., Goldschmidt, M. H., and Abraham, D.,** Transplantation of adult *Dirofilaria immitis* into Lewis rats: parasitologic and serologic findings, *J. Parasitol.*, 71, 391, 1985.
15. **Lee, C. C.,** *Dirofilaria immitis:* ultrastructural aspects of oocyte development and zygote formation, *Exp. Parasitol.*, 37, 449, 1975.
16. **Hawking, F.,** The 24-hr periodicity of microfilariae: biological mechanisms responsible for its production and control, *Proc. R. Soc. London Ser. B*, 169, 59, 1967.
17. **Mantovani, A.,** Transplacental transmission of microfilariae of *Dirofilaria immitis* in the dog, *J. Parasitol.*, 52, 116, 1966.
18. **Atwell, R. B.,** Prevalence of *Dirofilaria immitis* microfilariaemia in 6- to 8-week-old pups, *Aust. Vet. J.*, 57, 479, 1981.
19. **Todd, K. S., Jr., and Howland, T. P.,** Transplacental transmission of *Dirofilaria immitis* microfilariae in the dog, *J. Parasitol.*, 69, 371, 1983.
20. **Johnson, K. H. and Bemrick, W. J.,** *Dirofilaria immitis* microfilariae: electron microscopic examination of cuticle and muscle cells, *Am. J. Vet. Res.*, 30, 1443, 1969.
21. **Kozek, W. J.,** Ultrastructure of the microfilaria of *Dirofilaria immitis*, *J. Parasitol.*, 57, 1052, 1971.
22. **McLaren, D. J.,** Ultrastructural studies on microfilariae (Nematoda: Filarioidea), *Parasitology*, 65, 317, 1972.
23. **Taylor, A. E. R.,** Studies on the microfilariae of *Loa loa, Wuchereria bancrofti, Brugia malayi, Dirofilaria immitis, D. repens* and *D. aethiops, J. Helminthol.*, 34, 13, 1960.

24. **Cherian, P. V., Stromberg, B. E., Weiner, D. J., and Soulsby, E. J. L.,** Fine structure and cytochemical evidence for the presence of polysaccharide surface coat of *Dirofilaria immitis* microfilariae, *Int. J. Parasitol.*, 10, 227, 1980.
25. **O'Leary, T. P., Bemrick, W. J., and Johnson, K. H.,** Serial section analysis of cells in the microfilariae of *Dirofilaria immitis* containing neurosecretory-like granules, *J. Parasitol.*, 59, 701, 1973.
26. **Kozek, W. J.,** Unusual cilia in the microfilaria of *Dirofilaria immitis*, J. Parasitol., 54, 838, 1968.
27. **Bradley, T. J., Sauerman, D. M., Jr., and Nayar, J. K.,** Early cellular responses in the malpighian tubules of the mosquito *Aedes taeniorhynchus* to infection with *Dirofilaria immitis* (Nematoda), *J. Parasitol.*, 70, 82, 1984.
28. **Taylor, A. E. R.,** The development of *Dirofilaria immitis* in the mosquito *Aedes aegypti*, *J. Helminthol.*, 34, 27, 1960.
29. **Kartman, L.,** On the growth of *Dirofilaria immitis* in the mosquito, *Am. J. Trop. Med. Hyg.*, 2, 1062, 1953.
30. **Christensen, B. M.,** Laboratory studies on the development and transmission of *Dirofilaria immitis* by *Aedes trivittatus, Mosq. News*, 37, 367, 1977.
31. **Kutz, F. W. and Dobson, R.C.,** Effects of temperature on the development of *Dirofilaria immitis* (Leidy) in *Anopheles quadrimaculatus* Say and on vector mortality resulting from this development, *Ann. Entomol. Soc. Am.*, 67, 325, 1974.
32. **Christensen, B. M. and Hollander, A. L.,** Effect of temperature on vector-parasite relationships of *Aedes trivittatus* and *Dirofilaria immitis, Proc. Helminthol. Soc. Wash.*, 45, 115, 1978.
33. **Ando, K.,** Development of *Dirofilaria immitis* larvae without blood meal in *Aedes togoi* mosquito, *Mie Med. J.*, 33, 357, 1984.
34. **Nayar, J. K. and Sauerman, D. M., Jr.,** Physiological basis of host susceptibility of Florida mosquitoes to *Dirofilaria immitis, J. Insect Physiol.*, 21, 1965, 1975.
35. **Hamilton, D. R. and Bradley, R. E.,** Observations on the early death experienced by *Dirofilaria immitis*-infected mosquitoes (Diptera: Culicidae), *J. Med. Entomol.*, 15, 305, 1979.
36. **Kershaw, W. E., Lavoipierre, M. M. J., and Chalmers, T. A.,** Studies on the intake of microfilariae by their insect vectors, their survival, and their effect on the survival on their vectors. I. *Dirofilaria immitis* and *Aedes aegypti, Ann. Trop. Med. Parasitol.*, 47, 207, 1953.
37. **Grassi, B. and Noè, G.,** The propagation of the filariae of the blood exclusively by means of the puncture of peculiar mosquitoes, *Br. Med. J.*, 2, 1306, 1900.
38. **McGreevy, P. B., Theis, J. H., Lavoipierre, M. M. J., and Clark, J.,** Studies on filariasis. III. *Dirofilaria immitis:* emergence of infective larvae from the mouthparths of *Aedes aegypti, J. Helminthol.*, 48, 221, 1974.
39. **Kume, S. and Itagaki, S.,** On the life-cycle of *Dirofilaria immitis* in the dog as the final host, *Br. Vet. J.*, 111, 16, 1955.
40. **Kotani, T. and Powers, K. G.,** Developmental stages of *Dirofilaria immitis* in the dog, *Am. J. Vet. Res.*, 43, 2199, 1982.
41. **Hendrix, C. M., Wagner, M. J., Bemrick, W. J., Schlotthauer, J. C., and Stromberg, B. E.,** A scanning electron microscopic study of third-stage larvae of *Dirofilaria immitis, J. Parasitol.*, 70, 149, 1984.
42. **Lichtenfels, J. R., Pilitt, P. A., Kotani, T., and Powers, K. G.,** Morphogenesis of developmental stages of *Dirofilaria immitis* (Nematoda) in the dog, *Proc. Helminthol. Soc. Wash.*, 52, 98, 1985.
43. **Lichtenfels, J. R., Pilitt, P. A., and Wergin, W. P.,** Fine structure of the cuticle during development of the heartworm, *Dirofilaria immitis*, in dogs, *Proc. Helminthol. Soc. Wash.*, in press.
44. **Bancroft, T. L.,** Some further observations on the life-history of *Filaria immitis*, Leidy, *Br. Med. J.*, 1, 822, 1904.
45. **Webber, W. A. F. and Hawking, F.,** Experimental maintenance of *Dirofilaria repens* and *D. immitis* in dogs, *Exp. Parasitol.*, 4, 143, 1955.
46. **Newton, W. L.,** Experimental transmission of the dog heartworm, *Dirofilaria immitis*, by *Anopheles quadrimaculatus, J. Parasitol.*, 43, 589, 1957.
47. **Newton, W. L.,** Longevity of an experimental infection with *Dirofilaria immitis* in a dog, *J. Parasitol.*, 54, 187, 1968.
48. **Underwood, P. C. and Harwood, P. D.,** Survival and location of the microfilariae of *Dirofilaria immitis* in the dog, *J. Parasitol.*, 25, 23, 1939.
49. **Carlisle, C. H.,** The incidence of *Dirofilari immitis* (Heartworm) in dogs in Queensland, *Aust. Vet. J.*, 45, 535, 1969.
50. **Kazacos, K. R.,** The prevalence of heartworms *(Dirofilaria immitis)* in dogs from Indiana, *J. Parasitol.*, 64, 959, 1978.
51. **Lindemann, B. A. and McCall, J. W.,** *Dirofilaria immitis* in stray dogs from Richmond County, Georgia, *J. Parasitol.*, 67, 746, 1981.

52. **Tulloch, G. S., Pacheco, G., Casey, H. W., Bills, W. E., Davis, I., and Anderson, R. A.,** Prepatent clinical, pathologic, and serologic changes in dogs infected with *Dirofilaria immitis* and treated with diethylcarbamazine, *Am. J. Vet. Res.,* 31, 437, 1970.

53. **Guest, M. F., Wong, M. M., and Lavoipierre, M. M. J.,** Infectivity of irradiated larvae of *Dirofilaria immitis, Trans. R. Soc. Trop. Med. Hyg.,* 65, 403, 1971.

54. **Wong, M. M., Guest, M. F., and Lavoipierre, M. M. J.,** *Dirofilaria immitis:* fate and immunogenicity of irradiated infective stage larvae in beagles, *Exp. Parasitol.,* 35, 465, 1974.

55. **Kershaw, W. E., Lavoipierre, M. M. J., and Beesley, W. N.,** Studies on the intake of microfilariae by their insect vectors, their survival, and their effect on the survival of their vectors. VII. Further observations on the intake of the microfilariae of *Dirofilaria immitis* by *Aedes aegypti* in laboratory conditions: the pattern of the intake of a group of flies, *Ann. Trop. Med. Parasitol.,* 49, 203, 1955.

56. **Christensen, B. M.,** *Dirofilaria immitis:* effect on the longevity of *Aedes trivittatus, Exp. Parasitol.,* 44, 116, 1978.

57. **Fowler, J. L., Young, J. L., Sterner, R. T., and Fernau, R. C.,** *Dirofilaria immitis:* lack of correlation between numbers of microfilaria in peripheral blood and mature heart worms, *J. Am. Anim. Hosp. Assoc.,* 9, 391, 1973.

58. **Otto, G. F., Jackson, R. F., Bauman, P. M., Peacock, F., Hinrichs, W. L., and Maltby, J. H.,** Variability in the ratio between the numbers of microfilariae and adult heartworms, *J. Am. Vet. Med. Assoc.,* 168, 605, 1976.

59. **Otto, G. F.,** The significance of microfilaremia in the diagnosis of heartworm infection, in *Proceedings of the Heartworm Symposium '77,* Otto, G. F., Ed., Veterinary Medicine Publ., Bonner Springs, Kan., 1978, 22.

60. **Wong, M. M.,** Studies on microfilaremia in dogs. I. A search for the mechanisms that stabilize the level of microfilaremia, *Am. J. Trop. Med. Hyg.,* 13, 57, 1964.

61. **Pacheco, G.,** Relationship between the number of circulating microfilariae and the total population of microfilariae in a host, *J. Parasitol.,* 60, 814, 1974.

62. **Hinman, E. H., Faust, E. C., and DeBakey, M. E.,** Filarial periodicity in the dog heartworm, *Dirofilaria immitis* after blood transfusion, *Proc. Soc. Exp. Biol. Med.,* 31, 1043, 1934.

63. **Underwood, P. C. and Wright, W. H.,** Observations on the periodicity of *Dirofilaria immitis* larvae in the peripheral blood of dogs, *J. Parasitol.,* 20, 113, 1934.

64. **Hinman, E. H.,** Attempted reversal of filarial periodicity in *Dirofilaria immitis, Proc. Soc. Exp. Biol. Med.,* 33, 524, 1935.

65. **Kartman, L.,** Factors influencing infection of the mosquito with *Dirofilaria immitis* (Leidy, 1856), *Exp. Parasitol.,* 2, 27, 1953.

66. **Tongson, M. S. and Romero, F. R.,** Observations on the periodicity of *Dirofilaria immitis* in the peripheral circulation of the dog, *Br. Vet. J.,* 118, 299, 1962.

67. **Sawyer, T. K. and Weinstein, P. P.,** Experimentally induced canine dirofilariasis, *J. Am. Vet. Med. Assoc.,* 143, 975, 1963.

68. **Gubler, D. J.,** A comparative study on the distribution, incidence and periodicity of the canine filarial worms *Dirofilaria immitis* Leidy and *Dipetalonema reconditum* Grassi in Hawaii, *J. Med. Entomol.,* 3, 159, 1966.

69. **Church, E. M., Georgi, J. R., and Robson, D. S.,** Analysis of the microfilarial periodicity of *Dirofilaria immitis, Cornell Vet.,* 66, 333, 1976.

70. **Angus, B. M.,** Periodicity exhibited by microfilariae of *Dirofilaria immitis* in South East Queensland, *Aust. Vet. J.,* 57, 101, 1981.

71. **Kume, S.,** Experimental observations on seasonal periodicity of microfilariae, in *Proceedings of the Heartworm Symposium '74,* Morgan, H. C., Ed., Veterinary Medicine Publ., Bonner Springs, Kan., 1975, 26.

72. **Sawyer, T. K.,** Seasonal fluctuations of microfilariae in two dogs naturally infected with *Dirofilaria immitis,* in *Proceedings of the Hearworm Symposium '74,* Morgan, H. C., Ed., Veterinary Medicine Publ., Bonner Springs, Kan., 1975, 23.

73. **Wong, M. M.,** Experimental occult dirofilariasis in dogs with reference to immunological responses and its relationship to tropical eosinophila in man, *Southeast Asian J. Trop. Med. Public Health,* 5, 480, 1974.

74. **Wong, M. M., Suter, P. F., Rhode, E. A., and Guest, M. F.,** Dirofilariasis without circulating microfilariae: a problem of diagnosis, *J. Am. Vet. Med. Assoc.,* 163, 133, 1973.

75. **Rawlings, C. A., Dawe, D. L., McCall, J. W., Keith, J. C., and Prestwood, A. K.,** Four types of occult *Dirofilaria immitis* infection in dogs, *J. Am. Vet. Med. Assoc.,* 180, 1323, 1982.

76. **Custer, J. W. and Pence, D. B.,** Dirofilariasis in wild canids from the Gulf Coastal prairies of Texas and Louisiana, U.S.A., *Vet. Parasitol.,* 8, 71, 1981.

77. **Thornton, J. E., Bell, R. R., and Reardon, M. J.,** Internal parasites of coyotes in southern Texas, *J. Wildl. Dis.,* 10, 232, 1974.

78. **Graham, J. M.,** Filariasis in coyotes from Kansas and Colorado, *J. Parasitol.,* 61, 513, 1975.

79. **Franson, J. C., Jorgenson, R. D., and Boggess, E. K.,** Dirofilariasis in Iowa coyotes, *J. Wildl. Dis.,* 12, 165, 1976.
80. **Crowell, W. A., Klei, T. R., Hall, D. I., Smith, N. K., and Newsom, J. D.,** Occurrence of *Dirofilaria immitis* and associated pathology in coyotes and foxes from Louisiana, in *Proceedings of the Heartworm Symposium '77,* Otto, G. F., Ed., Veterinary Medicine Publ., Bonner Springs, Kan., 1978, 10.
81. **Kazacos, K. R. and Edberg, E. O.,** *Dirofilaria immitis* infection in foxes and coyotes in Indiana, *J. Am. Vet. Med. Assoc.,* 175, 909, 1979.
82. **Pence, D. B. and Meinzer, W. P.,** Helminth parasitism in the coyote *Canis latrans,* from the rolling plains of Texas, *Int. J. Parasitol.,* 9, 339, 1979.
83. **Weinmann, C. J. and Garcia, R.,** Coyotes and canine heartworm in California, *J. Wildl. Dis.,* 16, 217, 1980.
84. **Acevedo, R. A. and Theis, J. H.,** Prevalence of heartworm *(Dirofilaria immitis* Leidy) in coyotes from five northern California counties, *Am. J. Trop. Med. Hyg.,* 31, 968, 1982.
85. **Agostine, J. C. and Jones, G. S.,** Heartworms *(Dirofilaria immitis)* in coyotes *(Canis latrans)* in New England, *J. Wildl. Dis.,* 18, 343, 1982.
86. **King, A. W. and Bohning, A. M.,** The incidence of heartworm *Dirofilaria immitis* (Filarioidea), on the wild canids of Northern Arkansas, *Southwest. Nat.,* 29, 89, 1984.
87. **Kagei, N., Shiomi, H., Sugaya, H., and Akiyama, H.,** On the helminths from raccoon dogs in Japan, *Jpn. J. Parasitol.,* 32, 367, 1983.
88. **Mulley, R. C. and Starr, T. W.,** *Dirofilaria immitis* in red foxes *(Vulpes vulpes)* in an endemic area near Sydney, Australia, *J. Wildl. Dis.,* 20, 152, 1984.
89. **Hubert, G. F., Jr., Kick, T. J., and Andrews, R. D.,** *Dirofilaria immitis* in red foxes in Illinois, *J. Wildl. Dis.,* 16, 229, 1980.
90. **Simmons, J. M., Nicholson, W. S., Hill, E. P., and Briggs, D. B.,** Occurrence of *Dirofilaria immitis* in gray fox *(Urocyon cinereoargenteus)* in Alabama and Georgia, *J. Wildl. Dis.,* 16, 225, 1980.
91. **Carlson, B. L. and Nielsen, S. W.,** *Dirofilaria immitis* infection in a gray fox, *J. Am. Vet. Med. Assoc.,* 183, 1275, 1983.
92. **Okada, R., Imai, S., and Ishii, T.,** Clouded leopard *Neofelis neburosa,* new host for *Dirofilaria immitis,* *Jpn. J. Vet. Sci.,* 45, 849, 1983.
93. **White, G. L.,** *Dirofilaria immitis* and heartworm disease in the California sea lion, *J. Zool. Anim. Med.,* 6, 23, 1975.
94. **Medway, W. and Wieland, T. C.,** *Dirofilaria immitis* infection in a harbor seal, *J. Am. Vet. Med. Assoc.,* 167, 549, 1975.
95. **Williams, J. F. and Dade, A. W.,** *Dirofilaria immitis* infection in a wolverine, *J. Parasitol.,* 62, 174, 1976.
96. **Foil, L. and Orihel, T. C.,** *Dirofilaria immitis* (Leidy, 1856) in the beaver, *Castor canadensis, J. Parasitol.,* 61, 433, 1975.
97. **Miller, W. R. and Merton, D. A.,** Dirofilariasis in a ferret, *J. Am. Vet. Med. Assoc.,* 180, 1103, 1982.
98. **Parrott, T. Y., Greiner, E. C., and Parrott, J. D.,** *Dirofilaria immitis* infection in three ferrets, *J. Am. Vet. Med. Assoc.,* 184, 582, 1984.
99. **Narama, I., Tsuchitani, M., Umemura, T., and Kamiya, H.,** Pulmonary nodule caused by *Dirofilaria immitis* in a laboratory rabbit *(Oryctolagus cuniculus domesticus), J. Parasitol.,* 68, 351, 1982.
100. **Harwell, G. and Craig, T. M.,** Dirofilariasis in a red panda, *J. Am. Vet. Med. Assoc.,* 179, 1258, 1981.
101. **Narushima, E., Hashizaki, F., Kohno, N., Saito, M., Tanabe, K., Hayasaki, M., and Ohishi, I.,** *Dirofilaria immitis* infections in lesser pandas, *(Ailurus fulgens)* in Japan, *Jpn. J. Parasitol.,* 33, 475, 1984.
102. **Johnson, C. A.,** *Ursus americanus* (Black bear) a new host for *Dirofilaria immitis, J. Parasitol.,* 61, 940, 1975.
103. **Crum, J. M., Nettles, V. F., and Davidson, W. R.,** Studies on endoparasites of the black bear *(Ursus americanus)* in the southeastern United States, *J. Wildl. Dis.,* 14, 178, 1978.
104. **Klein, J. B. and Stoddard, E. D.,** *Dirofilaria immitis* recovered from a horse, *J. Am. Vet. Med. Assoc.,* 171, 354, 1977.
105. **Thurman, J. D., Johnson, B. J., and Lichtenfels, J. R.,** Dirofilariasis with arteriosclerosis in a horse, *J. Am. Vet. Med. Assoc.,* 185, 532, 1984.
106. **Baskin, G. B. and Eberhard, M. L.,** *Dirofilaria immitis* infection in a rhesus monkey *(Macaca mulatta),* *Lab. Anim. Sci.,* 32, 401, 1982.
107. **Donahoe, J. M. R.,** Experimental infection of cats with *Dirofilaria immitis, J. Parasitol.,* 61, 599, 1975.
108. **Wong, M. M., Pedersen, N. C., and Cullen, J.,** Dirofilariasis in cats, *J. Am. Anim. Hosp. Assoc.,* 19, 855, 1983.
109. **Campbell, W. C. and Blair, L. S.,** *Dirofilaria immitis:* experimental infections in the ferret *(Mustela putorius furo), J. Parasitol.,* 64, 119, 1978.
110. **Campbell, W. C., Blair, L. S., and McCall, J. W.,** *Brugia pahangi* and *Dirofilaria immitis:* experimental infections in the ferret, *Mustela putorius furo, Exp. Parasitol.,* 47, 327, 1979.

111. **Johnsen, D. O., DePaoli, A., Tanticharoenyos, P., Diggs, C. L., and Gould, D. J.**, Experimental infection of the gibbon *(Hylobates lar)* with *Dirofilaria immitis, Am. J. Trop. Med. Hyg.*, 21, 521, 1972.

112. **Wong, M. M.**, Experimental dirofilariases in macaques. Susceptibility and host responses to *Dirofilaria immitis,* the dog heartworm, *Trans. R. Soc. Trop. Med. Hyg.*, 68, 479, 1974.

113. **Wong, M. M. and Lim, L. C.**, Development of intraperitoneally inoculated *Dirofilaria immitis* in the laboratory Mongolian jird *(Meriones unguiculatus), J. Parasitol.*, 61, 573, 1975.

114. **Ohgo, T.**, Basic studies on the *Dirofilaria immitis* and the experimental infection with larval filariae in mice, *J. Osaka City Med. Cent.*, 29, 745, 1980.

115. **Delves, C. J. and Howells, R. E.**, Development of *Dirofilaria immitis* third stage larvae (Nematoda: Filarioidea) in micropore chambers implanted into surrogate hosts, *Trop. Med. Parasitol.*, 36, 29, 1985.

116. **Mann, P. H. and Fratta, I.**, Transplantation of adult heartworms, *Dirofilaria immitis,* into dogs and cats, *J. Parasitol.*, 39, 139, 1953.

117. **Carlisle, C. H.**, The experimental production of heartworm disease in the dog, *Aust. Vet. J.*, 46, 190, 1970.

118. **Rawlings, C. A. and McCall, J. W.**, Surgical transplantation of adult *Dirofilaria immitis* to study heartworm infection and disease in dogs, *Am. J. Vet. Res.*, 46, 221, 1985.

119. **Zielke, E.**, Preliminary studies on the transplantation of adult *Dirofilaria immitis* into laboratory rodents, *Ann. Trop. Med. Parasitol.*, 71, 243, 1977.

120. **April, M., D'Antonio, R., Scott, A. L., Malick, A., Roberts, E. P., and Levy, D. A.**, Development of a chemotherapeutic model for microfilaricidal drugs to *Dirofilaria immitis, Acta Trop.*, 41, 383, 1984.

121. **Mann, P. H. and Fratta, I.**, Survival of microfilariae of *Dirofilaria immitis* in rats and mice, *Science,* 117, 18, 1953.

122. **Mann, P. H. and Fratta, I.**, Observations on the experimental transfer of the microfilariae of *Dirofilaria immitis* to the mouse, rat, and the chick embryo, *J. Parasitol.*, 41, 218, 1955.

123. **Sakamoto, M., Kimura, E., Aoki, Y., and Nakajima, Y.**, Subcutaneous and intraperitoneal inoculation of *Dirofilaria immitis* microfilariae into mice, *Jpn. J. Parasitol.*, 33, 415, 1984.

124. **Zielke, E.**, On the longevity and behaviour of microfilariae of *Wuchereria bancrofti, Brugia pahangi* and *Dirofilaria immitis* transfused to laboratory rodents, *Trans. R. Soc. Trop. Med. Hyg.*, 74, 456, 1980.

125. **Grieve, R. B. and Lauria, S.**, Periodicity of *Dirofilaria immitis* microfilariae in canine and murine hosts, *Acta Trop.*, 40, 121, 1983.

126. **Earl, P. R.**, Filariae from the dog in vitro, *Ann. N.Y. Acad. Sci.*, 77, 163, 1959.

127. **Ando, K. and Kitamura, S.**, Osmotic pressure-dependent development of micorifilariae of *Dirofilaria immitis in vitro, Jpn. J. Parasitol.*, 31, 219, 1982.

128. **Sawyer, T. K. and Weinstein, P. P.**, The *in vitro* development of microfilariae of the dog heartworm *Dirofilaria immitis* to the "sausage-form", *J. Parasitol.*, 49, 218, 1963.

129. **Ando, K., Chinzei, Y., and Kitamura, S.**, Conditions of *in vitro* culture for development of microfilariae of *Dirofilaria immitis, Jpn. J. Parasitol.*, 29, 483, 1980.

130. **Yoeli, M., Alger, N., and Most, H.**, Studies on filaraisis. I. The behavior of microfilariae of *Dirofilaria immitis* in the wax-moth larva *(Galleria mellonella), Exp. Parasitol.*, 7, 531, 1958.

131. **Klein, J. B. and Bradley, R. E.**, Induction of morphological changes in microfilariae from *Dirofilaria immitis* by *in vitro* culture techniques, *J. Parasitol.*, 60, 649, 1974.

132. **Devaney, E. and Howells, R. E.**, Culture systems for the maintenance and development of microfilariae, *Ann. Trop. Med. Parasitol.*, 73, 139, 1979.

133. **Bender, A. P., Angrick, E., Dorn, C. R., and Donahoe, J. P.**, Microfilaremia in dogs, *Vet. Rec.*, 108, 41, 1981.

134. **Taylor, A. E. R.**, Maintenance of filarial worms *in vitro, Exp. Parasitol.*, 9, 113, 1960.

135. **Sawyer, T. K. and Weinstein, P. P.**, Morphologic changes occurring in canine microfilariae maintenance in whole blood cultures, *Am. J. Vet. Res.*, 24, 402, 1963.

136. **Sneller, V. P. and Weinstein, P. P.**, *In vitro* development of *Dirofilaria immitis* microfilariae: selection of culture media and serum levels, *Int. J. Parasitol.*, 12, 233, 1982.

137. **Devaney, E.**, The development of *Dirofilaria immitis* in cultured malpighian tubules, *Acta Trop.*, 38, 251, 1981.

138. **Yoeli, M., Upmanis, R. S., and Most, H.**, Studies on filariasis. III. Partial growth of the mammalian stage of *Dirofilaria immitis in vitro, Exp. Parasitol.*, 15, 325, 1964.

139. **Sawyer, T. K.**, *In vitro* culture of third-stage larvae of *Dirofilaria immitis, J. Parasitol.*, 49(Suppl. 2), 59, 1963.

140. **Sawyer, T. K.**, Molting and exsheathment *in vitro* of third-stage *Dirofilaria immitis, J. Parasitol.*, 51, 1016, 1965.

141. **Lok, J. B., Mika-Grieve, M., Grieve, R. B., and Chin, T. K.**, *In vitro* development of third- and fourth-stage larvae of *Dirofilaria immitis:* comparison of basal culture media, serum level and possible serum substitutes, *Acta. Trop.*, 41, 145, 1984.

142. **Wong, M. M., Knighton, R., Fidel, J., and Wada, M.,** *In vitro* culture of infective-stage larvae of *Dirofilaria immitis* and *Brugia pahangi. Ann. Trop. Med. Parasitol.,* 76, 239, 1982.
143. **Devaney, E.,** *Dirofilaria immitis:* the moulting of the infective larva in vitro, *J. Helminthol.,* 59, 47, 1985.
144. **Sawyer, T. K. and Weinstein, P. P.,** Third molt of *Dirofilaria immitis in vitro* and *in vivo, J. Parasitol.,* 51 (Suppl. 2), 48, 1965.

Chapter 3

THE BIOCHEMISTRY OF *DIROFILARIA IMMITIS*

Christopher Bryant

TABLE OF CONTENTS

I. INTRODUCTION

Knowledge of the biochemistry of the filarial worms is scantier than that of the other helminth groups. One reason is the apparent intransigency of the parasites as experimental material. A second is the apparently uncomplicated nature of their respiratory metabolism, for the filariae have long been thought to depend entirely on glycolysis for energy metabolism. New techniques of culture and enzyme assay have removed the first obstacle and new studies have shown that the second is an unjustified assumption. These are timely discoveries that, I hope, will stimulate biochemical work on the seven species of filarial worms that are parasitic in humans.

As Barrett[1] points out, filarial worms are species specific for their definitive hosts and, of the ones that infect humans, only *Brugia malayi* can be cultured in the laboratory in an alternative host. This difficulty has diverted attention to other filariae as possible models for human filariasis, and *Dirofilaria immitis* has received its share of attention. There are many advantages in working with *D. immitis*, prominent among them is the ready availability of ample material from dog pounds. The disadvantages it shares with the other Filarioidea are the difficulties of carrying out biochemical work on the microfilariae and the almost complete inaccessibility of larval stages in the mosquito. The biochemical work described in this chapter reflects this faithfully: a lot is known about the adults, little is known about the microfilariae, and almost nothing is known about the first three larval stages in the mosquito.

In writing this chapter, I have been extremely grateful to have Barrett's excellent review[1] at hand. This is a comprehensive summary of the biochemical literature on all the Filaroidea, and the interested reader is recommended to turn to it for comparative information. It has enabled me in this limited space to concentrate almost exclusively on the dog heartworm.

II. UPTAKE OF NUTRIENTS

Adult *D. immitis*, like other filarial worms, has a functional digestive tract, which has been observed to contain host erythrocytes. However, the parasite did not ingest trypan blue from the medium during in vitro incubation. In this it is similar to *B. pahangi*, although the latter has been shown to take up the dye in vivo. Adult filariae can absorb nutrients through the cuticle, and *D. immitis* utilized D-glucose and adenosine, presumably for RNA synthesis, even when its mouth was isolated from the nutrient solution.[2,3] It also took up uridine and uracil.[4] Uptake is a selective process, for L-glucose, sucrose, and thymidine are excluded, but the details of the transport mechanism are unknown. The presence in the cuticle of a number of enzymes that, in other tissues, are associated with nutrient uptake and digestion is consistent with an absorptive role. High levels of acid phosphatase, capable of hydrolyzing hemoglobin, are present in the hypodermis of *D. immitis*.

Jaffe and Doremus[5] first reported that microfilariae of *D. immitis* consume large quantities of glucose during in vitro incubation. They found that the rate of glucose consumption by the microfilariae increased with increasing glucose concentration until a maximum rate of uptake of 4 μmol/mℓ was reached. When ^{14}C-glucose was the substrate, 40% of the radioactivity was recovered in CO_2, 43% in the trichloracetic acid soluble fraction (comprising largely intermediates of glycolysis and the tricarboxylic acid cycle), 8% in RNA, 6% in lipids, 1% in glycogen, and 0.2% in DNA. The microfilariae are capable of incorporating adenine, adenosine, uracil, and uridine into RNA, but not thymidine. They are apparently unable to carry out the *de novo* synthesis of purines from glycine and formate.

Microfilariae do not possess a functional intestine, and so the utilization of glucose, amino acids, uracil, uridine, adenine, and adenosine is probably achieved by the cuticular routes.[5,6] As in the adult, cuticular uptake by microfilariae is selective because glutamate, histidine,

Table 1
PERCENTAGES OF
AMINO ACIDS
REMOVED FROM
MEDIUM BY *D. IMMITIS*
DURING INCUBATION
FOR 8 DAYS[6]

Amino acid	% Utilized
Isoleucine	7.7
Serine	5.8
Valine	4.1
Phenylalanine	3.0
Tyrosine	4.2
Methionine	1.8
Cystathionine	3.7
Aspartate	15.2[a]
Leucine	11.2[a]
Glutamine	43.3[a]
Threonine	32.4[a]
Proline	89.8[a]

Note: 6×10^4 microfilariae/ml TC-199 medium + 10% fetal bovine serum.

[a] Statistically significant uptake.

lysine, alanine, tryptophan, cystine, ornithine, and arginine are not absorbed from the medium during in vitro culture. Some uptake of isoleucine, serine, valine, phenylalanine, tyrosine, methionine, and cystathionine occurs,[6] while the concentrations of aspartate, leucine glutamine, threonine, and proline in the incubation medium are markedly reduced during 8 days of incubation (Table 1). The amount of glucose present fell by 40% during the same period.

III. RESPIRATORY METABOLISM

Until recently, the respiratory metabolism of parasitic helminths has been thought to conform to three main patterns (Figure 1). The first of these, supposedly well exemplified by the filarial worms, is homolactate fermentation. In homolactate fermentation, carbohydrate is converted exclusively to lactic acid by glycolysis. Lactic acid is excreted and energy generation is thus wholly anaerobic. A second type of metabolism, exemplified by *Hymenolepis diminuta*, follows the glycolytic route to the level of phosphoenolpyruvate (PEP). A subsequent carbon dioxide-fixation step yields oxaloacetate and malate. The latter then enters the mitochondrion for further metabolism — via a dismutation — to succinic, propionic, and acetic acids. The acids are excreted. The two diagnostic enzymes for such metabolism are phosphoenolpyruvate carboxykinase and fumarate reductase. A third type of metabolism, particularly well studied in *Ascaris lumbricoides*, but not illustrated here, extends this pathway by a number of redox and condensation reactions. The simple anaerobic picture of types one and two is complicated by recent experiments which suggest that pyruvate, as well as malate, can act as a mitochondrial substrate and by the superimposition of varying degrees of aerobic metabolism involving tricarboxylic acid cycle activity.[7]

It is extremely doubtful if any parasitic helminth exhibits homolactate fermentation, *sensu stricto*. There are several studies that now point to this conclusion, and there is considerable

A. Homolactate fermentation:

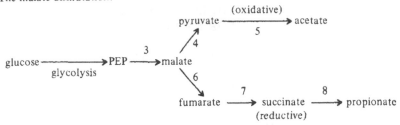

B. The malate dismutation:

FIGURE 1.   The path of carbon in helminth respiration. Important enzymes: (1) pyruvate kinase; (2) lactate dehydrogenase; (3) phosphoenolpyruvate carboxykinase; (4) "malic enzyme"; (5) the pyruvate dehydrogenase complex; (6) fumarase; (7) the fumarate reductase complex; (8) the succinate decarboxylase complex.

evidence of a substantial aerobic component in the respiration of the filarial worms. In the trematode *Schistosoma mansoni* (another supposed homolactate fermenter) even low tricarboxylic acid cycle activity, as demonstrated by the generation of small amounts of carbon dioxide under aerobic conditions, can contribute appreciably to energy metabolism because of the 18-fold greater yield of adenosine triphosphate (ATP) per mole of substrate oxidized compared with glycolysis.[8] Similar amounts of tricarboxylic acid cycle activity, also detected by carbon dioxide generation during aerobic incubation in vitro, have long been observed for filarial worms. *B. pahangi* and *Dipetalonema viteae* possess cristate mitochondria, usually indicative of aerobic capacity, and their oxygen uptake and motility are inhibited by cyanide.[1] The adults of both parasites have significant oxygen consumption, and rotenone and antimycin A (inhibitors of complex I and III in the mammalian electron-transport system [Figure 2], respectively) suppress this by about 80%. Uncouplers of oxidative phosphorylation stimulate endogenous oxygen consumption in *B. pahangi* microfilaria, suggesting that electron-transport processes are linked to ATP synthesis.[9] Furthermore, subcellular fractions prepared from these worms oxidized α-glycerophosphate, succinate, and malate, suggesting that they possess branched electron-transfer pathways that bifurcate on the substrate side of complex III. This is similar to the condition in other helminths, and rates of substrate oxidation were found to be comparable with those reported for other nematode parasites. Finally, as yet another complicating factor, submitochondrial particles from *B. pahangi* and *Di. viteae* possess a fumarate reductase associated with complex I which may be important in energy metabolism under anaerobic conditions.[10]

In another study, on *Litomosoides carinii*, the uptake of glucose during in vitro incubation was significantly higher when oxygen was present.[11] Anaerobically, there was an almost quantitative conversion of glucose to lactic acid, but aerobically appreciable quantities of acetate, acetoin (acetylmethylcarbinol), and carbon dioxide were formed in addition to lactate. Acetate and acetoin formation are often symptomatic of the activity of the pyruvate dehydrogenase complex. This complex catalyzes an important reaction that prepares carbon from glycolysis for entry into the tricarboxylic cycle so it is presumed that much of the carbon dioxide was derived from the cycle.

L. carinii possesses the enzymes of glycolysis, at much higher activities (excepting phosphofructokinase) than occur in rat liver, and those of the tricarboxylic acid cycle. The specific activities of the cycle enzymes are, when compared on the basis of mitochondrial protein, similar to those from rat liver. The results obtained from three different and independent experiments show that adult *L. carinii* is capable of oxidizing substrates completely to carbon

Complex I

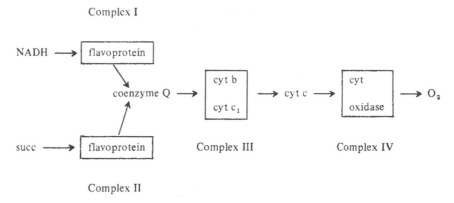

FIGURE 2. The mammalian electron-transport system briefly summarized. NADH, nicotinamide adenine dinucleotide (reduced); succ, succinate; cyt, cytochrome.

dioxide and water, using molecular oxygen. The first observation is that the $^{14}CO_2$ production from 6-^{14}C-glucose increases under aerobic conditions. The second is that pyruvate labeled at the methyl group gives $^{14}CO_2$. Both of these are symptomatic of tricarboxylic acid cycle activity. Third, carbon balance studies show that the rate of glucose-dependent carbon dioxide production exceeds that which can be accounted for by the formation of acetate and acetoin. In summary, the data suggest that 2% of utilized carbohydrate may have undergone complete oxidation to carbon dioxide and water. Starting from 100 mol of glucose, the energy yield of 98 mol converted to lactate is 196 mol ATP. The energy yield of 2 mol converted to carbon dioxide and water by the tricarboxylic acid cycle is 72 mol ATP. Under the conditions of the experiment, 27% of the ATP production of *L. carinii* is thus aerobic.

There is no evidence that the respiratory metabolism of *D. immitis* is markedly different from that of other filarial nematodes. *D. immitis* remains active during anaerobic in vitro incubation in a simple salt solution with only glucose for a substrate for up to 24 hr although oxygen is necessary for long term in vitro survival, which supports the view that the parasite has some aerobic capacity.[12]

During aerobic incubation with 1-^{14}C-glucose, incorporation into lactate accounted for 55% of the labeled carbon utilized. However, the fate of the 6-carbon was not determined so that no measure of the aerobic component of respiratory metabolism was obtained. About 2% of the label from the 1-carbon of glucose appeared in carbon dioxide which, with the observation that the enzymes glucose-6-phosphate dehydrogenase and 6-phosphophogluconate are present in extracts of the worm, supports the view that *D. immitis* possesses pentose phosphate pathway activity.[13,14] In addition, "transketolase system" (comprising phosphopentose isomerase, pentose epimerase, and transketolase) and transaldolase activities are present in preparations from the cytosol. The addition of methylene blue to the worm incubation media, which effectively acts to increase the amount of nicotinamide adenine dinucleotide available for reduction, increased the recovery of xylulose-5-phosphate. Finally, the incorporation of carbon from labeled glucose into nucleic acids also occurs.[15] These observations, taken together, provide considerable circumstantial evidence for the operation of the pentose phosphate pathway in *D. immitis*. However, most of the evidence is derived from isotopic studies and the definitive enzymology has yet to be done.

D. immitis possesses all the enzymes of glycolysis and the tricarboxylic acid cycle, although the activities of aconitase and isocitrate dehydrogenase are low and probably rate limiting. This, and the failure to demonstrate malate dehydrogenase activity in the direction of oxaloacetate formation, suggest that the cycle has very low activity.[16] However, it should be noted that the equilibrium of the malate dehydrogenase reaction is normally heavily in favor of the formation of malate and that recent measurements in a more modern assay system show that reversal is, in fact, possible.[48]

Table 2
LIPID COMPOSITION OF
D. IMMITIS[18]

Lipid class	Percent
Triglycerides	0.4
Diglycerides	2.7
Phosphoglycerides	21.9
Cholesterol	40.9
Cholesterol esters	28.7
Free Fatty acids	5.4

Note: Total lipid 5.22 μmol/g wet wt.

Walter and Albiez[17] have partially purified an NADP-dependent malic enzyme from *D. immitis* homogenates, whereas Turner and Hutchinson[18] have described an NAD-dependent one from mitochondria. It has been suggested that the oxidative decarboxylation of malate and of 6-phosphogluconate in the pentose phosphate pathway may account for the observed carbon dioxide generation during in vitro incubation of the worms. It certainly may in part. However, many experiments were conducted under aerobic conditions with uniformly labeled glucose, so that a contribution from tricarboxylic acid cycle activity cannot be excluded.

Phosphoenolpyruvate carboxykinase has been found in *D. immitis*.[19] Its presence, and that of malic enzyme, and the absence of evidence for any of the pathways leading to the excretion of succinate, propionate, or acetate have led to the suggestion that pyruvate might be derived, at least in part, from the cleavage of malate.[20]

$$PEP \rightarrow oxaloacetate \rightarrow malate \rightarrow pyruvate$$

The enzymes concerned are, in order, phosphoenolpyruvate carboxykinase, malate dehydrogenase, and "malic enzyme". However, as Barrett[1] points out, there is no experimental evidence to support this scheme, and other functions for the enzymes, such as glyconeogenesis, NADPH generation for synthetic reactions, or the transport of acetyl CoA across mitochondrial membranes, may be more important. The last may be particularly significant in helminths that excrete acetate.

Many organisms are capable of using glycerol for energy generation, oxidizing it, after an initial phosphorylation step, via the latter half of the glycolytic pathway. However, there is little evidence for such a process in *D. immitis*, which uses glucose some 200 times faster than it uses glycerol. A small amount of radiocarbon from labeled glycerol was incorporated into phosphoglycerides and diglycerides.[21]

IV. LIPIDS AND LIPID METABOLISM

It is something of a curiosity that, while studies of the general metabolism of filarial worms have tended to lag, much more is known about their lipid metabolism than that of other helminth groups.

A lipid analysis[22] of *D. immitis* shows that the major lipid classes are the sterols and their esters. Total phospholipid was present in the concentration range 5 to 7 μg/mg wet weight of worm and, of this, phosphatidylcholine and phosphatidylethanolamine made up 85%. In the fatty-acid fraction 16 and 18 carbon chains predominate and the ratio of saturated to unsaturated fatty acids is about 2:1. Parallel studies of the lipids in dog plasma show little correlation with those of *D. immitis*. The lipid pattern of the parasite more nearly resembles those of the other filariae. A summary of the lipid composition of *D. immitis* is given in Table 2.

Glycerol uptake in *D. immitis* was studied first of all from the point of view of energy metabolism, and it was established that it plays little part.[21] Almost as an afterthought, the same report noted that, while little glycerol was utilized for respiration by worms in vitro, the little that was taken up was incorporated into phosphoglycerides, diglycerides, and free cholesterol. The uptake of ^{14}C-labeled glycerol, acetate, and oleic acid by *D. immitis* and their incorporation into the various lipid classes was therefore studied in more detail.[18] Acetate and oleic acid are more readily used than glycerol, and radiocarbon was found in phospholipids, cholesterol and its esters, free fatty acids, di- and triglycerides, and sterol esters. The free fatty-acid fraction was very poorly labeled, and from the data presented it is doubtful that fatty-acid synthesis is a significant function of the parasite. The diglyceride and phosphoglyceride fractions were rapidly labeled irrespective of precursor. Glycerol and oleic acid are apparently capable of being incorporated unchanged into glycerides, while acetate is incorporated into the acid component of glycerides and into free fatty acids. This is in line with observations on other helminths, none of which is capable of synthesising fatty acids *de novo*, but may be capable of a limited amount of chain lengthening.

Three plasmalogens have also been detected in *D. immitis* and their syntheses have been examined in some detail.[23,24] Phosphatidylserine synthesis occurs in extracts from adult female *D. immitis* but only by way of an exchange reaction, catalyzed by a Ca^{2+}-dependent enzyme. In this reaction, L-serine replaces one of the bases in preformed phospholipid. The rate of incorporation is low (about 44 pmol/min/mg protein), and other animals, such as the rat, have alternative pathways. Phosphoethanolamine synthesis also occurs by way of an exchange reaction in which ethanolamine replaces choline or serine in phosphatidylcholine or phosphatidylserine. However, two other mechanisms for the formation of phosphatidyl-ethanolamine were observed. One depends on the simple decarboxylation of phosphatidyl-serine. The other involves the Mg^{2+}-dependent enzyme, ethanolamine phosphotransferase, and proceeds according to the following sequence of reactions:

1. ethanolamine + ATP → phosphoethanolamine + ADP

2. phosphoethanolamine + CTP → CDP-ethanolamine + PP_i

3. CDP-ethanolamine + 1,2-diacylglycerol → phosphatidyl-
    ethanolamine + CMP

where ATP and ADP = adenosine tri- and diphosphate; CTP, CDP, and CMP = cytidine tri-, di-, and monophosphate; and PP_1 = inorganic phosphate.

In *D. immitis*, formation of phosphatidylethanolamine by decarboxylation is about 30 times faster than the above pathway and about 3000 times more active than the base exchange mechanism. Phosphatidylcholine can be synthesized by an analogous pathway and also by an *S*-adenosylmethionine-mediated methylation of phosphatidylethanolamine.[25]

The synthesis of isoprenoid derivatives by filarial worms has recently received intensive study. There are many compounds in living organisms whose structures are apparently based on isoprene (2-methylbutadiene):

$$CH_2=\overset{\displaystyle \overset{CH_2}{|}}{C}-CH=CH_2$$

These compounds contain multiples of five carbon atoms so arranged that they can be notionally separated into isoprene-like fragments. Such compounds are called terpenes, and are found in oils like camphor and geraniol, in rubber, and in carotenes from plants. They are also present as side chains in various vitamins and cofactors for biochemical reactions, such as the family of ubiquinones (coenzymes Q) that are important in the respiratory electron-transport system in mitochondria, and folate metabolism.[26] A ubiquinone has the structure

$$CH_2$$
$$|$$
$$quinone-(CH_2-CH=C-CH_2)_nH$$

where n refers to the number of isoprene units in the side chain. Ubiquinones of different lengths are referred to as ubiquinone 9 or 10 depending on the number of units that they possess. There are many compounds with such side chains present in nematodes. They include ubiquinone itself, cholesterol, rhodoquinone, farnesol-like compounds, and the polypropanol solanesol.

One important metabolic pathway involving isoprenoid compounds leads to the synthesis of sterols. The starting point is mevalonic acid (3,5-dihydroxy-3-methylvaleric acid):

$$H_3C \qquad OH$$
$$HOOC-CH_2-C-CH_2-CH_2-OH$$

The sequence is

mevalonic acid → 5-phosphomevalonic acid → 5-pyrophosphomevalonic

acid → isopentenyl pyrophosphate → 3,3-dimethylallyl pyrophosphate

→ farnesyl pyrophosphate → presqualene pyrophosphate → squalene

→ lanosterol

Adult *D. immitis* cannot synthesize squalene or sterols *de novo*, although it can synthesize ubiquinone 9, a number of dolichol isoprenologs, and, predominantly, a short-chain isoprenoid alcohol, geranyl geraniol (Table 3). Neither can it add the isoprenoid tail to menadione in the synthesis of vitamin K_2 from menadione.[27]

These finding are not wholly consistent with the earlier record[18] that a small amount of cholesterol was formed after incubation with labeled glycerol or acetate, but are more consistent with a number of other observations. For example, there was a barely detectable incorporation of radiocarbon from mevalonate, the preferred substrate, into squalene and cholesterol.[27] There is also a marked ability to synthesize isoprenoids, which is symptomatic of the inability to make sterols.[28] It is therefore probable that *D. immitis* does not carry out the *de novo* synthesis of sterols.

Members of an interesting group of compounds, the ecdysteroids, have recently been detected in *D. immitis*.[29] In insects, molting and metamorphosis is controlled by ecdysteroid hormones. Nematodes also undergo several molts during their development, but it is not clear whether ecdysone is involved in nematode molting, or even whether the Insecta can be used as a good analogy for nematode endocrinology. However, both male and female *D. immitis* contained ecdysone, 20-hydroxyecdysone, 20,26-dihydroxyecdysone, and possibly ponasterone A. Generally, the concentrations were lower than those normally encountered in insects.

The discovery of dolichols in *D. immitis* is not surprising as they are very widely distributed in invertebrates. Dolichols are implicated as cofactors in the transfer of sugars to glycoproteins and proteoglycans and, as microfilariae have a surface coat that is mainly glycoprotein,[30] may be of great importance in the synthesis of cuticular components. It is therefore significant that the isolation of a glycosyl transferase, which utilizes dolichol phosphate to promote the

Table 3
DISTRIBUTION OF RADIOACTIVITY FROM DL-¹⁴C-MEVALONATE AMONG THE LIPIDS OF MALE ADULT *D. IMMITIS* AFTER INCUBATION IN VITRO[27]

Lipid fraction	Total incorporation (%)
Phospholipids[a]	21.4
Neutral lipids	40.5
Free fatty acids	2.6
Cholesterol	0.1
Geranyl geraniol	11.4
Dolichols	1.4
Ubiquinones	2.9
Squalene	0.03
Stearyl palmitate	1.0

Note: The total used was less than 1% of the available mevalonate; female worms were not significantly different from males.

[a] Includes phosphatidylethanolamine, lysophosphatidylcholine, phosphatidylcholine, sphingomyelin, and phosphatidylserine in the approximate ratios of 5:5:1:3:5.

transfer of glycosyl residues from sugar nucleotides in the formation of glycoproteins, has recently been reported.[31] This enzyme was located in the microsome fraction derived from homogenates of *D. immitis,* and is able to transfer mannose from guanosine diphosphomannose to dolichol monophosphate in the presence of a divalent cation. The microsomes also promote the synthesis of a glycoprotein from mannose and α-lactalbumin.

V. FOLIC ACID METABOLISM

The folic acid derivatives are extremely important in metabolism because they assist in the transfer of one-carbon groups to acceptor molecules during biochemical syntheses. In particular, they are concerned with purine and pyrimidine metabolism, and therefore have attracted considerable interest as potential targets for antiparasite drugs. It seems likely that adult filariae do not synthesize dihydrofolate, but there is also no evidence that they can utilize dihydrofolate from nutrient medium during in vitro incubation. The various interactions of the folate analogs are shown in Figure 3.[1,4,32]

Although the properties of the enzymes are generally similar to the corresponding mammalian enzymes as well as those in *Aedes aegypti,* there are a number of differences, particularly in sensitivity to drugs, that are of interest to pharmacologists. There is also some evidence that the functional role of methylenetetrahydrofolate reductase in adult *D. immitis* differs from its counterparts in nonhelminths in that the filarial enzyme preferentially catalyzes the oxidation of N^5-methyltetrahydrofolate to N^5,N^{10}-methyltetrahydrofolate which can, in turn, be converted to other tetrahydrofolate cofactors.

In summary, then, it has been demonstrated that *D. immitis* possesses the complete suite of enzymes to catalyze the formation of tetrahydrofolate and the interconversion of all known single-carbon unit-donating tetrafolate cofactors. That they function in this way has been shown by incubating *D. immitis* in a purine-free medium that contained radioactive N^5-methyltetrahydrofolate as the only folate source. A purine ribonucleotide was extracted from

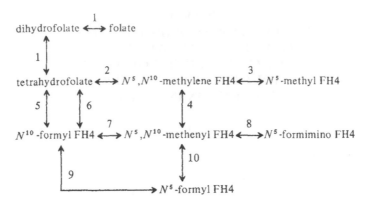

FIGURE 3.   The metabolism of folic acid (FH4 = tetrahydrofolate). All of the enzymes involved in these interconversions have been demonstrated in *D. immitis*.[1,4,32]   They are (1) dihydrofolate reductase; (2) serine hydroxymethyltransferase; (3) N^5,N^{10}-methylenetetrahydrofolate reductase; (4) N^5,N^{10}-methylenetetrahydrofolate dehydrogenase; (5) N^{10}-formyltetrahydrofolate synthetase; (6) N^{10}-formyltetrahydrolate dehydrogenase; (7) N^5,N^{10}-methenyltetrahydrofolate cyclohydrolase; (8) N^5-formyl, N^{10}-formyltetrahydrofolate mutase; (10) N^5-formyltetrahydrofolate cyclohydrase.

the parasite that was labeled in positions 2 and 8 on the purine ring. It seems likely that N^{10}-formyltetrahydrofolate and N^5,N^{10}-methenyltetrahydrofolate are first synthesized from the precursor, and that they donate carbons 2 and 8, respectively.[33,34]

VI. OTHER ENZYMES STUDIED

A. Phosphatases

A range of parasitic nematodes has been shown to possess high acid and low alkaline phosphatase activities.[35] In adult *D. immitis,* both the intestine and the hypodermis are positive. The presence of the enzymes in the hypodermis is probably a correlate of the ability of the worm to transport nutrients across the cuticle as acid phosphatase is involved in the functions of absorption, secretion, and excretion.[36]

There has been great interest in the possible use of the phosphatases as a diagnostic tool to distinguish between microfilariae of various species.[37] There are completely distinct patterns of acid phosphatase activity in *D. immitis* and *Dipetalonema reconditum* even when infections are mixed.[38] The histochemical distribution of acid phosphatase has since been determined in a number of nematodes, including several species of filarial worms, and differs consistently between species. There are also apparently different patterns of activity associated with each developmental stage in the mosquito. The first larval stage of *Dirofilaria immitis* shows maximum acid phosphatase activity in the anal vesicle and the buccal cavity. In the second stage, the enzyme is distributed throughout the alimentary canal and by the time the infective third stage has been reached the whole body of the worm is positive.[39]

B. Acid Protease

Acid proteases are widespread in helminths, irrespective of host or site of predilection, and also occur in *D. immitis*.[40] It seems that the enzyme levels are much higher in blood nematodes than in those inhabiting the gastrointestinal tract. In *D. immitis,* the greatest activity is found in the intestine. The enzyme has a pH optimum lying between 3.1 and 4.5, and hemoglobin appears to be the preferred substrate. This is consistent with the habitat of the parasite and the fact that it has been observed to ingest erythrocytes although other proteins, such as myoglobin, casein, and albumin are also hydrolyzed.[41]

Table 4
K$_m$s (mM) FOR HEXOKINASE
FROM MALES AND FEMALES OF
TWO SPECIES OF *DIROFILARIA*[49]

	D. immitis		D. roemeri	
	ATP	glucose	ATP	glucose
Male	2.06	0.024	3.07	0.66
Female	0.85	0.100	0.37	0.37

There are at least two acid proteases although the possibility that they are comprised of the same subunits (molecular weight 23,000) cannot be ignored.[42] Two acid proteases have been partially purified from soluble extracts of *D. immitis* and designated F$_p$I and F$_p$II. The molecular weight of F$_p$I is about 170,000, with a pH optimum between 4.6 and 5.8. It is obviously a sulfhydryl enzyme as its activity in enhanced by the protective effects of such reagents as mercaptoethanol and cysteine while it is inhibited by the sulfhydryl inhibitors like iodoacetate. F$_p$II has a molecular weight of 48,000, a pH optimum lying between 2.6 and 3.4, and is highly sensitive to pepstatin (80% inhibition at 10 nM) suggesting that, like cathepsin, it is a carboxyl protease.[43].

C. Enzymes of Respiratory Metabolism

Lactate dehydrogenase from *D. immitis* has been partially purified.[44] It has a molecular weight of about 130,000, similar to that of other helminths. The equilibrium is strongly in favor of lactate formation. The K$_m$s (substrate concentration giving half maximum activity) for NADH and NAD are 0.28 and 0.9 mM, and for pyruvate and lactate, 0.3 and 11 mM, respectively. The enzyme is noteworthy because it is inhibited by suramin (0.006 mM gives 50% inhibition), a compound which has been suggested for use against human filariases.

"Malic enzyme" (malate dehydrogenase, decarboxylating) has also been partially purified from *D. immitis*.[17] It catalyzes the reaction

$$malate + NADP \rightarrow pyruvate + NADPH_2 + CO_2$$

Equilibrium is greatly in favor of pyruvate formation, and K$_m$s for malate and pyruvate are 0.85 and 5.0 mM, respectively. It is very sensitive to suramin, a concentration of 0.015 mM giving 50% inhibition.[44] The emphasis on decarboxylation offers support for a role of malic enzyme in maintaining cytosolic pyruvate pools for lactate dehydrogenase (see Section III, Respiratory Metabolism).

Pyruvate kinase from *D. immitis* is similar to that of other invertebrates, having a molecular weight of about 237,000. It occurs in two forms; an L-form which is strongly regulatory and is activated by fructose-1,6-biphosphate, and an M-form.[19,45]

Hexokinase exhibits one peculiarity that makes it worthy of mention, a marked sexual dimorphism. It has a pH optimum between 7.8 and 8.2, occurs in at least three isozymic forms, and may exhibit variation between strains. Hutchison et al.[46] report K$_m$s of 0.32 mM (glucose), 0.86 mM (fructose), and 0.39 mM (ATP), while Flockhart and Kohlhagen[49] have noted much lower K$_m$ values for glucose and much higher ones for ATP (Table 4). The sexual dimorphism is apparent in the K$_m$s for both *D. immitis* and *D. roemeri,* and implies that the enzyme from the female is much more sensitive to fluctuations in internal ATP levels than that from the male.

D. Thymidine Kinase

Cytosolic thymidine kinase is an important enzyme in the group of enzymes that salvage pyrimidines. It catalyzes the reaction

$$\text{thymidine} + \text{ATP} \rightarrow \text{thymidine--5--phosphate} + \text{ADP}$$

It has been partially purified (molecular weight 180,000) and is dependent on Mg^{2+}, with a pH optimum of 7.0. K_ms for thymidine and ATP are 60 μm and 1.6 μM, respectively.[47]

VII. CONCLUSIONS

There is much yet to be learned about the biochemistry of *D. immitis*. Present evidence suggests that the parasite is capable of both aerobic and anaerobic metabolism. The concept of homolactate fermentation, of which the filariae were once the great exemplars, must be rejected. As well as glycolysis, the filariae possess tricarboxylic acid cycle and fumarate reductase activities and, although the latter has not yet been reported in *D. immitis,* there is no evidence to show that, biochemically, *D. immitis* is different from the other filariae. However, as a cautionary inscription in my laboratory continually reminds all who work in it, "absence of evidence is not evidence of absence".

Much more is known about the lipid and folate metabolism of filariae than of any other helminth group. Once again, there is little to single out *D. immitis* from the others. It appears to be capable of a little fatty acid chain lengthening and not much else. It is probably not capable of synthesizing sterols *de novo,* and, although it has a full array of enzymes for the interconversion of the folic acid derivatives, it remains to be confirmed that it is capable of synthesizing purines and pyrimidines, although there is some evidence that it possesses the metabolic equipment for their salvage from the medium.

Almost nothing is known about protein synthesis or nucleic acid metabolism. The mechanisms of transcuticular nutrient uptake remain a mystery and the biochemical regulation of development has not even been considered. There is clearly much to be done, and it is dangerous to rely too heavily on parallel studies in other helminths to fill the void. It is a paradox of comparative biochemistry that, as our knowledge of helminth metabolism increases, we learn that it is often invalid to extrapolate from one organism to another.

REFERENCES

1. **Barrett, J.,** Biochemistry of filarial worms, *Helminthol. Abstr. Ser. A.,* 52, 1, 1983.
2. **Yanagisawa, T. and Koyama, T.,** (Quoted by Barrett, J., in *Helminthol. Abstr. Ser. A,* 52, 1, 1983.).
3. **Chen, S. N. and Howells, R. E.,** The uptake *in vitro* of monosaccharides, disaccharides and nucleic acid precursors by adult *Dirofilaria immitis, Ann. Trop. Med. Parasitol.,* 75, 329, 1981.
4. **Jaffe, J. J., McCormack, J. J., and Meymarian, E.,** Comparative properties of schistosomal and filarial dihydrofolate reductases, *Biochem. Pharmacol.,* 21, 719, 1972.
5. **Jaffe, J. J. and Doremus, H. M.,** Metabolic patterns of *Dirofilaria immitis* microfilariae *in vitro, J. Parasitol.,* 56, 254, 1970.
6. **Ando, K., Mitsuhashi, J., and Kitamura, S.,** Uptake of amino acids and glucose by microfilariae of *Dirofilaria immitis in vitro, Am. J. Trop. Med. Hyg.,* 29, 213, 1980.
7. **Bryant, C. and Flockhart, H. A.,** Biochemical strain variation in parasitic helminths, *Adv. Parasitol.,* 25, 276, 1986.
8. **Van Oordt, B. E. P., Van den Heuvel, J. M., Tielens, A. G. M., and Van den Bergh, S. G.,** The energy production of the adult *Schistosoma mansoni* is for a large part aerobic, *Mol. Biochem. Parasitol.,* 16, 117, 1985.
9. **Mendis, A. H. W. and Townson, S.,** Evidence for the occurrence of respiratory electron transport in adult *Brugia pahangi* and *Dipetalonema viteae, Mol. Biochem. Parasitol.,* 14, 337, 1985.

10. **Fry, M. and Brazeley, E. P.,** Mitochondrial NADH-fumarate reductase in adult *Brugia pahangi, Tropenmed. Parasitol.,* 36 (Suppl.), 25, 1985.
11. **Ramp. T. and Kohler, P.,** Glucose and pyruvate catabolism in *Litomosoides carinii, Parasitology,* 89, 229, 1984.
12. **Hutchinson, W. F. and McNeill, K. M.,** Glycolysis in the adult dog heartworm, *Dirofilaria immitis, Comp. Biochem. Physiol.,* 35, 721, 1970.
13. **Earl, P. R.,** Filariae from the dog *in vitro, Ann. N.Y. Acad. Sci.,* 77, 163, 1959.
14. **Hutchison, W. F. and Turner, A. C.,** Glycolytic end products of the adult dog heartworm, *Dirofilaria immitis, Comp. Biochem. Physiol.,* 62B, 71, 1979.
15. **Turner, A. C. and Hutchison, W. F.,** Oxidative decarboxylation reactions in *Dirofilaria immitis* glucose metabolism, *Comp. Biochem. Physiol.,* 73B, 331, 1982.
16. **McNeill, K. M. and Hutchison, W. F.,** The tricarboxylic acid cycle enzymes in the adult dog heartworm, *Dirofilaria immitis, Comp. Biochem. Physiol.,* 38B, 493, 1971.
17. **Walter, R. D. and Albiez, E. J.,** Inhibition of NADP-linked malic enzyme from *Onchocerca volvulus* and *Dirofilaria immitis* by suramin, *Mol. Biochem. Parasitol.,* 4, 53, 1981.
18. **Turner, A. C. and Hutchison, W. F.,** Lipid synthesis in the adult dog heartworm, *Dirofilaria immitis, Comp. Biochem. Physiol.,* 64B, 403, 1979.
19. **Brazier, J. B. and Jaffe, J. J.,** Two types of pyruvate kinase in schistosomes and filariae, *Comp. Biochem. Physiol.,* 44B, 145, 1973.
20. **McNeill, K. M. and Hutchison, W. F.,** Carbohydrate metabolism of *Dirofilaria immitis,* in *Canine Heartworm Disease: The Current Knowledge,* Bradley, R. E., Ed., University of Florida, Gainesville, Fla., 1972, 51.
21. **Hutchison, W. F. and Turner, A. C.,** Glycerol metabolism in the adult dog heartworm, *Dirofilaria immitis, Comp. Biochem. Physiol.,* 64B, 399, 1979.
22. **Hutchison, W. F., Turner, A. C., Grayson, D. P., and White, H. B.,** Lipid analysis of the adult dog heartworm, *Dirofilaria immitis, Comp. Biochem. Physiol.,* 53B, 495, 1976.
23. **Srivastava, A. K. and Jaffe, J. J.,** Phosphatidylcholine synthesis in adult *Dirofilaria immitis* females, *Int. J. Parasitol.,* 15, 27, 1985.
24. **Srivastava, A. K. and Jaffe, J. J.,** Phosphatidylserine synthesis in adult *Dirofilaria immitis* females, *Int. J. Parasitol.,* 16, 9, 1986.
25. **Srivastava, A. K., Jaffe, J. J., and Lambert, R. A.,** Phosphatidylethanolamine synthesis in adult *Dirofilaria immitis* females, *Int. J. Parasitol.,* 15, 429, 1985.
26. **Comley, J. C. W., Jaffe, J. J., Chrin, L. R., and Smith, R. B.,** Synthesis of ubiquinone 9 by adult *Brugia pahangi* and *Dirofilaria immitis:* evidence against its involvement in the oxidation of 5-methyltetrahydrofolate, *Mol. Biochem. Parasitol.,* 2, 271, 1981.
27. **Comley, J. C. W. and Jaffe, J. J.,** Isoprenoid biosynthesis in adult *Brugia pahangi* and *Dirofilaria immitis, J. Parasitol.,* 67, 609, 1981.
28. **Pennock, J. F.,** Terpenoids in marine invertebrates, in *International Review of Biochemistry: Biochemistry of Lipids,* Vol. 14, Goodwin, T. W., Ed., University Park Press, Baltimore, 1977, 153.
29. **Mendis, A. H. W., Rose, M. E., Rees, H. H., and Goodwin, T. W.,** Ecdysteroids in adults of the nematode, *Dirofilaria immitis, Mol. Biochem. Parasitol.,* 9, 209, 1983.
30. **Cherian, P. V. R., Strombert, B. E., Weiner, D. J., and Soulsby, E. J. L.,** Fine structure and cytochemical evidence for the presence of a polysaccharide surface coat of *Dirofilaria immitis* microfilariae, *Int. J. Parasitol.,* 10, 227, 1980.
31. **Comley, J. C. W., Jaffe, J. J., and Chrin, L. R.,** Glycosyl transferase activity in homogenates of adult *Dirofilaria immitis, Mol. Biochem. Parasitol.,* 5, 19, 1982.
32. **Jaffe, J. J. and Chrin, L. R.,** Folate metabolism in filariae: enzymes associated with 5,10-methylene-tetrahydrofolate, *J. Parasitol.,* 66, 53, 1980.
33. **Jaffe, J. J. and Chrin, L. R.,** Involvement of tetrahydrofolate cofactors in *de novo* purine ribonucleotide synthesis by adult *Brugia pahangi* and *Dirofilaria immitis, Mol. Biochem. Parasitol.,* 2, 259, 1981.
34. **Jaffe, J. J., Chrin, L. R., and Smith, R. B.,** Folate metabolism in filariae. Enzymes associated with 5,10-methenyltetrahydrofolate and 10-formyltetrahydrofolate, *J. Parasitol.,* 66, 428, 1980.
35. **Maki, J. and Yanagisawa, T.,** Histochemical studies on acid phosphatase of the body wall and intestine of adult filarial worms in comparison with that of other parasitic nematodes, *J. Helminthol.,* 54, 39, 1980.
36. **Barka, T.,** Cellular localisation of acid phosphatase activity, *J. Histochem. Cytochem.,* 10, 231, 1962.
37. **Acevedo, R. A., Theis, J. H., Kraus, J. F., and Longhurst, W. M.,** Combination of filtration and histochemical stain for detection and differentiation of *Dirofilaria immitis* and *Dipetalonema reconditum* in the dog, *Am. J. Vet. Res.,* 42, 537, 1981.
38. **Chalifoux, L. and Hunt, R. D.,** Histochemical differentiation of *Dirofilaria immitis* and *Dipetalonema reconditum, J. Am. Vet. Med. Assoc.,* 158, 601, 1971.
39. **Omar, M. S.,** Distribution of acid phosphatase activity in the larval stages of *Wuchereria bancrofti, Brugia malayi, Brugia pahangi* and *Dirofilaria immitis* in the mosquito, *Tropenmed. Parasitol.,* 28, 100, 1977.

40. **Sato, K., Takahashi, J. and Sawada, T.,** Studies on the globinolytic enzymes of *Dirofilaria immitis, Schistosoma japonicum* and *Clonorchis sinensis, Jpn. J. Parasitol.,* 25, 8, 1976.
41. **Maki, J., Furuhashi, A., and Yanagisawa, T.,** The activity of acid proteases hydrolysing haemoglobin in parasitic helminths with special reference to interspecific and intraspecific distribution, *Parasitology,* 84, 137, 1982.
42. **Sato, K. and Suzuki, M.,** Studies on an acid protease from *Dirofilaria immitis. Jpn. J. Parasitol.,* 32, 89, 1983.
43. **Swamy, K. H. S. and Jaffe, J. J.,** Isolation, partial purification and some properties of two acid proteases from adult *Dirofilaria immitis, Mol. Biochem. Parasitol.,* 9, 1, 1983.
44. **Walter, R. D.,** Inhibition of lactate dehydrogenase activity from *Dirofilaria immitis* by suramin, *Tropenmed. Parastiol.,* 30, 463, 1979.
45. **Munday, K. A., Giles, I. G., and Poat, P. C.,** Review of the comparative biochemistry of pyruvate kinase, *Comp. Biochem. Physiol.,* 67, 403, 1980.
46. **Hutchison, W. F., Turner, A. C., and Oelshlagel, F. J.,** Hexokinase of the adult dog heartworm, *Dirofilaria immitis, Comp. Biochem. Physiol.,* 58B, 131, 1977.
47. **Jaffe, J. J., Comley, J. C. W., and Chrin, L. R.,** Thymidine kinase activity and thymidine salvage in adult *Brugia pahangi* and *Dirofilaria immitis, Mol. Biochem. Parasitol.,* 5, 361, 1982.
48. **Kohlhagen, S.,** personal communication, 1985.
49. **Flockhart, H. A. and Kohlhagen, S.,** unpublished observations, 1985.

Chapter 4

CLINICAL SIGNS AND DIAGNOSIS OF CANINE DIROFILARIASIS

Richard B. Atwell

TABLE OF CONTENTS

I. CLINICAL HISTORY

Generally, the epidemiological information associated with an infected dog relates to lack of or inadequate use of diethylcarbamazine (DEC) and access to infected mosquitos, both in the immediate vicinity and while dogs are in transit through infected areas. The inadequate use of DEC includes the following: too little, too infrequent, started too late, did not allow for growth of the pup.

Dogs presenting with mild heartworm infection usually have few clinical signs.[1,2] The infection is usually diagnosed by routine or elective testing rather than being based on particular history or on the presentation of specific clinical signs. Moderate to severe cases of heartworm disease have a history of loss of condition and body weight, dry harsh hair coat, a chronic cough, and occasional inappetance. More severe cases exhibit exercise intolerance, syncope, and signs of right-sided congestive heart failure (RSCHF). Hemoptysis is sometimes seen in severe cases where owners report blood loss from the head usually associated with exercise and/or bouts of coughing.[3,4] There is no specific breed susceptibility although the Bull Terrier breed seems to be over-represented in the more severely affected cases seen in Australia.[5]

The history associated with cases of caval syndrome is covered in Chapter 11 and of the DEC reaction in Chapter 9.

The history associated with unusual cases of the disease relates to the particular body area involved. Posterior paralysis and other central nervous system (CNS) signs are seen with epidural and/or cerebral location of the parasite, and muscle and skin necrosis with systemic arterial embolization.[6]

Sometimes ectopic parasites are located by accident, (e.g., at laparotomy) or are associated with gangrenous wounds thought originally to be unassociated with the parasite (e.g., toe necrosis due to femoral artery obstruction and periphreal ischemic necrosis with worms protruding from the wound).[5]

II. CLINICAL SIGNS

It needs to be emphasized that even with all the available (and sometimes costly) diagnostic support technology, there is no excuse for an incomplete and poorly performed physical examination.[7] Examination should consist of a distant inspection as the dog walks to the table, observing gait, respiratory effort, degree of cachexia, abdominal distension, etc. as well as the dog's reactions to its environment. Close physical examination should be thorough — examining all the body areas: head and neck, thorax, abdominal cavity, and inguinal and peripheral areas. A thorough cardiorespiratory examination is also essential if maximal clinical information is to be obtained with each case.

A. Usual Presentations

Usual clinical signs in each individual case are dependent on the extent of pulmonary disease and its effect on the general health of the dog (e.g., secondary metabolic anemia), as well as the specific effects from the cor pulmonale lesions. These lesions include increments to the afterload of the right ventricle due to progressive pulmonary arterial and related pulmonary parenchymal disease.[2,3]

Mild cases have few if any clinical signs and are diagnosed with laboratory tests. The danger in such cases is that the veterinarian may be misled into thinking that other nonrelated clinical signs are associated with a microfilaremia from a mild infection, thus possibly disguising another disease or problem in the dog.

Moderately affected cases have a chronic cough, weight and condition loss, and hair coat changes. These latter changes are not believed to be directly associated with the disease but are more likely due to general metabolic stress. Mucosae may be pale due to both a peripheral perfusion deficiency resulting from cor pulmonale and to anemia. The latter is associated with metabolic stress and low-grade hemolytic anemia, caused by increased mechanical fragility of erythrocytes.[8] This mucosal paleness may be accentuated with exercise, which tends to give a rough guide to the extent of compromise of the dog's pulmonary arterial perfusion capacity. The right cardiac apex beat may be more obvious due to right ventricular dilation, but care is needed with this interpretation in deep chested dogs. Hepatomegaly and ascites are not usually seen in moderate cases. Auscultation is usually normal. Sometimes friction sounds can be heard due to pleural adhesions which are not an uncommon sequela to peripheral pulmonary hemorrhage in this disease.[9] An accurate description of abnormal respiratory sounds of heartworm cases has not been reported and with the recent change in nomenclature of respiratory sounds some confusion does exist in this area. However, accentuated crackles are sometimes heard in moderate cases presumably associated with areas of extensive hemorrhage and inflammation.

Severe cases usually have tachypnea and tachycardia with the animal having no exercise tolerance, and perhaps syncope, even with mild exercise. Dyspnea with exercise is also apparent and reflects the extent of the pulmonary arterial disease. Arrhythmias are unusual which is surprising considering the right side of the heart is so involved. In particular, the right atrium is affected by stretching and dilation in cases of right-sided congestive heart failure (RSCHF) and extensive right atrial involvement in cases of caval syndrome. Mucosae are pale, and sometimes cool, dry, and sticky. These findings are caused by both anemia and peripheral perfusion deficits. The inability to perfuse damaged pulmonary arteries, effectively creating underloading of the left ventricle and thus reducing cardiac output, leads to reduced peripheral perfusion. Cyanosis is rare in severe cases. However, it can be seen when other respiratory disease is present in infected dogs, resulting in diagnostic confusion.[5]

A systolic jugular pulse (believed to be associated with tricuspid regurgitation) is usually present with cases of RSCHF in which the right apex beat is usually pronounced. The liver is palpably enlarged, particularly when the abdomen at the left costal arch is gently palpated. Ascites is often very pronounced but is sometimes difficult to detect, particularly in the older, obese dog. The amplitude of the femoral artery pulse is usually lower than normal with the rate sometimes increased. The pulse rhythm is usually normal but pulse deficits can be present in severe cases of RSCHF and in those cases that have arrhythmias. Peripheral cooling can also be present in severe cases. Edema of the back legs or subcutaneous edema is unusual in cases of RSCHF and, if present, is a poor prognostic sign.

Auscultation in severe cases will occasionally reveal a systolic murmur over the pulmonic area (believed to be due to worm-induced turbulence in the pulmonary outflow tract seen in angiograms)[5] and over the tricuspid area, due to both worm-induced turbulence and regurgitation secondary to dilatation of the valvular annulus. The second heart sound will be louder and will sometimes appear split, presumably due to delayed closure of the pulmonic valve associated with the reduced rate of pulmonary blood flow and pulmonary hypertension.[3] Sometimes precordial thrills are felt, in particular, over the tricuspid area when worm entanglement and/or turbulence is present, or in cases of caval syndrome. Although not pathognomonic, a marked thrill over the tricuspid area in a dog with acute RSCHF is suggestive of severe worm entanglement.[5] Pulmonic valve incompetence due to inefficient valve closure resulting from distortion or worm presence is not common.[3]

Pulmonary auscultation is difficult in these cases due to the extent of cardiorespiratory distress. Friction sounds and crackles can be heard, but, as with cardiac auscultation, what is heard will depend on the patient, the extent of disease, and on the auscultatory skills of the veterinarian.

B. Unusual Presentations

Table 1 outlines other less common presentations and their associated clinical signs. It is

Table 1
RARE/UNUSUAL PRESENTATIONS ASSOCIATED WITH CANINE DIROFILARIASIS

Condition	Clinical Signs	Ref.
Ischemic myopathy	Posterior weakness, peripheral hypothermia, low amplitude femoral pulse due to arterial thrombosis	10
Peripheral ischemia/necrosis	Lameness, gangrenous lesions, painful foci, similar to ischemic myopathy	6
Paradoxical embolism	Associated with right to left intracardiac shunts; dependent on area of embolism; lameness, pain, necrosis, etc.; similar to peripheral ischemia, mostly embolizes hind legs and kidney	11
Filariae in CNS/spinal cord	Ataxia, hemiplegia, depression, lethargy, seizures	12
Cerebral infarction	Circling, rolling, ataxia, quadriplegia, dysphagia, blindness, coma	13
Worm entanglement of the tricuspid valve	Acute RSCHF, shock, collapse, tachycardia, tachypnea, dyspnea, hemoglobinuria, marked thrill and murmur over tricuspid area, jugular pulse	
Arrhythmias	Unusual, usually superventricular tachycardia or atrial fibrillation	
DIC	Unusual bleeding, thrombocytopenia and signs of the primary cause of DIC	14
Defibrination syndrome	Unusual bleeding, jaundice, hypofibrinogenemia	15
DIC and associated polyneuropathy	Weakness, muscle atrophy, poor peripheral sensory response	16
Pneumothorax	Respiratory distress proportional to loss of lung function	17
Hemothorax	As for pneumothorax and the secondary effects of blood loss	
Filariae free in abdominal or pleural cavity	Usually no direct clinical signs except in the pleural cavity where filariae usually enter the cavity from lung lesions	
Filariae in anterior chamber	Uveitis, hypopyon, hyphema, mobility of worm is visible with penlight and slit lamp	
Cutaneous granulomas	Histology needed to diagnose, nonspecific grossly	18

obvious that combinations of some of these presentations can exist in a single case making diagnosis more difficult and prognosis more severe. Table 2 refers to some of the more common drug-associated clinical signs which may become apparent and may confuse the clinical assesment of a particular case during the course of treatment.

C. Other Specific Presentations and Suggested Treatment
1. Occult Disease

This presentation has probably been overemphasized in the literature because of the overreliance in the past for diagnosis based only on tests to confirm microfilarial presence and because one of the forms of occult disease is usually a severe clinical presentation, the so-called immune-mediated occult presentation.[2]

The occult condition (an infected case without the presence of circulating microfilariae) is caused by the following situations: (A) inability to produce microfilariae as with (1) single-sex or (2) single-worm infections, (3) prepatent, (4) immature, or (5) geriatric infections, or (6) ectopic infections; (B) infections where any microfilariae produced are (7) destroyed by host defense mechanisms in the spleen, liver, and lung (so-called immune-mediated occult); and (C) where microfilarial production can be (8) reduced or stopped due to the sterilizing effects of drugs (e.g., levamisole,[19] ivermectin[20]) or by immunological mechanisms[3]

Table 2
POSSIBLE DRUG-RELATED PRESENTATIONS IN
CLINICAL CASES

Condition	Clinical signs
Thiacetarsamide	Depression, anorexia, vomiting, jaundice, toxic hepatitis/nephritis, coma, death
Levamisole[a]	
Gut	Vomiting, diarrhea, inappetance, anorexia
CNS	Depression, lethargy, excitability (hallucinations?), ataxia (hind legs particularly), behavioral changes, temperament changes, seizures, death
Skin	Dry and scaly coat, alopecia, toxic epithelionecrolysis (TEN)
Blood	Hemolytic anemia, thrombocytopenia, bone marrow depression
Systemic	Hypo- and hyperthermia
Diathiazanine iodide	Vomiting, anorexia
Ivermectin	Shock, depression, lethargy, inappetance
Post-adulticidal therapy complications	Due to inflammation and thrombosis, thromboembolism, coughing, fever, depression, anorexia, hemoptysis, dyspnea, pale mucosa, thrombocytopenia, bronchial-arterial fistula, exsanguination
Phlebitis (thiacetarsamide)	Pain, swelling, necrosis, slough at I/V site
Aspirin side effects	Vomiting, depression, bleeding (gastrointestinal in particular), melena, death due to exsanguination
Prednisolone side effects	Polydipsia, polyuria, polyphagia, steroid-induced hepatopathy

Note: Refer to Chapter 8 for more detail.

[a] See Reference 2.

directed at the uterus of the worm. This last form (C) is not well documented in the literature. Of these eight forms of occult disease, the immune-mediated occult case is the most clinically significant presentation. In these cases severe respiratory distress is evident in the dogs (polypnea, dyspnea) associated with the other signs of heartworm disease — namely coughing, sometimes blood-tinged sputum, weight loss, depression, exercise intolerance, etc.[3] These cases also usually have more severe ausculatory changes (increased crackles), a responsive anemia, eosinophilia (peripherally and with bronchial aspiration), protein-losing nephropathy, and hypergammaglobulinemia.[21,22] The pneumonitis of heartworm disease is believed to be Type 3 antigen deposition-hypersensitivity.[5] The immune-mediated occult case is an extreme example of an over-reaction to deposited antigen (killed and dying microfilariae) in the capillaries and parenchyma of the lung.

Basically then, these cases have severe clinical signs with no circulating microfilariae with a predominance of respiratory signs at presentation. Radiography confirms the respiratory involvement with extensive prednisolone-responsive alveolar patterns overlying the more chronic interstitial vascular heartworm pattern. The response to prednisolone therapy is usually effective and spectacular improvement can be seen in some cases.

2. Pulmonary Artery-Bronchial Fistula

These can occur in severe cases where extensive pulmonary pathology is present. The weakened arterial walls can either rupture into the closely associated airway or rupture into the pulmonary parenchyma.[6] In the latter case bleeding is usually minimal and few directly associated clinical signs are seen apart from coughing. However, if bleeding occurs into the airways and then into the dependent parenchyma, dogs present with acute cardiopulmonary distress associated with fresh blood being seen about the head, exiting via the mouth or nose from a source of bronchial bleeding in the lungs. Any immediate treatment is always associated with a grave prognosis as a "bleed out" can easily occur into the bronchial tree and dependent pulmonary tissue. However, the following suggestions may be of use in a less severe case: absolute rest and stressless manipulation, sedation, oxygen therapy (provided it is given without stress), antitussive therapy (as chronic coughing is believed to be a cause of sudden pulmonary arterial hypertension that may lead to arterial rupture), hypotensive therapy to lower systemic blood pressure and hopefully reduce pulmonary arterial pressure, and assessment of the need for fluid or whole blood replacement.[2] Obviously, these situations are critical and any therapy that may reduce the possibility of further bleeding due to blood-pressure effects in the pulmonary arteries should be instigated. The immediate outcome in such cases is probably more dependent on the size of the vascular defect and resultant fistula and on the clotting status of the dog rather than on specific drug therapy. Once stabilized and a focal bleeding site is established by radiography, lobectomy can be considered.[3] However, such cases usually have substantial disease and anesthesia and surgery are usually difficult and not without risk.

In any severe case, consideration should be given to this occurence as a potential development during or after treatment and is yet another reason for rest following adulticidal therapy. Similarly, exercise (even months after therapy) that is likely to lead to pulmonary hypertension should be avoided. Paroxymal coughing should also be controlled in the severely affected but successfully treated dog, as arterial damage produces areas of weakness that can persist even after the dog appears clinically stabilized.[5]

3. Severe Pulmonary Thromboembolism

These dogs are presented with extreme respiratory distress, pale mucosa and other signs of shock, no exercise tolerance at all, and signs of severe heartworm disease usually 7 to 17 days following arsenical therapy.[22] However, spontaneous cases can be seen in which the extent of thrombosis can be extremely severe.[5] The signs have usually developed over 24 to 48 hr and the absence of hemoglobinuria differentiates cases of caval syndrome. Diagnosis is supported by tests to confirm intravascular coagulation (e.g., fibrin degradation products [FDP], coagulogram abnormalities, thrombocytopenia) and the presence of widely dilated, truncated, opaque, pulmonary arteries, particularly about the carina.

Cage rest, prednisolone, fluid therapy, terbutaline, and antibiotics are administered. Comsumptive thrombocytopenia can be seen simultaneously with signs of melena, epistaxis, hemoptysis, petechial hemorrhage, and sometimes hematuria. Treatment involves heparin 150 to 250 units/kg three times a day, platelet-rich plasma, and vincristine (0.4 mg/m^2 surface area) if the platelet count is very low. Aspirin needs to be used very cautiously once the coagulogram and platelet numbers are considered normal.

III. DIAGNOSTIC PROCEDURES*

To confirm the suspected diagnosis of dirofilariasis based on clinical signs and historical factors in each case, a series of diagnostic tests are used.[24] While not absolute, the following tests and their sequential use are based on ease of use, costs, accuracy, reliability, and the associated benefits of prognostic assessment with some of these tests.

Microfilaremias are evident in a variable percentage of dogs depending on the infection

---

* This section has been written chiefly for the practicing veterinarian who has limited access to the published material, to some of the research-oriented technology used in diagnosis, and to other assessment procedures at institutions.

rate in each geographic area, on the local prevalence of occult disease, and on the local infection rate of *Dipetalonema reconditum*.

Distinguishing different types of microfilariae is based on motility on a thick smear examination, and more accurately, on measurements of length and width, and on head and tail characteristics. Different types of microfilariae have been reported (see Chapter 1), but the chief need is to distinguish between microfilariae of *Dirofilaria immitis* and of *Di. reconditum*. It must be noted that variable handling of blood samples can alter dimension measurements of microfilariae via a calibrated ocular micrometer.[3]

A. Tests for Detection and Identificaton of Microfilariae
1. Wet Smear (Thick Smear)
This is a simple, easy, and cheap test. A drop of blood (ethylenediamine tetraacetic acid [EDTA] or heparin) is placed on a glass slide and a coverslip applied. This avoids the effects of drying and heating from any light source. Under low power, microfilariae are detected most easily by their movement, particularly about the edge of the coverslip. *D. immitis* microfilariae tend to have stationary motility whereas those of *Di. reconditum*, which are usually in very low numbers, usually move across the field of view. Obviously if there are no microfilariae evident, the infection may be occult or the level of microfilariae may be too low to be seen in the small sample of blood used, particularly in the cold winter months when peripheral microfilarial counts can be reduced (e.g., 93.5% down to 75.5%).[25]

2. Microcapillary Hematocrit[26]
This technique is not used widely, but involves microscopic examination for motile microfilariae in and above the buffy coat of the prepared microhematocrit tube placed horizontally on the microscope stage. The tube needs to be rotated to examine all areas and the focus racked up and down to examine different levels of the tubes.

3. Concentration and Stain Tests[26]
The modified Knott test involves the use of 1 mℓ of the patient's blood, collected in EDTA, which is then added to 9 mℓ 2% formalin in a 15-mℓ centrifuge tube, mixed, and then centrifuged at 1000 to 1500 rpm for 5 min. The supernatant is decanted to leave 1 mℓ of sediment to which is added 0.5 to 1.0 mℓ of either 1/500 methyl green, or 1/1000 or 1/2000 methylene blue. Several slides are then prepared to examine all of the stained sediment. This technique allows fixation of microfilariae, staining of microfilariae, and concentrates the 1-mℓ sample improving the accuracy of the test. In fact, larger quantities of blood can be used if further accuracy is required in dogs with very low counts. Additionally, the sediment can be placed in a standard laboratory counting chamber and accurate counts made based on the known blood-sample volume.

As the microfilariae are fixed accurate measurements can also be made and the head shape and tail characteristics examined more closely. The test is inexpensive, provided access to a centrifuge is available. Disadvantages of the technique are possible cross contamination if centrifuge tubes are reusable and incompletely washed, and the adherance of microfilariae to the walls of the centrifuge tubes during preparation and transference to the counting chamber or slides.[27]

Various types of lysing fluid can be used with concentration and with filtration tests. One such solution, Lyse Alyve Solution® (Henry Schein, U.S.,) allows microfilariae to remain viable and therefore more easily seen on the filter or in the sediment.

4. Filter and Stain Test[28]
These are widely used in practice as a rapid, simple, less expensive test. One mℓ of blood collected in EDTA or fresh blood is mixed with 10 mℓ of lysing solution, usually com-

mercially supplied (e.g., Difil®, Evsco, Buena, N.J.) The filter (preferably 5 μm pore size) is carefully placed in the holder (see guidelines by supplier) without wrinkling and the filter holder assembled. The diluted and lysed solution is then syringed through the filter followed by 10 mℓ of distilled water which helps to avoid excess debris on the membrane. Variations exist at this point. Either the sample is stained while the filter is assembled[26] (5 to 10 mℓ of 1/10,000 toluidine blue followed by 10 mℓ distilled water) or after the filter has been placed on a slide using a commercial test stain (e.g. Difil®). The filter is then covered with a coverslip and, after allowing 2 to 5 min for staining to become effective, is examined for microfilariae. Advantages of the test are ease of use and short preparation time. However, cross contamination can occur from the lysing solution and if the filter holders are not washed thoroughly.[27] Morphology of the microfilariae is difficult to assess and sometimes they are hard to see due to background debris (e.g., unlysed cells). Microfilarial distortion can also occur, with the microfilariae usually becoming shorter due to shrinkage. Counting is possible but more difficult than with the modified Knott test. Microfilariae can also be lost due to wrinkling of the filter, through the pores of the filter, and in excess fluid at the edge of the slide when the coverslip is applied.[27]

In general, if morphology and counts are necessary the modified Knott test is used, whereas for speed and comparable accuracy, where the cost of the commercial kit is not considered significant, the filtration test is used.

5. Histochemical Tests[29]

A further test exists (Apfil®, Apex Labs, Sydney, Australia) which helps to histochemically differentiate microfilariae.[30] This is similar to the filter tests except that additional steps based on acid phosphate staining are used to produce differential staining patterns (*Di. reconditum* — overall red stain; proximal 1/3 pale red, distal 2/3 red; *D. immitis* — distinct enzyme bodies stained red, excretory and anal pores). The test does not appear to be widely used, particularly where the prevalence of *Di. reconditum* is low and the directions, as with all commercial tests, need to be carefully followed to produce reliable results.

6. Summary

Thus, in using concentration and/or filtration tests most cases with microfilariae will be detected. However, the wet smear and the capillary tube are more likely to produce false-negative results as the blood sample is small, the technique simplistic, and no objective differentiation of microfilariae is possible.

False-negative results with any test are, however, possible if the test procedures are not adhered to, if the microfilarial count is very low, if technical errors occur, or if microfilarial seasonal or daily periodicity is a factor in the geographic locality involved. Obviously, occult disease will never be diagnosed with the above tests.

If cost comparisons are excluded and the tests used correctly, the microfilarial detection results of the concentration tests and the filter tests are of the same order and seem to be the best "first order" procedure to adopt in practice. Most veterinarians perform a wet smear as a crude screening test and, if negative, proceed to concentration or filtration before using other tests.

The published results of comparative experiments of several tests are variable due to developmental stages of the tests and probably also due to the operators involved. For example, one study[31] showed the wet smear detected 79.8% of 119 mixed microfilarial samples, 89% with the modified Knott test, and 97.5% with the filter test (Difil®), whereas a review[32] of the results of other researchers seemed to suggest that the accuracies of the concentration and filter test were not practically different.

Microfilarial counts do not relate to the extent of pathology, to adult numbers, or to the so-called severity of infection.[33] Actual counts become important in two areas. Dogs are more likely to show adverse reactions to DEC if the microfilarial count is high (see Chapter 9), and dogs with very high microfilarial counts are more likely to be those that react to ivermectin therapy being used as a microfilaricide.[34]

Chapter 1 outlines some of the confusion over microfilarial variation. However, many factors will affect what peripheral count is obtained. It will depend on season and time of day to a different degree in different parts of the world, perhaps on the degree of excitement at collection time, the production rate of fertile filariae, the accuracy of the test you choose when used correctly, and most of all, on within-dog variation and within-sample variation. In fact, if dogs are closely monitored sometimes extreme within-dog variation of microfilarial numbers will be recorded.[35]

False-positive diagnosis of *D. immitis* infection, apart from incorrect microfilarial differentiation, can occur with regard to juvenile microfilaremias, which are usually very low and usually terminated early in life.[36] These dogs may, however, be potential occult cases due to preexisting immune recognition of microfilariae. The other possible cause is with dogs given infected blood during transfusion. Microfilariae so infused usually are removed quickly,[37] but survival times have been prolonged (over 2 years) in some cases. In such cases antigen tests could identify the case as negative for adult antigen suggesting that adult filariae are not present.[38]

The other use for tests for microfilariae is with the annual checkup prior to DEC usage. In heavily infected areas, dogs should be examined annually to ensure that infection (usually due to inadequte DEC coverage) has not occurred. This is particularly important if the dog has been off DEC during winter, with the possibility of a DEC reaction occurring if a microfilaremia has developed and DEC usage is recommenced automatically without checking.

Inadequate clearance of microfilariae after adulticide therapy can also be confusing in subsequent diagnostic tests as they may last 12 to 18 months after adulticide therapy.[39]

B. Tests to Diagnose Infection and to Assess Infection and Disease Status

1. Serological Tests

These tests are not absolute, should not be interpreted in isolation, and do not offer a universal diagnostic tool for all cases of heartworm infection. There have been numerous reports of the use of commercial serology to aid diagnosis over the last 3 to 5 years. Chapter 10 covers other aspects of serology and some of the earlier work using other methods. Confusion has been apparent with the introduction of some of these tests and their accuracy has varied tremendously. There has also been a basic change in the expectations of these tests from originally aiding diagnosis in occult cases to being expected to diagnose all infections in all cases, as well as detecting all types of occult disease, a situation which is probably unrealistic.

a. Indirect Fluorescent Antibody (IFA) Test

This test, for immune-mediated occult disease, is effective as it does detect IgG antibodies (in the patient's sera) in 80 to 95% of cases.[40]

A positive result suggests antibodies to microfilariae do exist in the blood sample from the case. Obviously this test is performed only in the well-established laboratory with the availability of a fluorescence microscope. Again, as with all tests, interpretation of results must be in the light of all other clinical information and not just absolutely on one laboratory finding. As an example, Grieve and colleagues[41] showed that the IFA had a positive predictive value of 60% and a negative predictive value of 37.5%. That is, it was used to detect occult disease, but as immuno-mediated occult disease is only one of eight basic types these low values would be expected. Thus, for the diagnostician such a test, where there is no hope of prematurely knowing the type of occult disease, is virtually useless if used in isolation. This test does not detect the other types of occult disease, the most common cause of which are single sex, single worm, and immature infections (57 to 83% of cases).[33] In fact, in some areas the occult rate is low and so the need to use these sereological tests is lessened, with reliance being more on concentration and filtration-staining techniques.

b. Other Antibody Detection Tests

With the develpment of further tests, attempts have been made to improve both sensitivity and specificity so that all infections can be detected, i.e., all occult cases and microfilaremic cases as well. In actual fact, while such a test would be advantageous, economics and the efficiency of the simple first-order screening tests encourage practitioners to adopt the practical approach of wet smear, followed by a concentrate/filter test, and subsequently to use the newer serological tests. So in essence, there have been a lot of demands placed on these tests to be "correct in all situations" when their original aim was to help accurately diagnose, not the microfilariae-positive dogs, but the occult case.

Serological tests were mostly based on enzyme immunoassay (EIA) systems (often collectively called enzyme-linked immunosorbent assay [ELISA] tests) to detect antibody to *D. immitis* in the circulation. Numerous tests were available (e.g., Dirokit®, MAbCO, Brisbane, Australia; Dirotect®, Mallinckrodt, St. Louis, Mo.), but most are now not in use or have been withdrawn. They will still, however, have a use in detecting prepatent or immature infections where antibody levels are usually raised or increasing and where detectable antigen is most likely not present.

The chief problems with these tests were low accuracy and cross reactivity with other parasites, in particular *Di. reconditum*. Their chief design for use was to detect the occult case which then avoided the overreliance on the laboratory-based IFA test. However, these tests were then produced for general use and were released, possibly too early, into a market that was not fully aware of the limitation of such tests; the result being that they were used on all sorts of cases and so inaccuracies were reported.

The other obvious defect of such technology is that antibody presence (apart from the possibility of cross reactivity) does not necessarily infer that infection is present (e.g., following adulticidal therapy with dead filariae but a marked antibody titer) nor does it equate with the severity of infection. The immune-responsiveness of the individual dog to variable stages of the life cycle and to variable adult burdens is not predictable and, therefore, a 3 + as compared to a 1 + reaction is meaningless in terms of what it represents in adult parasite burden or associated pathology. Similarly, it is possible to have an active infection but a negative antibody response or in some cases even immuno-suppression of the host.

Thus, the tests designed to detect antibody should not be used to diagnose clinical cases of the disease except, perhaps, in suspected prepatent/immature cases or in the immune-mediated occult case where the dog's immune system is hyperresponsive to heartworm infection and so should have a higher antibody titer. Table 3 outlines some of the tests that have been or still are available. As with all tests, directions supplied by the companies must be rigidly adhered to, particularly with regard to temperature, product and sample storage, and time intervals during the tests. Expiry dates also need to be checked. Concurrent anti-inflammatory therapy may also affect the results of antibody detection tests.

Other technologies[42] for antibody detection do exist and two tests are available at the laboratory level. One, Modified ELISA (J. M. Veterinary Laboratories, Chula Vista, Calif.), detects IgG and IgM to antigen derived from adults and microfilariae. The other, Track XI® (Daryl Laboratories, U.S.), is an IFA system whereby colloid polymer columns containing adult worm antigen are exposed to test samples and then to a fluorescence reagent and a fluorescence reader. Both are designed to detect antibody to *D. immitis*, but, as with all antibody tests, care is needed in their interpretation in each clinical case.

A latex agglutination test is also available (Dirocult®, Synbiotics, San Diego, U.S.) that detects antibody when antigen-coated microscopic beads are agglutinated by the cross linking

Table 3
SEROLOGICAL TESTS TO
DETECT ANTIBODY THAT HAVE
BEEN USED IN DIAGNOSIS

Test	Type	Source
Dirocult®	LA	Synbiotics
Dirokit®	EIA	MAbCO
Dirotect®	EIA (ELISA)	Mallinckrodt
Track XI®	IFA	Daryl Labs

Note: LA = latex agglutination, EIA = enzyme immunoassay, ELISA = enzyme-linked immunosorbent assay, IFA = indirect fluorescent antibody test.

of specific heartworm antibody in the test sample, which must be fresh and unfrozen. As with all latex agglutination tests, experience is required with their interpretation and good lighting and background are required to accurately see the agglutination.[43]

Recent papers have shown how inaccurate these tests can be. Sisson and colleagues[44] showed a low range of sensitivities (39 to 74%) in occult cases. In some calculations doubtful cases were excluded, but if doubtful results exist then there is even more confusion for the clinician with the case at hand needing diagnosis.[44] Similarly, the work by Greene and colleagues,[45] comparing filtration with tests for antibody based on predictive values (the actual reliablity of a positive or negative result), showed that the predictive value of a positive test for occult cases was 52 to 55% and that for a negative test was 66 to 71%, with test efficiency being of the order of 62%. Thus, based on these two recent papers using the tests to help diagnose all types of occult cases less than acceptable results were obtained.

Similarly, research evaluating an antibody test found positive predictive values averaging 47.5%, but in some age groups the test results were very poor.[41] However, the negative predictive value was 100% suggesting that such a test, coupled with microfilarial detection and tests for antigen, would be excellent in selecting dogs for research purposes where heartworm-free cases are essential.

Another problem is confusion over the effect of the death of infective larvae following DEC therapy and concurrent antibody titers. Grieve et al.[46] showed that in experimental infections, the DEC-treated group of dogs did not differ serologically from the non-DEC group 11 weeks postinfection, but began to differ at 13 weeks post-treatment. So it is conceivable that dogs being given DEC (and so prophylactically protected), in endemic areas, could still produce false-positive serology with larval death.

Lombard[47] summarizing published data stated that published results of antibody tests in occult cases had given 5 to 88% false-positive results and 5 to 68% false-negative results. As always, to be more interpretive, the types of occult cases should be detailed, but the trend is obvious; antibody tests used in isolation are not accurate and should be used only with specific cases and situations.

c. Antigen Detection Tests
i. Diagnosis

These tests[48-50] are now considered to be the diagnostic tests of choice to detect circulating antigen after initial screening has been performed to detect microfilariae. In addition to diagnosis, prognostic assessment will be possible in the future involving estimation of adult worm burdens and assessment of worm mortality following adulticide therapy. In addition to this, the possibility of these tests also being suitable for predicting the potential extent of

Table 4
RESULTS OF COMMERCIAL TESTS[a] TO DETECT ANTIGEN IN CIRCULATION

Test	n	Sensitivity	Specificity	Accuracy	Ref.
Difil II® (LA)	97	70	90	80	Courtney[67]
Dirochek®	174	90.3	99.1	94.8	Synbiotics[67]
EIA-ELISA	NK	33.0	84.0	58.5	Grieve[67]
Greyhound	426	89	97	93	Courtney et al. 1986
		(+ PV = 96%; − PV = 92%)			unpublished
Greyhound	294	89	95	91	Courtney et al. 1986
					unpublished
		(+ PV = 97%; − PV = 82%)			
Dirokit latex® (LA)	304	84.3	88.8	86.7	MAbCO[b] (Agen)
	57	85.3	95.7	89.5	Atwell[67]
Diromail®	181	92.2	97.4	94.5	MAbCO (Agen)
EIA					Atwell[67]
Fresh samples	69	100	100	100	48
Stored Samples	18	78	96	92	48
Filarochek®	72	98	100	99	Weil[67]
EIA, ELISA, Macrobed	314	97.3	98.2	97.7	Mallinckrodt
	NK	97.0	85.0	91.0	Courtney[67]
	NK	67.0	90.0	78.5	Grieve[67]
Greyhound	294	91	98	94	Courtney et al. 1986
		(+ PV = 99%; − PV = 86%)			unpublished
Greyhound	NK	96	98	97	Courtney et al. 1986
					unpublished
		(+ PV = 98%; − PV = 97%)			

Note: LA = latex agglutination, EIA = enzyme immunoassay, ELISA = enzyme-linked immuno-sorbent assay, PV = predictive value, (+) = positive, (−) = negative, NK = not known.

[a]   Data on Cite® (Agritech, U.S.) and Clin Ease® not available during preparation of this text.
[b]   Based on 5 sets of results from Drs. Atwell, Vankan, Pope (Australia), Courtney, and Smith (U.S.).

thromboembolism following adulticide and worm death is under investigation.[38] However, it is stressed that ectopic infections will probably not be detected, the worms needing to be in the circulation for antigen levels to be detected.

Table 4 outlines the various tests and their reported accuracies. Predictive values allowing for local prevalence rates have now been employed which is a further aid to practitioners in assessing the real practical value of the results from these tests in their own locality.

The one disadvantage of these tests is their cost. Another potential problem is the psychological effect that new "state of the art" technology can have on the inexperienced user where new and changing concepts tend to be more readily rejected.

As far as the research dog is concerned it is imperative that the animal be free of the disease prior to costly experimentation. A combination of tests to detect microfilariae (e.g., modified Knott test), antibody (for prepatent infections), and antigen (for occult infections) will ensure the most accurate assessment available. Other tests, e.g., radiography, are only of use once infection is established sufficiently to produce detectable lesions. Arteriography is of use earlier than plain radiography and will show worm presence, but is invasive, time consuming, and expensive.

Compared to the antibody tests, there is little if any cross reactivity to *Di. reconditum* with the antigen tests.[38] However, there are no "absolutely always accurate" antigen detection tests available. Some are more sensitive (correctly detecting infected cases), or others more specific (correctly detecting noninfected cases), and all tests should have good sample reproducibility. It is emphasized that they are not screening tests for all dogs. Screening cases should be selected (e.g., suspect occult cases or particular problem cases where a "rule out" of heartworm infection may be needed). Predictive values (PV) based on known prevalence rates should also be used to enable the diagnostician to be able to assess each individual case with a sense of applied practicality; e.g., a PV of a positive result of 69% suggests that when a test result is positive, it has a 69% chance of being correct. Of course, we should apply these same PV to all laboratory tests to ensure more realistic interpretation.[51]

Generally speaking, the latex agglutination tests are less accurate but easier to perform and less complicated than the EIA tests (see Table 4). For example the Dirochek® test (Synbiotics, U.S.) is reported to be easier and faster (20 min as opposed to 40 min) to use than the Filarochek® test (Mallinckrodt, U.S.), in which the color reactions can be sometimes difficult to interpret. The Diromail® test (Agen, N.J., U.S.; MAbCO, Australia) is a laboratory-based EIA using air-dried samples sent by mail from the practitioner.

All such tests* may not detect low numbers of worms (<5) which are more likely to be immature, small, and nongravid.[38,52]

ii. Assessment of Worm Numbers

Weil and colleagues[50] showed correlation of antigen levels with adult worm numbers ($r = 0.82$; $p < 0.001$) as did the work of Atwell and others[52] with a similar correlation of $r = 0.8$ where they related antigen levels to filarial mass. It was felt that this mass correlation would probably be more of an indicator of potential thromboembolic problems rather than just a guide to the number of worms. It may be a more suitable guide as the antigen is believed to be produced by both male and female worms. Courtney[38] also has shown correlation of antigen levels with dog-weight-adjusted worm numbers.

These data could also be used to predict which dogs with their respective worm burdens would develop thromboembolic complications. Of course, the extent to which thromboembolism will develop is not directly related to just worm numbers, but it is more likely that the greater the mass of dead material the greater is the chance that thrombosis and thromboembolism will occur. Thus, there is the potential to use the antigen tests to estimate worm numbers and perhaps assess or predict thromboembolic potential. However, just as the antigen test may not diagnostically detect low worm numbers or immature worms, the use of the test to assess thromboembolic potential would be similarly limited. However, such information if available to practitioners would enable better assessment of each individual infection and of the possibility of complications.

iii. Estimation of Live/Dead Ratios Following Adulticide Therapy

Research work presented at the Heartworm Symposium in 1986 (Figure 1) showed that antigen tests could be used to assess adulticide therapy.[52,53] These results were based on a small number of dogs (n = 16), but indications are that antigen levels do fall to significantly lower levels (when compared to preadulticide levels) after adult filariae are killed. Weil[53] showed that by 8 weeks postadulticide 93% of total detectable antigen was reduced, and at 12 weeks no antigen was detected. Separation was apparent between cleared (treated) dogs (n = 4) and the untreated infected control dogs (n = 4). Based on the work of Atwell and colleagues,[52] it was shown that estimations based on the ratios of the preadulticide and of the 60 day postadulticide antigen levels could be predictive of dogs with significant live worm burdens (>1 live worm) 60 days after adulticide therapy.

Thus, it is possible that in the near future a service will be available to indicate whether the dog is cleared of adults or still harbors a significant adult worm burden. As with diagnosis and with worm number estimation, it will be unlikely that immature filariae will be detected by such tests as they most likely secrete too little detectable antigen.

* At least seven antigen tests are now commercially available and local distributors should be contacted as to the most reliable cost-effective test.

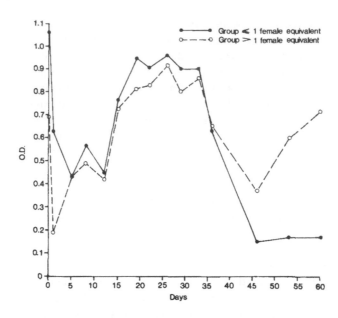

FIGURE 1.   Mean optical density (OD) values (representing antigen levels) of infected (>1 female equivalent) and noninfected (<2 female equivalents) dogs during the period following adulticide therapy enabling prediction of infection status at day 60. (From Atwell, R. B., et al., *Proceedings of the Heartworm Symposium '86*, Otto, G. F., Ed., American Heartworm Society, Washington, D.C., 1986, 71. With permission.)

iv. Summary

In summary, antigen tests are much more accurate diagnostically than the previous serological tests. They should generally be used only after routine tests for the presence of microfilariae have been performed in dogs in which the disease is suspected based on epidemiological, environmental, and historical factors, and/or associated clinical signs. Based on the current data on the use of antigen tests, these tests should not be used as "first order" routine diagnostic screening tests on all dogs, even more so, in areas where the prevalence of the disease is low.

They also have the additional benefits of giving prognostic information about worm numbers, and therefore, perhaps, the thrombotic/thromboembolic potential, and about the effectiveness of adulticide therapy with live/dead filarial estimations 8 to 12 weeks following therapy when antigen levels would be compared to those levels prior to therapy. It is conceivable that in the near future these tests could be used as much, if not more, for their prognostic benefits than for diagnosis, particularly if costs are minimized.

C. Other Tests to Aid Diagnosis and to Assess Disease Status
1. Radiography*
2. Arteriography[54]

This is a useful test to further assess the degree of vascular compromise, even in very early infections, and to see filariae *in situ* as negative shadows in the flow of contrast agent. It is used experimentally to visualize early arterial disease, e.g., 3 months onward, to ensure cases are clear of infection and disease before experiments and, perhaps, in diagnostically confusing cases when luminal detail is required. As with radiography, true functional assessment of the pulmonary arterial tree cannot be made except that blood flow rates and

* See Chapter 5, Radiology of Heartworm Disease.

obstructed, turbulent flow patterns can be visualized. Angiography (and also scintillography)[55] allows the clinician to assess disease status particularly with regard to arterial compromise, but does not indicate functionally the extent of the effect of pathology on pulmonary hemodynamics, particularly the status of the potentially highly resistant, peripheral, microscopic vascular bed of the pulmonary artery tree. However, this technology is usually only available at tertiary institutions and the literature does cover the topic.[3,4]

3. Circulation Time

This technique[7] has been used with sodium fluorescein, but seems to be now out of favor. It involved the injection of a dye into a peripheral vein and recording the appearance of fluorescence observed on the mucosae. No accurate estimation of pathology could be assessed and variation occurred due to variable dog size and degree of excitement. However, as most of our diagnostic data do not give us true functional organ assessment perhaps this simple technique should again be studied and standardized for bolus site and rate of injection, heart rate, dog weight, etc. It should be of great use if it would give an estimation of the perfusion capability of each patient and, therefore, of the potential risks associated with an estimated dead worm burden and associated thromboembolism. Another partial circulation time was that using arteriography observing the time taken for blood to move from the distal bed of the caudal lobes back to the left atrium, the normal being $2^1/_2$ sec.[54] Once again, the procedure was invasive, but did give an estimate of microvascular function rather than of the obvious large proximal artery pathology seen with radiography and arteriography.

4. Electrocardiography

This is useful as a prognostic indicator. If there is evidence of right ventricular hypertrophy (at least three right ventricular hypertrophy [RVH] criteria, e.g., S wave enlargements on leads CV6LL, CV6RU, 1, and 2; MEA frontal plane greater than 103°) a poorer prognosis needs to be given.[4] Knight[3] showed in 23 heartworm dogs with anatomical RVH that S > 0.8 mV (CV6LL) was seen in 91% of cases; S > 0.7 mV (CV6LU), 78% of cases; S > 0.05 mV (lead 1), 74%; frontal plane MEA > 103° (1 and AVF), 70%; R/S ratio (CV6LU) < 0.87, 61%; and S > 0.35 mV (lead 2) to be present in 61% of cases. The enlarged S waves occurred in CV6LL and LU in 7.1 and 2.9% of 70 control dogs. Thus, the selection of three criteria should enable a confident diagnosis of RVH to be made and a relevant prognosis given.

It is known that right ventricular dilation (RVD) occurs before RVH.[56] Analysis of 137 infected dogs showed that the severe RVD and RVH were equated with RSCHF, that RVD and RVH usually occurred together, and that RSCHF was seldom seen without RVD or RVH.[57] Specific RVH criterion were a reduced R/S on CV6RU, a positive T wave on V_{10}, an increased MEA on the tranverse plane, and an enlarged S wave on lead 2, while an enlarged S wave on lead CV6LU was sensitive but not specific. Such results would indicate that the dog has a severely dilated and hypertrophied right ventricle and was in RSCHF or, with exercise, thrombosis, or stress, could easily decompensate into RSCHF. It would also suggest that systolic pulmonary arterial pressure was at least 50 mmHg and mean pressure at least 30 mmHg.[58,59] Figure 2 shows a typical case in RSCHF with RVH.

With three criteria evident, the information is of the order of 90% reliable.[4] However dogs can still have severe pulmonary hypertension without RVH if the pathology, e.g., thrombosis, has occurred recently.[3]

It is stressed that the electrocardiography (EKG) is not necessarily diagnostic of heartworm disease, but if the dog has the disease, and alterations to its EKG pattern suggesting RVH are present, then the information is supportive of the diagnosis and is prognostic, as severe arterial disease is usually the cause of the right ventricular pressure overload and so of the hypertrophic response. It also reinforces the need for cage rest in such cases as exercise not

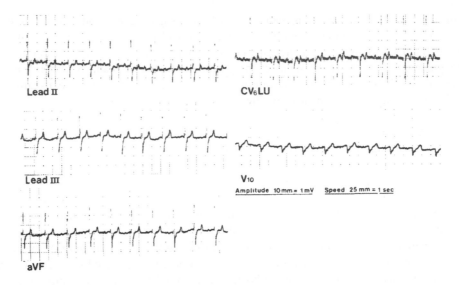

Lead II   CV₆LU

Lead III   V₁₀
Amplitude 10 mm = 1 mV   Speed 25 mm = 1 sec

aVF

FIGURE 2.   EKG from a dog with right-sided congestive heart failure and right ventricular hypertrophy associated with severe dirofilariasis. (From Atwell, R. B., *Vet. Rec.*, 104, 114, 1978. With permission.)

only increases pulmonary artery pressures but also reduces renal perfusion and so encourages fluid and salt retention. Both of these exercise-associated hemodynamic factors place increased load on the right side further adding, with time, to the hypertrophic stimulus which, with continued demand, will eventually lead to ischemic cardiomyopathy and so to failure.

As discussed under clinical signs, arrhythmias are uncommon and, if they occur, are usually associated with RSCHF or caval syndrome. If the arrhythmia is detremental to ventricular function, digoxin, used as an antiarrhythmic agent, is indicated.

5. Echocardiography[59]

This technology allows visualization of the dilated right side and of actual worm presence as echo-dense shadows. Objective data can be obtained from the M mode and the right ventricular internal diastolic diameter seems a reliable guide; dogs with cor pulmonale being clearly different to control animals.[60] It is also possible to see abnormal tricuspid valve movement and septal mobility, possibly due to volume and pressure overload seen in severe cases of heartworm disease once the tricuspid value is incompetent. This technique can also be used to guide intracardial forceps[61] and to help appreciate parasite movement between the right atrium and ventricle. Worm entanglement of the tricuspid valve can also be visualized as can suspect thrombosis in the cardiac chambers or on the valve itself.

6. Pulmonary Artery Hemodynamics

Several papers[62,63] documenting hemodynamic studies have helped to understand the hemodynamics of this disease. There is no doubt that pulmonary arterial hypertension (and elevated right-sided pressures) does exist, that it is elevated with exercise, and that arterial rupture can occur, particularly during hypertensive periods associated with exercise or coughing. The real prognostic indicator of this disease would be a test to assess the extent of pulmonary perfusion deficit induced at the distal vascular (possibly arteriolar) level by thromboembolism and the distal arteritis involved in the pathophysiology of this disease. Static pressures are not a true guide to the extent of compromise as pressure needs to be recorded under exercise stress to be truly indicative of the extent of pulmonary hypertension. In addition, it is not just the actual pressure but the ability of the pulmonary arterial tree to be perfused (i.e., pulmonary flow rate, cardiac output, pulmonary vascular resistance, and pressure) that needs to be assessed. Unfortunately, a noninvasive, simplistic system does not exist to obtain prognostic information on the extent of the functional deficit of the pulmonary arterial tree.

Another pressure factor once suggested to be prognostic was the rate of change of pressure in the pulmonary artery (dP/dt).[64] Once again, it involves invasive procedures, but in theory seems to be related to the extent to which the pulmonary arterial pathology reduces the elastic (easily and rapidly expanded) nature of the arterial tree, in particular the larger elastic proximal arteries, and so would seem to be a good prognostic indicator.

7. Other Laboratory Tests

There are few other tests that are diagnostic or supportive of canine dirofilariasis. Clinical pathology is supportive (e.g., eosinophilia, basophilia), but it is not diagnostic or prognostic. Plasma biochemistry changes are similarly neither diagnostic nor prognostic. They may be supportive of concurrent disease assessment in older dogs as an aid to overall assessment prior to adulticide therapy (e.g., hemogram, liver and kidney profile, and urinalysis), but are not predictive of potential toxicity to arsenicals.[65]

In the interest of the middle-aged to older patient, where owners wish to have clinical pathological tests performed (and to pay for these tests), and in the interest of being as thorough as possible with every case, the use of extensive clinical pathology testing can be justified. Obviously, in indicated cases (e.g., caval syndrome, severe anemia, drug toxicity, a geriatric case with RSCHF, etc.) and in a potential litigation situation there is no excuse not to use such tests; failing to peform such tests would most likely be seen as incompetent. However, in the healthy, mildly infected case of canine dirofilariasis, the use of routine screening panels needs to be assessed on the basis of costs, cost effectiveness, the actual attainment of clinically useful information,[65], client resistance or insistence, the probability of litigation, and on the individual approach to each type of owner and clinical case by individual veterinarians. Simply, a balance needs to be established between what can be done and what needs to or should be done in each case.

In summary, it would seem that clinical pathology, apart from serological tests as discussed earlier, is supportive in mildly affected dogs, can offer essential information in selected cases (e.g., drug side-effects or severe disease), and is not predictive of which dogs will react adversely to adulticide therapy. Table 5 records some of the abnormal (reported) findings in heartworm cases.

Table 6 gives a summary of a practical approach to diagnosis, prognostic assesment of the severity of the disease, and evaluation of the success of therapy. Other more expensive, impractical, invasive tests are available and futuristic assessment procedures are discussed in Chapter 7.

Table 5
ABNORMAL CLINICAL PATHOLOGY REPORTED IN CASES OF DIROFILARIASIS[3,4,21,66]

Condition	Associated clinical data
Anemia	Three types: (1) metabolic, (2) subclinical hemolytic, (3) hemolytic (caval syndrome); normochronic, normocytic — usually mild; usually responsive and more severe in occult cases; increased mechanical fragility
Eosinophilia	80—90% of cases
Basophilia	50% of cases
Neutrophilia	In more severe cases
Lymphopenia	Probably a stress factor in severe cases
DIC	Rarely reported — seen associated with postadulticide treatment
Thrombocytopenia	Seen in caval cases and with severe thromboembolism; ''activated'' platelets also seen in caval cases
Proteinuria	Particularly seen in severely affected dogs; seen in immune-mediated occult and RSCHF cases
Plasma ALT and SAP	Marked elevations in caval syndrome cases and with toxicity to adulticides
Plasma urea	Mild elevations in some RSCHF cases; usually prerenal elevations seen in severe RSCHF; marked changes in caval cases
Hyperglobulinemia	Usually beta is elevated in immune-mediated occult cases; usually seen with more severe disease
Hypoalbuminemia	Seen with high globulin and with protein-losing nephropathy
Acidosis	Mild in severe cases
Urinalysis	Casts (hemoglobin) seen in caval cases; proteinuria — albumin; hemoglobin in caval cases; other concurrent disease findings; useful with assessment of post adulticide toxicity

Table 6
SUMMARY OF DIAGNOSTIC AND PROGNOSTIC PROCEDURES

Diagnostic sequence

Wet smear	Microfilarial presence
Concentration/Filtration	Microfilarial presence
Antigen detection tests	Adult presence
Radiography	Extent of disease

Prognostic assessment (noninvasive)

Severity of clinical signs	
Exercise tolerance and development of pale mucosae and dyspnea	
Antigen tests	Worm number; thromboembolic potential?
Radiography	Severity of arterial/parenchymal disease, dilation of main arteries as guide to loss of downstream perfusion
Electrocardiography	Development of right ventricular hypertrophy
Echocardiography	Ventricular dilation and hypertrophy; worm presence
Therapy response	Clinical improvement with cage rest and prednisolone/aspirin therapy
Futuristic	Thrombotic and thromboembolic potential assessment; vasodilatory capacity of the pulmonary vasculature; standardized exercise test; standardized circulation time or other indicator of pulmonary perfusion

Adulticide therapy assessment

Clinical response	Short term coughing, respiratory signs — suggestive of death of adults; long term improvement in exercise and general health
Antigen tests	Before and 8—12 weeks following adulticide therapy to assess worm mortality

Table 6 (continued)
SUMMARY OF DIAGNOSTIC AND PROGNOSTIC PROCEDURES

Microfilarial clearance	Not necessarily indicative of clearance of all adults, e.g., a residual population of female filariae may remain without microfilarial production
Radiography	Clearance of alveolar patterns and of periarterial opacity; interstitial patterns usually remain

REFERENCES

1. **Atwell, R. B.,** Treatment of severe canine dirofilariasis and associated cardiac decompensation, *Aust. Vet. Practit.*, 12, 132, 1982.
2. **Atwell, R. B.,** Canine dirofilariasis — clinical update, *Vet. Ann.*, 26, 250, 1986.
3. **Knight, D. H.,** Heartworm disease, in *Textbook of Veterinary Internal Medicine*, 2nd ed., Ettinger, S. J., Ed., W. B. Saunders, Philadelphia, 1983, 1097.
4. **Calvert, C. A. and Rawlings, C. A.,** Pulmonary manifestations of heartworm disease. Symposium on respiratory disease, *Vet. Clin. North Am: Small Animal Pract.*, 15, 991, 1985.
5. **Atwell, R. B.,** unpublished data, 1984.
6. **Kotani, T.,** Pathological studies on canine dirofilariasis, *Bull. Univ. Osaka Prefect. Ser. B*, 35, 93, 1983.
7. **Jackson, W. F.,** Management of the symptomatic patient, in *Proceedings of the Heartworm Symposium '74*, Morgan, H. C., Ed., Veterinary Medicine Publ., Bonner Springs, Kan., 1975, 56.
8. **Ishihara, K., Kitagawa, H., Ojima, M., Yagata, Y., and Suganuma, U.,** Clinicopathological studies on canine dirofilarial hemoglobinuria, *Jpn. J. Vet. Sci.*, 40, 525, 1978.
9. **Atwell, R. B.,** Aspects of Natural and Experimental Canine Dirofilariasis, Ph.D. thesis, University of Queensland, Brisbane, Australia, 1983.
10. **Stuart, B. P., Hoss, H. E., Root, C. R., and Short, T.,** Ischemic myopathy associated with systemic dirofilariasis, *J. Am. Anim. Hosp. Assoc.*, 14, 36, 1978.
11. **Tomimura, T., Funahashi, N., Kotani, T., Oka, T., Mochizuki, H., Noda, S., and Nomura, K.,** Studies of paradoxical embolism with *D. immitis* in dogs, *Jpn. J. Parasitol.*, 18, 265, 1969.
12. **Kotani, T., Tomimura, T., Ogura, M., Yoshida, H., Mochizuki, H., and Koreeda, T.,** Cerebral infarction caused by *D. immitis* in three dogs, *Jpn. J. Vet. Sci.*, 37, 379, 1975.
13. **Carlisle, M. S., Webb, S. M., Sutton, R. H., Hampson, E. C., and Blum, A. J.,** Case report — adult *D. immitis* in the brain of a dog, *Aust. Vet. Practit.*, 14, 10, 1984.
14. **Kociba, G. J. and Hathaway, J. E.,** Disseminated intravascular coagulation associated with heartworm disease in the dog, *J. Am. Anim. Hosp. Assoc.*, 10, 373, 1974.
15. **Spaulding, G. L., Wilkins, R. J., Tilley, L. P., Johnson, G. F., Kay, W. J., Indrieri, R. J.,** Grand rounds conference — defibrination syndrome following therapy for heartworm disease, *J. Am. Anim. Hosp. Assoc.*, 11, 310, 1975.
16. **Dillon, A. R. and Braund, K. G.,** Distal polyneuropathy after canine heartworm disease therapy complicated by disseminated intravascular coagulation, *J. Am. Vet. Med. Assoc.*, 181, 239, 1982.
17. **Saheri, Y., Ishitari, R., and Miyamoto, Y.,** Acute fatal pneumothorax in canine dirofilariasis, *Jpn. J. Vet. Sci.*, 43, 315, 1981.
18. **Scott, D. W.,** Nodular skin disease associated with *D. immitis* in the dog, *Cornell Vet.*, 69, 233, 1979.
19. **Tulloch, G. S.,** A new approach to the treatment of heartworm disease in dogs, in *Proceedings of the Heartworm Symposium, '74*, Morgan, H. C., Ed., Veterinary Medicine Publ., Bonner Springs, Kan., 1975, 85.
20. **Anantaphruti, M., Kino, H., Terada, M., Ishii, A. I., and Sano, M.,** Studies on chemotherapy of parasitic helminths. XIII. Efficacy of ivermectin on circulating microfilaria and embryonic development in the female worm of *D. immitis*, *Jpn. J. Parasitol.*, 31, 517, 1982.
21. **Atwell, R. B. and Buoro, I. B. J.,** Clinical presentations of canine dirofilariasis with relation to their hematological and microfilarial status, *Res. Vet. Sci.*, 35, 364, 1983.
22. **Calvert, C. A. and Thrall, D. E.,** Treatment of canine heartworm disease coexisting with right-side heart failure, *J. Am Vet. Med. Assoc.*, 180, 1201, 1982.
23. **Calvert, C. A.,** Treatment of heartworm disease with severe pulmonary arterial disease, in *Proceedings of the Heartworm Symposium '86*, Otto, G. F., Ed., American Heartworm Society, Washington, D.C., 1986, 43.

24. **Whitlock, H. V.,** Laboratory procedures for the detection and identification of microfilariae in blood canine heartworm disease, *Symp. 40th Proc. Postgraduate Committee Vet. Sci.*, University of Sydney, Australia, 1978, 29.

25. **Kume, S.,** Experimental observations on seasonal periodicity of microfilariae, in *Proceedings of the Heartworm Symposium '74*, Morgan, H. C., Ed., Veterinary Medicine Publ., Bonner Springs, Kan., 1975, 26.,

26. **Altman, N. H.,** Laboratory diagnosis of *D. immitis:* evaluation of current tests, in *Canine Heartworm Disease: The Current Knowledge*, Bradley, R. E., Ed., University of Florida, Gainesville, Fla., 1972, 87.

27. **Calvert, C. A. and Rawlings, C. A.,** Diagnosis and management of canine heartworm disease, in *Current Veterinary Therapy*, Vol. 8, Kirk, R. W., Ed., W. B. Saunders, Philadelphia, 1983, 348.

28. **Wylie, J. P.,** Detecting microfilariae, *J. Am. Med. Vet. Assoc.*, 157, 512, 1970.

29. **Chalifoux, L. and Hunj, R. D.,** Histochemical differentiation of *D. immitis* and *Dipetalonema reconditum*, *J. Am. Vet. Med. Assoc.*, 158, 601, 1971.

30. **Whitlock, H. V., Poster, C. J., and Kelly, T. D.,** The PKW acid phosphatase modificaion for the recovery and histochemical identification of microfilariae of *D. immitis* in blood, *Aust. Vet. Practit.*, 8, 201, 1978.

31. **House, C. and Glover, F.,** Evaluation of an improved filter test for microfilariae detection, in *Proceedings of the Heartworm Symposium '74*, Morgan, H. C., Ed., Veterinary Medicine Publ., Bonner Springs, Kan., 1975, 19.

32. **Jackson, R. F. and Otto, G. F.,** Detection and differentiation of microfilariae, in *Proceedings of the Heartworm Symposium '74*, Morgan, H. C., Ed., Veterinary Medicine Publ., Bonner Springs, Kan., 1975, 21.

33. **Otto, G. F.,** The significance of microfilaremia in the diagnosis of heartworm infection, in *Proceedings of the Heartworm Symposium '77*, Otto, G. F., Ed., Veterinary Medicine Publ., Bonner Springs, Kan., 1978, 22.

34. **Jackson, R. F., Seymour, W. G., and Beckett, R. S.,** Routine use of 0.05 mg/kg of ivermectin as a microfilaricide, in *Proceedings of the Heartworm Symposium '86*, Otto, G. F., Ed., American Heartworm Society, Washington, D.C., 1986, 41.

35. **Boreham, P. F. L. and Atwell, R. B.,** unpublished data, 1982.

36. **Atwell, R. B.,** Prevalence of *Dirofilaria immitis* microfilariae in 6- to 8- week-old pups, *Aust. Vet. J.*, 57, 579, 1981.

37. **Carlisle, C. and Atwell, R. B.,** unpublished data, 1981.

38. **Courtney, C. H.,** Comparison of tests for immunodiagnosis of canine dirofilariasis, in *Proceeding of the Heartworm Symposium '86*, Otto, G. F., Ed., American Heartworm Society, Washinton, D.C., 1986, 77.

39. **Jackson, R. F.,** Treatment of the asymptomatic dog, in *Proceedings of the Heartworm Symposium '74*, Morgan, H. C., Ed., Veterinary Medicine Publ., Bonner Springs, Kan., 1975, 51.

40. **Wilkins, R. J., Hurvitz, A. I., Gula, I., and Tilley, L. P.,** Clinical experience with fluorescent antibody test in dogs with occult dirofilariasis, *Proceedings of Heartworm Symposium '77*, Otto, G. F., Ed., Veterinary Medicine Publ., Bonner Springs, Kan., 1978, 45.

41. **Grieve, R. B., Glickman, L. T., Bater, A. K., Mika-Grieve, M., Thomas, C. B., and Patronek, G. J.,** Canine *D. immitis* infection in a hyperenzootic area: examination by parasitologic findings at necropsy and by two serodiagnostic methods, *Am. J. Vet. Res.*, 47, 329, 1986.

42. **Corwin, R. M., Pratt, S. E., and Wagner, C. A.,** Diagnosis and treatment of canine heartworm disease — an update, *Mod. Vet. Pract.*, 66, 548, 1985.

43. **Atwell, R. B. and Rezakhani, A.,** Preliminary assessment of Dirocult®, a latex agglutination test for detection of *Dirofilaria immitis* infection in the dog, *Aust. Vet. J.*, 63, 127, 1986.

44. **Sisson, D., Dilling, G., Wong, M. M., and Thomas, W. P.,** Sensitivity and specifity of the indirect-fluorescent antibody test and two enzyme-linked immunosorbent assays in canine dirofilariasis, *Am. J. Vet. Res.*, 46, 1529, 1985.

45. **Greene, R. T., Bennett, R. A., Woody, D., and Troy, G. C.,** Evaluation of a microfilter technique and two serological tests used in the diagnosis of canine heartworm disease, *J. Am. Anim. Hosp. Assoc.*, 22, 153, 1986.

46. **Grieve, R. B., Mika-Johnson, M., Jacobson, R. H., and Cypess, R. H.,** Enzyme linked immunosorbent assay for measurement of antibody responses to *D. immitis* in experimentally infected dogs, *Am. J. Vet. Res.*, 42, 66, 1981.

47. **Lombard, C.,** Summary of the heartworm symposium, Paper presented at the Am. Acad. of Vet. Cardiol., Am. Anim. Hosp. Assoc. Meeting, New Orleans, March 1986.

48. **Rylatt, D. B., Atwell, R. B., Watson, A. R. A., Jenkins, A., Blake, A. S., Cottis, L. E., and Bundesen, P. G.,** A second generation heartworm test, *Aust. Vet. Practit.*, 14, 174, 1984.

49. **Weil, G. J., Malane, M. S., and Powers, K. G.,** Detection of circulating parasite antigens in canine dirofilariasis by counterimmunoelectrophoresis, *Am. J. Trop. Med. Hyg.*, 33, 425, 1984.

50. **Weil, G. J., Malane, M. S., Powers, K. G., and Blair, L. S.,** Monoclonal antibodies to parasite antigens found in the serum of *D. immitis*-infected dogs, *J. Immunol.*, 134, 1185, 1985.

51. **Gertsman, B. B. and Cappucci, D. T.**, Evaluating the reliability of diagnostic test results, *J. Am. Vet. Med. Assoc.*, 188, 248, 1986.

52. **Atwell, R. B., Vankan, D. M., Cotis, L. E., Blake, A. S., Rylatt, D. B., Watson, A. R. A., and Bundeson, P. G.**, The use of an antigen test for diagnosis as an indicator of filarial numbers and for assessing filarial mortality following Caparsolate® therapy, in *Proceedings of the Heartworm Symposium '86*, Otto, G. F., Ed., American Heartworm Society, Washington, D.C., 1986, 71.

53. **Weil, G. J.**, The practical significance of parasite antigen detection for canine dirofilariasis, in *Proceedings of the Heartworm Symposium '86*, Otto, G. F., Ed., American Heartworm Society, Washington, D.C., 1986, 95.

54. **Bisgard,G. E. and Lewis, R. E.**, *In vivo* arteriography in canine heartworm disease, in *Canine Heartworm Disease: The Current Knowledge*, Bradley, R. E., Ed., University of Florida, Gainesville, Fla., 1972, 117.

55. **Cornelissen, J. M. M., Wolvekamp, W. T. C., Stokhof, A. A., van den Brom, W. E., and de Vries, H. W.**, Primary occlusive pulmonary vascular disease in a dog diagnosed by a lung perfusion scintigram, *J. Am. Anim. Hosp. Assoc.*, 21, 293, 1985.

56. **Rawlings, C. A. and Lewis, R. E.**, Right ventricular enlargement in heartworm disease, *Am. J. Vet. Res.*, 38, 1801, 1977.

57. **Calvert, C. A., Losonsky, J. M., Brown, J., and Lewis, R. E.**, Comparisons of radiographic and electrocardiographic abnormalities in canine heartworm disease, *Vet. Radiol.*, 27, 2, 1986.

58. **Hill, J. D.**, Electrocardiographic diagnosis of right ventricular enlargement in dogs, *J. Electrocardiol. San Diego*, 4, 347, 1971.

59. **Knight, D. H.**, Heartworm heart disease, *Adv. Vet. Sci. Comp. Med.*, 21, 107, 1977.

60. **Lombard, C. W., and Buergelt, G. D.**, Echocardiographic and clinical findings in dogs with heartworm induced cor pulmonale, *Compen. Contin. Educ.*, 5, 971, 1983.

61. **Ishihara, K., Sasaki, Y., and Kitagawa, H.**, Development of a flexible alligator forceps, *Jpn. J. Vet. Sci.*, 48, 989, 1986.

62. **Knight, D. H.**, Effect of chronic *D.immitis* infestations on cardio-pulmonary diagnosis and right ventricular activation in the dog, in 18th Gaines Veterinary Symposium, Guelph, Ontario, Canada, 1968, 15.

63. **Rawlings, C. A.**, Cardiopulmonary function in the dog with *D. immitis* infection during infection and after treatment, *Am. J. Vet. Res.*, 41, 319, 1980.

64. **Musselman, E.**, Pulmonary vascular hemodynamic studies in canine dirofilariasis, in *Proceedings of the Heartworm Symposium '74*, Morgan, H. C., Ed., Veterinary Medicine Publ., Bonner Springs, Kan., 1975, 43.

65. **Hribernik, T. N., and Hoskins, J. D.**, Laboratory data and its predictive value in the treatment of heartworm cases, in *Proceedings of the Heartworm Symposium '86*, Otto, G. F., Ed., American Heartworm Society, Washington, D.C., 1986, 107.

66. **Buoro, I. B. J. and Atwell, R. B.**, Urinalysis in canine dirofilariasis with emphasis on proteinuria, *Vet. Rec.*, 112, 252, 1983.

67. **Otto, G.F., Ed.**, *Proceedings of the Heartworm Symposium '86*, American Heartworm Society, Washington, D.C., 1986.

Chapter 5

RADIOLOGY OF HEARTWORM DISEASE

Carol H. Carlisle

TABLE OF CONTENTS

I. INTRODUCTION

Radiography is valuable for determining the presence and severity of heartworm disease. However, it is not a shortcut to diagnosis and should be used as an ancillary aid to a thorough clinical examination.[1]

II. INDICATIONS FOR RADIOLOGY

Survey radiographs of the thorax provide valuable information on the extent and degree of change in the pulmonary parenchyma, pulmonary arteries, and the heart. Radiographs are useful in that:

1. They can provide evidence to confirm suspected heartworm disease.
2. They can provide valuable assistance with a diagnosis that has not been suspected clinically.
3. They provide information on the location and extent of the pulmonary vascular changes.
4. They provide a visual picture of the extent of the parenchymal lesions associated with the disease and which basic lung pattern predominates.
5. They supply information on the cardiopulmonary complications which can be associated with heartworm disease.
6. They may indicate additional radiographic procedures which may be useful.
7. They are very convenient for diagnosing occult dirofilariasis.
8. In association with the clinical signs and results of other diagnostic studies, radiographs can be useful in determining the treatment regime and prognosis of this disease.
9. The course of the disease, especially in response to treatment can be monitored.
10. They aid in differentiating heartworm disease from other thoracic diseases.

In summary, radiography provides information about the topographic location and extent of the thoracic disease[2] and assists in the planning and assessment of a treatment program.[3]

III. RADIOGRAPHIC POSITIONING OF THE ANIMAL

Two survey radiographs are necessary for the examination of the thorax of animals with heartworm disease. One is taken in lateral recumbency and the other in dorso-ventral recumbency. Although slight differences are seen between radiographs taken in right vs. left lateral recumbency,[2] these are not important for the assessment of heartworm disease. The main aim is to be consistent and take all radiographs on either one side or the other. The dorso-ventral position, with the animal crouched on the cassette, is selected in preference to the ventro-dorsal position because the position of the heart changes to a lesser extent and so retains a more consistent shape.[4,5] Most of the information required to diagnose the disease can be obtained from these two films. Oblique views may be useful to differentiate pulmonary consolidation from rib or paraspinal tissues. Radiographs taken with a horizontal beam may be useful to demonstrate free fluid in the thorax. To achieve maximal detail radiographs should be taken with the lungs fully inflated in inspiration.

IV. PRODUCTION OF SUCCESSFUL RADIOGRAPHS

Successful radiography of the thorax is limited by a number of factors and these fall into two groups. The first includes technical problems and requires a basic knowledge of the equipment available and in what circumstances it should be used. The second involves a knowledge of the problems associated with radiography of the thorax generally and in animals, in particular. There is marked variation in thoracic shapes and difficulty in controlling both respiration and movement can also be encountered. Many text books have outlined the principles of radiographic and darkroom techniques,[6-9] but as they are important they will be summarized briefly in this chapter.

A. Problems Related to Equipment

The technical qualities of a satisfactory radiograph have been summarized and include (1) adequate contrast; (2) correct density; (3) minimum magnification and distortion; (4) sharp delineation of detail; and (5) an absence of artifacts due to radiographic and processing faults.

(1) Poor visual contrast is commonly the result of under- or overexposure, under- or overdevelopment, or fog. It must be understood that even low-powered X-ray machines are capable of producing good quality thoracic radiographs, but it is preferable to have a more powerful unit with a higher tube-current output to permit short exposure times. Most problems are caused by a poor understanding of how to use the equipment, by the use of poor quality equipment, or, more commonly, by the use of inadequate darkroom facilities and techniques. One of the most frequent problems leading to poor contrast is under- or overdevelopment caused by inaccurate timing or, more frequently, underdevelopment from the use of exhausted developing solutions. The developer should be replenished or changed frequently. Fog results from a number of sources,[1] but the most common cause is scattered radiation which occurs with thoraxes 10 to 15 cm or more in diameter. The simplest way to reduce this scatter is by coning the X-ray beam. The coned field should start a few centimeters cranial to the first rib and extend at least to the caudal end of the costal arch or second lumbar vertebra and only a few centimeters beyond the body. A fine-line grid with 32 or more lines per centimeter and a ratio of at least 8:1 will also decrease scatter, but it has the disadvantage of increasing the tube current (ma) × times (s) (mas) by a factor of 3. This is undesirable in the unanesthetized dog because the exposure time is too slow to prevent blurring from respiratory movements. With low-powered X-ray units the practitioner often has to make the choice between using a grid or shorter exposure times. Fortunately, air in the lungs provides natural contrast, so, unless the dog is large and fat, the use of a grid is not absolutely necessary.

(2) Adequate density of a radiograph is essential for examination of thoracic views. High peak kilovoltage (kVP) values should be used for two reasons. Firstly, they provide a well-balanced, long, gray scale,[2] and secondly, they allow the mas to be reduced. This is an advantage when a grid has to be used or to reduce exposure for an unanesthetized dog. Judging the exposure factors is more difficult for the dorso-ventral projection, but it is considered adequate if the sternum and spine can be faintly but clearly seen through the heart shadow.[10]

(3) Magnification and distortion are minimized if the animal is positioned as close to the film as possible and at an adequate film-focal distance. The beam should be centered approximately 2 cm behind the end of the scapula for both projections. The center of the X-ray beam must be at right angles to the animal and the film and the thorax should be parallel to the film. However, despite awareness of all these points, it should be remembered that organ silhouettes are always larger than the organ itself and that some distortion will inevitably occur, particularly in the periphery of the radiograph.

(4) Poor definition or sharpness results from problems inherent in the equipment and inefficiency of the operator. An inherent problem is geometric unsharpness caused by having to use a large focal area on the anode to overcome the production of excessive heat. This results in a penumbra or nonsharp edges. If a small focal spot could be used, the edges of the object being X-rayed would be sharp. Operator problems include unsuitable film-focal distances (as unsharpness increases with a decrease in the film-focal distance), incorrect choice of intensifying screens and film, and movement unsharpness.

(5) It is important to exercise care when handling and processing films to eliminate artifacts.

B. Problems Related to the Patient

There are a number of important factors associated with the animal which can cause poor quality radiographs. These include (1) incorrect positioning of the animal; (2) unnecessary superimposition of structures over the thoracic organs; (3) movement; and (4) variations in the diameter and density of the thorax.

(1) For the lateral view, the animal should be positioned so the dorsal curvature of the ribs does not extend unilaterally above the spine. For the dorso-ventral view, the sternum and spine are superimposed. It is sometimes difficult to achieve good dorso-ventral radiographs if the patient is uncooperative. Even if it is a well-behaved animal, it is often hard to determine when the spine and sternum are superimposed.

(2) The forelimbs should be pulled well forward to avoid the detail of the vessels of the cranial mediastinum and pulmonary lobes being obscured by the biceps. The hind limbs should not be pulled too far backwards as this may rotate the caudal section of the thorax.

(3) Movement is a major problem with radiography of the thorax. It can be voluntary movement of the animal or involuntary from respiration and cardiac contractions. Ideally, the animal should be anesthetized and intubated, and radiographs taken with the lungs moderately inflated by gently squeezing the rebreathing bag. This is not always possible, and if the animal cannot be anesthetized it should be heavily sedated and completely restrained. In this case respiration may be controlled by gently occluding the nostrils or blowing on the nose. Some animals, however, resent this restraint and struggle.

The problem related to movement of the animal can be overcome by using higher kilovoltages and shorter exposure times. Times of 0.02 to 0.04 sec are adequate, but X-ray machines which generate 200 to 300 ma are necessary. If less powerful units are used, the ancillary equipment should be specially chosen to permit as short an exposure time as possible. Fast intensifying screens or rare earth screens together with fast X-ray films should be used and a grid used only when absolutely necessary.

(4) Variations in the density and depth of the thorax among dogs pose a problem. Often, in thick, muscular animals the cranial areas will be adequately exposed on the radiograph, but the caudal lobes will be overexposed. The use of high kVP resulting in a more penetrating beam is an advantage, but it is also useful to take advantage of the heel effect. In this effect more of the beam is absorbed at the anode side of the tube, resulting in a lower intensity at this end. The spine of the animal should be positioned parallel to the long axis of the X-ray tube with the denser cranial section of the animal placed under the cathode.

V. EXAMINATION OF RADIOGRAPHS

Radiographs should be examined under optimal conditions using suitable illuminators. Dark areas on the radiograph can be enhanced by a spotlight. Light in the viewing room should be minimal, and lights in viewing boxes which are not in use should be turned off. It is important to be consistent when placing the radiographs on the viewing box. In the dorso-ventral radiograph the left side of the dog is on the right side of the viewer, but in the lateral view the direction which the head faces is optional although it must be consistently facing the same way.

If it is necessary to examine the radiographs while they are wet, a second examination should be done after the films are dry because the layer of water may diminish the detail.

VI. INTERPRETATION OF RADIOGRAPHS

Having established that the radiographs are of satisfactory quality, interpretation is based

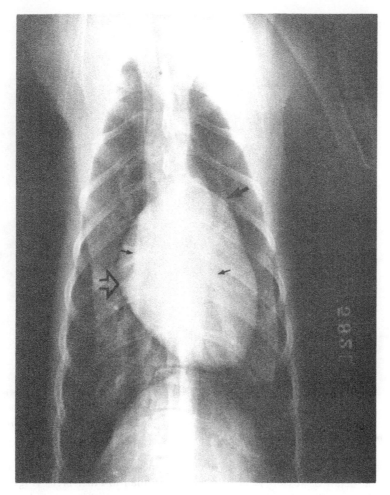

FIGURE 1. Dorso-ventral view. Normal 2-year-old male greyhound showing main pulmonary artery (large closed arrow), caudal pulmonary lobar arteries (small arrows), and the right side of the heart (open arrow).

on a sound knowledge of both the normal anatomy of the thorax and an understanding of radiographic signs of heartworm disease. A systematic approach is essential. The radiographs are evaluated for quality, the presence of artifacts, and for normal anatomical variations. The thorax is then examined carefully, starting at the periphery. The soft tissues, spine, sternebrae, and ribs are examined before individual organs such as the heart and lung. In heartworm disease, the pulmonary lobar arteries should be examined in detail. It has been suggested that the cranial lobar arteries should be used as a reference for determining if vessels are of normal size.[11] The normal artery and vein should be the same size at the level of the fourth rib. In this disease, however, this is only useful if the cranial arteries are involved, which is often not the case.[12]

Following the general examination, specific areas of the radiographs are examined thoroughly for signs of heartworm disease (Figures 1 and 2). These areas include enlargement of the pulmonary artery segment, the branches of the pulmonary lobar arteries, the right side of the heart, the pulmonary parenchyma for increased density particularly around the pulmonary arteries, the liver for enlargement, the pleural cavities for fluid accumulation, and, if included, the abdomen for presence of ascites associated with right-sided congestive heart failure.

The pulmonary artery segment is best evaluated from the dorso-ventral view, where it is found between twelve and two o'clock.[13] In the lateral view, the region around the carina

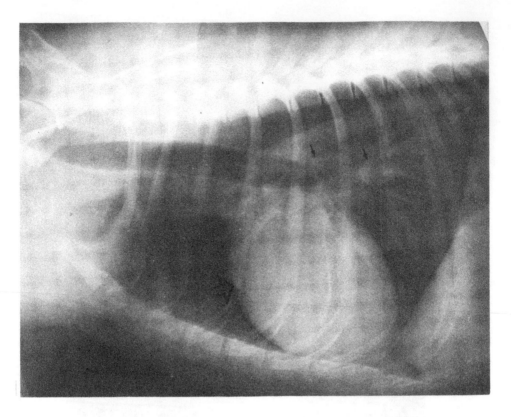

FIGURE 2.   Lateral view. Same dog as Figure 1 showing caudal pulmonary lobar arteries (small arrows) and the right side of the heart (open arrow).

should be examined carefully for signs of enlargement of the right and left caudal lobar arteries as the detail of vessels is often clearer here than in other areas (Figure 3). In the dorso-ventral view, the arterial changes may be seen where the vessels branch from the main pulmonary artery (Figure 4). However, in some instances the cardiac shadow may inhibit visibility, and parenchymal changes may mask the edges of the pulmonary vessels. Sometimes, the shape of the arteries can be seen clearly only over the diaphragm and liver (Figure 5). The whole length of each artery must be examined carefully for loss of arborization associated with enlargement, tortuosity, truncation, and sacculation, and, in rare cases, if the artery is occluded near its origin, for a decrease in size of the peripheral portions of the vessels. It must be emphasized that both views are necessary for evaluation of the heart, pulmonary arteries, and pulmonary parenchyma.

Cardiovascular changes associated with the disease have been described in clinical cases[14-17] and parenchymal changes have also been described in detail.[3] Angiocardiography or arteriography has been used in both clinical and experimental cases to show the presence and site of the adult worms and the shape and size of the heart and pulmonary lobar arteries.[12,18-24]

VII. RADIOGRAPHIC FEATURES OF HEARTWORM DISEASE

The severity of radiographic signs is, in general, correlated with the severity of the clinical signs. They are dependent on the number of heartworms, the duration of infestation, and the severity of the reaction to the presence of the worms.

A. Mild Dirofilariasis

In the early stages of the disease or in some cases where the number of worms are low, the radiographic signs on survey radiographs may be minimal or ambiguous. In some, no

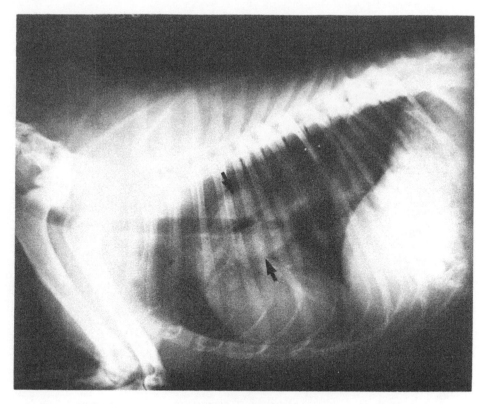

FIGURE 3.   Lateral view. Five-year-old male kelpie with heartworm disease showing marked enlargement of the caudal pulmonary lobar arteries around the carina (arrows).

radiographic signs may be evident at all. In dogs with low numbers of worms, the radiographs may remain ambiguous or negative for a long period. These radiographic changes have been described in both clinical and experimental cases.[2,3,12,13,25]

1. Peripheral Pulmonary Arterial Enlargement

The most common radiographic finding in mild cases is a dense focal area which may develop at the site of an obstruction to flow in the peripheral regions of the right caudal lobar artery;[26] some hemorrhage may occur around the site. This radiographic sign appears to be pathognomonic for mild heartworm disease.[12,13,20,25,26] It has been shown that heartworms have a predilection for localizing in the caudal lobar arteries with the right side being involved first,[20,21,27,28] and this enlargement may be the only sign of the disease (Figures 5 to 7). It is, therefore, important to examine the lateral radiograph around the carinal region of the dog, particularly below the end-on view of the right cranial lobar bronchus. In the dorso-ventral view, the right caudal branch should also be examined from its origin caudally.

It has been suggested that this predilection for the right branch may be caused by the larger size of the right compared with the left caudal lobar artery, as well as the curved pathway of the flow of blood. However, it has been shown that in the normal dog, the left caudal pulmonary artery is larger than the right.[29] In experimental studies, it has been confirmed that the right caudal lobar artery enlarged before the left caudal lobar artery regardless of the numbers of worms, and that enlargement of the arteries was related more to the duration of the infection than to the number of worms present.[12,24-26] Intimal proliferation is associated with the presence of live worms,[30-32] and as the arterial walls thicken[26] it is clear that the duration of the infection is important.

Enlargement and tortuosity of the peripheral portion of the caudal lobar arteries are common with both low and high numbers of heartworms, but they occur earlier and are more severe with greater numbers of worms.[24,25]

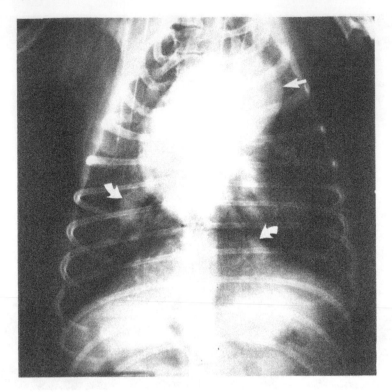

FIGURE 4.   Dorso-ventral view. Four-year-old female fox terrier with heartworm disease showing marked enlargement of the pulmonary segment (straight arrow) and enlarged tortuous caudal pulmonary lobar arteries (curved arrows).

2. Right Ventricular Enlargement

Right ventricular enlargement can occur within 6 months of infection and before the adults are sexually mature.[17,33] Its development appears to be related to the number of heartworms; it does not occur in dogs harboring less than 20 worms and in only 1 in 5 dogs harboring between 40 and 100 worms. This ventricular enlargement was once thought to be simple hypertrophy in response to resistance loading on the heart. It has now been shown that dilation occurs before hypertrophy develops, and this is thought to be the result of the interrelationship of volume and pressure loading[34] of the right ventricle in this disease.

3. Main Pulmonary Artery Enlargement

There was no increase in the size of the main pulmonary artery for 1 year after experimental infection of dogs with low numbers of worms (21 worms), but with larger numbers of worms there was a slight increase. In dogs injected with 50 larvae, the main pulmonary artery segment was enlarged 12 months after infection.[25]

4. Parenchymal Changes

In the experimental dogs studied for 1 year after experimental infection no parenchymal changes were found[25] while in longstanding clinical cases mainly chronic lung changes have been reported.[3]

5. Summary

In mild infections with low numbers of heartworm, arterial and parenchymal changes may be absent, or they may take a long time to develop and may not become particularly severe. With heavier infestations, however, only mild signs may be detected on radiographs

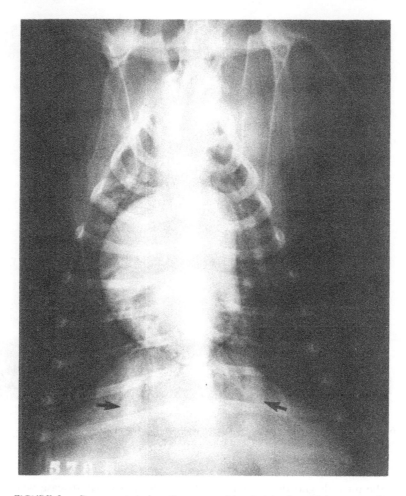

FIGURE 5. Dorso-ventral view. Seven-year-old male labrador with heartworm disease showing the enlarged caudal pulmonary lobar arteries visible when superimposed on the diaphragm (arrows).

when the disease is in the early stages. Specific areas of the thoracic radiographs should be examined carefully. These include the left and right caudal lobar arteries situated dorsal and ventral to the carina on the lateral projection and the pulmonary segment for enlargement of the main pulmonary artery on the dorso-ventral projection. The right side of the heart should be observed for signs of enlargement. Particular attention should be directed at the more peripheral sections of the caudal lobar arteries which have been shown to be the most common site of enlargement in early cases. In cases where changes on survey radiographs are absent or equivocal, angiography can be used to demonstrate heartworms as radiolucent streaks in the pulmonary arteries.[21,22] Angiography, however, is difficult to perform and requires special equipment. It cannot generally be recommended as a procedure for mild heartworm cases as clinical and serological examination can provide adequate diagnostic information.

B. Moderate to Severe Dirofilarisis

Usually more than one of the major radiographic signs of heartworm disease is present in moderate to severe cases. The most common signs in clinical cases are an enlarged right side of the heart in the shape of a reversed D (72% of 135 dogs), the enlargement of the pulmonary artery segment (71%), and enlargement of the branches of the pulmonary arteries (54%). At least one of these signs was present in 90% of affected dogs.[17]

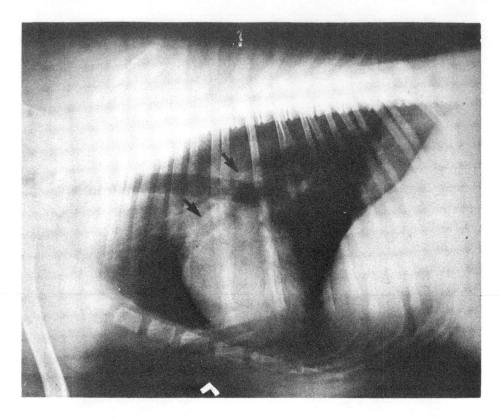

FIGURE 6.   Lateral view. Nine-year-old male standard poodle with mild heartworm disease showing enlargement of the left caudal and right cranial lobar arteries (arrows).

1. Peripheral Pulmonary Arterial Changes

These are the most diagnostic radiographic feature of heartworm disease when they are present. It has been found that increased pulmonary vascularity occurred in 54% of the naturally infected, randomly selected dogs and that this number would be expected to increase with progression of the disease.[17]

The diagnostic importance of these arterial changes depends on the stage of development of the disease. They are important in early infections (Figures 6 and 7), particularly where there is a large worm burden, as they may be the only diagnostic sign before changes in the pulmonary artery and heart develop. In advanced cases, the arterial changes give an indication of the severity of arterial pathology, but they may not be as useful a diagnostic aid as in the early stage of the disease because they may be masked by parenchymal lesions (Figure 8).

Changes seen in the lobar arteries include dilatation, sacculation, tortuosity, scalloping, serations, and loss of arborization (Figures 3 to 5), and, in some cases, the vessels may be small or totally disappear (Figures 5 and 6). These arterial changes may occur singly or, more commonly, several of these changes may be seen in the individual case. The number of and the degree of these changes increase with the severity of the disease.

The increase in the size of the arteries can be judged by comparing them with the veins or with the size of a rib. In normal animals, the arteries will be less than the width of a rib, whereas, in extreme cases of heartworm disease, they may be dilated to six times the size of a rib.[3] Tortuosity of lobar arteries is common. Saccular widening and serations of the peripheral section of the caudal lobar arteries often occurs (Figures 3 and 4) as does blunting or pruning towards the periphery of the arteries (Figure 4). In some cases the whole of a lobar artery may be obliterated due to severe truncation proximally so that none of the above signs can be detected.

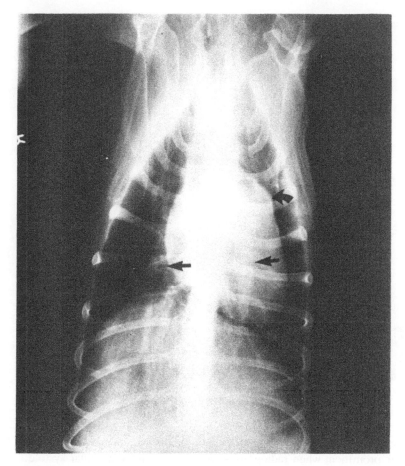

FIGURE 7. Dorso-ventral view. Same dog as in Figure 6 showing a slightly enlarged pulmonary segment (curved arrow) and enlarged caudal pulmonary lobar arteries (straight arrows).

In experimentally infected dogs with high worm burdens, some enlargement and tortuosity of the caudal pulmonary arteries were observed 4 months after infection, and these changes continued to increase in severity through the observation period of 12 months.[12,25] Angiograms revealed enlarged vessels which were tortuous and contained radiolucent-filling defects due to worms. In some instances, the caudal lobar arteries were already obstructed as indicated by a decreased flow of contrast medium in the affected vessels. This was usually most readily observed in the right caudal lobar arteries.

2. Right-Sided Cardiac Enlargement

The amount of enlargement varies from a minimal change to the typical reverse "D" shape, and the enlargement appears to develop slowly. In many clinical cases, the apex of the heart is displaced further to the left thoracic wall in the dorso-ventral view as the curvature of the right side increases. Occasionally, displacement to the right is seen.

3. Main Pulmonary Artery Enlargement

The change in the main pulmonary artery is easily observed on the dorso-ventral view. The size of this section can vary dramatically from being negligible to very prominent and may be an equivocal sign, in the absence of other signs, of heartworm disease (Figures 4 and 5). In some moderately affected dogs, however, it may be normal, whereas in dogs with congestive heart failure, the enlargement may be hidden by the severe cardiomegaly.

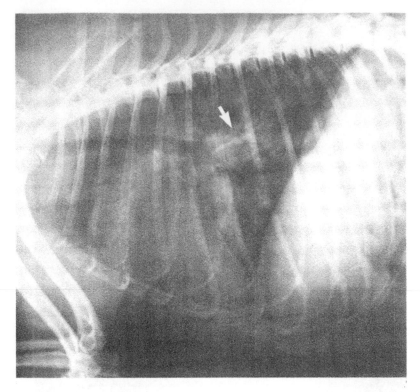

FIGURE 8.   Lateral view. Same dog as in Figure 4 showing obliteration (arrow) of the vessels by parenchymal changes.

In some breeds of dog, e.g., greyhounds, the pulmonary artery may normally be larger or it may appear prominent when radiographs are taken during ventricular systole.

4. Parenchymal Lesions

In moderate cases of heartworm disease the most productive region to examine is the peripheral parenchyma associated with the caudal lobar arteries.

a. Basic Patterns of Lung Disease in Moderate to Severe Dirofilariasis

As the disease progresses, the pulmonary parenchyma becomes more dense, particularly around the pulmonary arteries. This distribution of parenchymal changes is indicative of dirofilariasis. The caudal lobes are more frequently affected than the middle lobes, and the cranial lobes are generally least affected. Increased lung density, obliteration of vascular detail around the pulmonary arteries, loss of detail of the aorta, posterior vena cava, and some regions of the cardiac silhouette indicate the presence of pulmonary lesions, and the severity of the case can be judged from these changes.

Dirofilariasis produces mixed radiographic patterns of lung disease, but they should be classified somewhat arbitrarily according to the major structures involved, e.g., alveolar, interstitial, bronchial, or vascular.[3]

It is important to try to establish the predominant pulmonary tissue pattern as a basis for determining treatment and prognosis. The alveolar pattern is usually very dominant and may mask the presence of other lung tissue involvement, but careful examination of the radiograph may reveal the more chronic interstitial, bronchial, and vascular changes. Medical treatment will help to reduce the alveolar changes which then allows assessment of other tissues. A marked clinical and radiographic improvement should be seen before an adulticide is administered. Patchy, fluffy, coalescing densities with air bronchograms and air alveolograms indicate an alveolar pattern. These are created by suffusion of the tissue with hemorrhage, the migration of inflammatory cells into alveoli, and the compression of the alveolar lumen by thickened interalveolar septa.

Interstitial or bronchial lung patterns are indicative of more chronic disease. There may be some response to medical treatment directed against the inflammatory response, but the more chronic fibrotic changes will remain. A vascular lung pattern is indicative of the most severe form of pulmonary disease, and, generally, improvement does not occur without treatment with an adulticide. In the severe chronic cases there may be no change following treatment.

b. Other Parenchymal Changes

Hemorrhage into pulmonary tissue or intercurrent bronchopneumonia may cause blotchy focal pulmonary consolidations on the radiographs. Densities around arteries may be caused by inflammatory infiltration, hemorrhage or local edema in response to embolic material, the presence of live and dead worms and microfilariae, and, perhaps, hypersensitivity reactions to adult and immature worms. These reactions spread from the arteries into the surrounding parenchymal tissue. The region around the caudal arteries is a common site for increased density in moderate to severely affected cases.

Distinct focal consolidations which resemble metastatic lesions also occur. These may be caused by focal areas of hemorrhages or emboli, in which case the margin is usually not distinct. Granulomas may also develop and be almost impossible to distinguish from metastatic lesions, but can be differentiated from lesions due to hemorrhages or emboli by their clearly defined margins.

Accumulation of fluid in the pleural cavities may obscure the detail of the thoracic viscera in radiographs of dogs with congestive heart failure resulting from advanced heartworm disease. However, grossly distorted pulmonary arteries may still be visible through the fluid.[3] Hepatomegaly and ascites also occur terminaly in these cases as can pericardial effusion which is considered to be a rare occurrence.

Dirofilariasis is usually a slowly progressive disease, but sometimes the chronic course is interrupted by more acute symptoms. The most common acute injury to occur is erosion through a vessel into a bronchus. The final outcome of this depends on the size of the rupture and the vessel involved. If it is small, the blood is usually distributed throughout the bronchial tree to produce a dominent alveolar pattern and respiratory distress which subsides with time. An alveolar pattern is seen radiographically in the lungs. If it is large or continues to bleed, it may result in sudden death of the animal. Other complications such as bronchopneumonia and rupture of a dilated pulmonary artery with exsanguination into the pleural cavities have been observed.[35] Allergic pneumonitis[16] and multifocal pulmonary consolidations associated with disseminated intravascular coagulation have also been reported.[36]

5. Summary

In moderate to severe cases, radiographic changes are more easily recognized than in mild cases. The heart should be examined for right-sided cardiac and main pulmonary artery enlargement indicated by the reverse ''D'' shape. Parenchymal changes, particularly around the pulmonary arteries, and a mixed pattern are common in the lung tissue. Radiographic signs consistent with right-sided congestive heart failure may be seen.

VIII. CHANGES IN RADIOGRAPHIC FINDINGS ASSOCIATED WITH TREATMENT

Destruction of the adult worms with an adulticide produces significant changes in the lungs when they are examined radiographically. It has been shown experimentally[26] that the main pulmonary arteries and the cranial lobar arteries reach their maximum size 3 months after treatment. The right caudal lobar arteries are largest at 5 months and the right ventricle at 7 months. The right ventricle continued to enlarge throughout a 12-month period of observation. The main pulmonary artery and lobar arteries then decrease in size, but the right ventricle continues to enlarge for at least 12 months. The parenchymal lesions in the caudal lung are most evident during the first 6 months following treatment, but then resolve to some extent in dogs where all heartworms have been killed.[26] Resolution of the dilatation, tortuosity and obstruction of blood vessels and of the parenchymal lesions indicate successful treatment, but generally some arterial abnormalities persist, particularly in the peripheral regions of the lung. Prominent radiographic changes will persist if all the worms are not killed.

A marked increase in radiographic density of the lung parenchyma occurs after treatment due to a granulomatous response to the dying and dead adult worms. The location of the dead worms correlates well with the site of parenchymal changes[27] being more frequent in the right than in the left caudal lobes and more frequent in the caudal lobes than the middle or cranial lobes.

Radiographic examinations of experimental dogs which had been infected 27 or 31 months earlier with 100 larvae and then treated with an adulticide 3 weeks before arteriography revealed intraluminal obstruction to blood flow to the caudal lobes. At this time, dilatation was less in the main pulmonary arteries but worse in the peripheral arteries when compared with pretreatment X-rays. Focal areas of increased parenchymal density developed in the peripheral areas after treatment, but they had decreased in severity 5 weeks after treatment.[37]

The effect of aspirin and prednisolone in modifying the reaction of the lungs to dead worms has been studied.[38] It was found that the radiographic changes were less severe in dogs given aspirin (10 mg/kg) or prednisolone (1 mg/kg) from the first day of adulticide treatment for 4 weeks. Prednisolone appeared to be more effective than aspirin in relieving parenchymal changes.

REFERENCES

1. **Douglas, S. W. and Williamson, H. D.**, *Veterinary Radiological Interpretation*, 1st ed., William Heinemann, London, 1970.
2. **Suter, P. F. and Lord, P. F.**, *Thoracic Radiography: A Text Atlas of Thoracic Diseases of the Dog and Cat*, 1st ed., Weltswill, Switzerland, 1984.
3. **Carlisle, C. H.**, Canine dirofilariasis: its radiographic appearance, *Vet. Radiol.*, 21, 123, 1980.
4. **Carlisle, C. H. and Thrall, D. E.**, A comparison of normal feline thoracic radiographs made in dorsal versus ventral recumbency, *Vet. Radiol.*, 23, 3, 1982.
5. **Ruehl, W. W. and Thrall, D. E.**, The effect of dorsal versus ventral recumbency on the radiographic appearance of the canine thorax, *Vet. Radiol.*, 22, 10, 1981.
6. **Ticer, J. W.**, *Radiographic Technique in Small Animal Practice*, W. B. Saunders, Philadelphia, 1975.
7. **Gillette, E. L., Thrall, D. E., and Lebel, J. L.**, *Carlson's Veterinary Radiology*, Lea & Febiger, Philadelphia, 1977.
8. **Douglas, S. W. and Williamson, H. D.**, *Principles of Veterinary Radiography*, 3rd ed., Balliere Tindall, London, 1980.
9. **Ryan, G. D.**, *Radiographic Positioning of Small Animals*, Lea & Febiger, Philadelphia, 1981.
10. **Wyburn, R. S. and Lawson, D. D.**, Simple radiography as an aid to the diagnosis of heart disease in the dog, *J. Small Anim. Pract.*, 8, 163, 1967.
11. **Thrall, D. E. and Losonsky, J. M.**, A method for evaluating canine pulmonary circulatory dynamics from survey radiographs, *J. Am. Anim. Hosp. Assoc.*, 12, 457, 1976.
12. **Rawlings, C. A., Lewis, R. E., and McCall, J. W.**, Development and resolution of pulmonary arteriographic lesions in heartworm disease, *J. Am. Anim. Hosp. Assoc.*, 16, 17, 1980.
13. **Rawlings, C. A., McCall, J. W., and Lewis, R. E.**, The response of the canine's heart and lungs to *Dirofilaria immitis*, *J. Am. Anim. Hosp. Assoc.*, 14, 17, 1978.
14. **Jackson, W. F.**, Radiographic examination of the heartworm infected patient, *J. Am. Vet. Med. Assoc.*, 154, 380, 1969.

15. **Ettinger, S. J. and Suter, P. F.,** *Canine Cardiology,* W. B. Saunders, Philadelphia, 1970.

16. **Boring, J. G.,** Radiographic diagnosis of heartworm disease, in *Proceedings of the Heartworm Symposium '74,* Morgan, H. C., Ed., Veterinary Medicine Publ., Bonner Springs, Kan., 1975, 32.

17. **Lewis, R. E. and Losonsky, J. M.,** The frequency of roentgen signs in heartworm disease, in *Proceedings of the Heartworm Symposium '77,* Otto, G. F., Ed., Veterinary Medicine Publ., Bonner Springs, Kan., 1978, 73.

18. **Hobson, H. P.,** Angiocardiography in canine dirofilariasis. I. Preliminary studies, *J. Am. Vet. Med. Assoc.,* 135, 537, 1959.

19. **Hahn, A. W.,** Angiocardiography in canine dirofilariasis. II. Utilization of rapid film change technique, *J. Am. Vet. Med. Assoc.,* 136, 355, 1960.

20. **Liu, S. K., Yarns, D. A., Carmichael, J. A., and Tashjian, R. J.,** Pulmonary collateral circulation in canine dirofilariasis, *Am. J. Vet. Res.,* 30, 1723, 1969.

21. **Tashjian, R. J., Liu, S. K., Yarns, D. A., Das, K. M., and Stein, H. L.,** Angiocardiography in canine heartworm disease, *Am. J. Vet. Res.,* 31, 415, 1970.

22. **Bisgard, G. E. and Lewis, R. E.,** *In vivo* arteriography in canine heartworm disease, in *Canine Heartworm Disease: The Current Knowledge,* Bradley, R. E., Ed., University of Florida, Gainesville, Fla., 1972, 117.

23. **Kraczynski, J. and Daehler, M. H.,** Contrast radiography as an aid in diagnosing canine dirofilariasis, *J. Am. Vet. Med. Assoc.,* 162, 397, 1973.

24. **Thrall, D. E., Badertscher, R. R., McCall, J. W., and Lewis, R. E.,** The pulmonary arterial circulation in dogs experimentally infected with *Dirofilaria immitis:* its angiographic evaluation, *J. Am. Vet. Radiol. Soc.,* 20, 74, 1979.

25. **Thrall, D. E., Badertscher, R. R., Lewis, R. E., McCall, J. W., and Losonsky, J. M.,** Radiographic changes associated with developing dirofilariasis in experimentally infected dogs, *Am. J. Vet. Res.,* 41, 81, 1980.

26. **Rawlings, C. A., Losonsky, J. M., Lewis, R. E., and McCall, J. W.,** Development and resolution of radiographic lesions in canine heartworm disease, *J. Am. Vet. Med. Assoc.,* 178, 1172, 1981.

27. **Atwell, R. B. and Carlisle, C. H.,** The distribution of filariae, superficial lung lesions and pulmonary arterial lesions following chemotherapy in canine dirofilariasis, *J. Small Anim. Pract.,* 23, 667, 1982.

28. **Buoro, I. B. J., Atwell, R. B., and Heath, T.,** Angles of branching and the diameters of pulmonary arteries in relation to the distribution of pulmonary lesions in canine dirofilariasis, *Res. Vet. Sci.,* 35, 353, 1983.

29. **Atwell, R. B. and Buoro, I. B. J.,** Pulmonary arterial luminal diameters in relation to the distribution of the lesions associated with canine dirofilariasis, *Aust. Vet. J.,* 62, 29, 1985.

30. **Hennigar, G. R. and Ferguson, R. W.,** Pulmonary vascular sclerosis as a result of *Dirofilaria immitis* infection in dogs, *J. Am. Vet. Med. Assoc.,* 131, 336, 1957.

31. **Adcock, J. L.,** Pulmonary arterial lesions in canine dirofilariasis, *Am. J. Vet. Res.,* 22, 655, 1961.

32. **Simpson, C. F. and Jackson, R. F.,** Pathophysiology of heartworm disease, in *Proceedings of the Heartworm Symposium '74,* Morgan, H. C., Ed., Veterinary Medicine Publ., Bonner Springs, Kan., 1975, 38.

33. **Rawlings, C. A. and Lewis, R. E.,** Does heartworm disease produce dilation or hypertrophy of the right ventricle?, in *Proceedings of the Heartworm Symposium '77,* Otto, G. F., Veterinary Medicine Publ., Bonner Springs, Kan., 1978, 76.

34. **Rawlings, C. A. and Lewis, R. E.,** Right ventricular enlargement in heartworm disease, *Am. J. Vet. Res.,* 38, 1801, 1977.

35. **Giles, R. C., Jr. and Hildebrandt, P. K.,** Ruptured pulmonary artery in a dog with dirofilariasis, *J. Am. Vet. Med. Assoc.,* 163, 236, 1973.

36. **Kociba, G. J. and Hathaway, J. E.,** Disseminated intravascular coagulation associated with heartworm disease in the dog, *J. Am. Anim. Hosp. Assoc.,* 10, 373, 1974.

37. **Rawlings, C. A., Losonsky, J. M., Schaub, R. G., Greene, C. E., Keith, J. C., and McCall, J. W.,** Postadulticide changes in *Dirofilaria immitis*-infected beagles, *Am. J. Vet. Res.,* 44, 8, 1983.

38. **Rawlings, C. A., Keith, J. C., Lewis, R. E., Losonsky, J. M., and McCall, J. W.,** Aspirin and prednisolone modification of radiographic changes caused by adulticide treatment in dogs with heartworm infection, *J. Am. Vet. Med. Assoc.,* 182, 131, 1983.

Chapter 6

PATHOLOGY AND PATHOGENESIS OF DIROFILARIASIS

Richard H. Sutton

TABLE OF CONTENTS

I. INTRODUCTION

Dirofilariasis is a multisystemic disorder with the lungs, heart, liver, and kidneys being the main organs affected. While the pathogenesis, pathology, and consequences of the disease have been the subject of many reviews[1-6] most of the reports on the pathology of dirofilariasis refer to advanced natural disease. In an endeavor to define the pathogenesis, sequential studies on experimental disease have relied more on radiographic, particularly angiographic, methods with pathological study being confined to occasional observations at various stages of the disease.

The pulmonary arterial system is the prime site of the pathology associated with *Dirofilaria immitis* infection and, as the disease progresses, the effects are reflected by alterations both functionally and pathologically not only in the pulmonary vasculature but also in the interstitial lung tissue. As a consequence, the heart and the liver become affected, mainly by the development of pulmonary hypertension leading to right-sided circulatory failure. The kidney is also partially involved in this chain of events, but it appears that the principal effect on this organ is immunologically based and is, therefore, a separate facet of the disease.

II. THE PATHOLOGY OF THE PULMONARY ARTERIAL SYSTEM AND ITS DEVELOPMENT

A. Gross Pathology

The appearance of the lungs at post-mortem examination is variable, this being dependent on both the severity and the stage of the disease. In general, natural disease is well advanced before clinical signs appear. While there are similarities in pathological observations irrespective as to whether it is natural or experimental infection, the severity of the disease and the rate at which it develops are related to the magnitude of infection.[7-12] For example, dogs artificially infected by subcutaneous inoculation of infective larvae will develop moderate to severe arterial disease within 5 to 6 months whereas most natural infections progressively develop over several years.[2,5,9] In an average-sized dog (25 kg), mature parasites, which are usually found in the pulmonary arteries, may be found in the right ventricle of the heart if the number of worms exceeds about 25. A number in the order of 50 or more may lead to their presence in the right atrium and venae cavae.[4] Therefore, while the primary pathology is in the pulmonary vasculature, other changes in the interstitial tissue of the lungs, the presence of right-sided congestive heart failure, and the development of the liver failure or caval syndrome are all determined to some degree by the magnitude of the infection. This, in turn, will be reflected in the clinical and post-mortem findings. Accordingly, in the individual case the gross pathological appearance can show considerable variation.

In dirofilariasis the lungs may have a mottled appearance. This can be due to a number of factors. In one major study the most frequent pulmonary change observed was congestion[13] which ranged from being irregular in pattern to diffuse. This was most severe in the caudal lobes of the lung. Generally, however, the presence of pulmonary congestion is a nonspecific finding in that it is usually an indicator of circulatory failure irrespective of cause. Atelectasis may be present in the ventral aspects of the lung lobes if hydrothorax is present[14,15] (Figure 1). The latter, which is an infrequent occurrence in dirofilariasis, is sometimes found in association with ascites which results from right-sided congestive heart failure. Excess pericardial fluid can sometimes be present. Hydrothorax may also be seen in association with dilation of lymphatics on the surface of the pulmonary pleura.[16] The pathogenesis of hydrothorax, however, is not clear because it does not occur as part of a generalized edema. It is possible that it reflects specific venous stasis as a part of circulatory failure, but its inconsistent presence suggests that the reasons for its occurrence are complex.

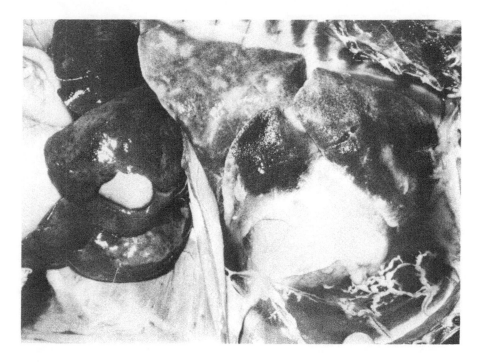

FIGURE 1.   Hydrothorax in dirofilariasis with ventral compression (atelectasis), as shown by the darker coloration of most of the ventral lungs. The lungs are mottled, due to congestion and hemosiderosis, the gall bladder is edematous, and the capsular surface of the liver is coated with fibrin.

Scattered, often well-circumscribed focal lesions are another common feature[11,16,17] and contribute to the mottled appearance of the lungs (Figures 1 and 2). These may vary from being intensely red, which is consistent with hemorrhage,[16] to being consolidated, irregular, grayish nodules.[17] While the focal hemorrhagic areas could suggest that infarction has occurred,[13] studies have shown that in moderate to severe heartworm disease there is adequate collateral circulation to prevent this.[18] For example, ligation of the pulmonary blood supply to one lung in normal dogs does not result in infarction.[19,20]

The focal lesions which are associated with parasitic thrombi in the branches of the pulmonary artery are usually more frequent in the caudal lobes. In addition to those of a hemorrhagic and nodular appearance, the lesions may have a pale fibrous center, surrounded by a zone of hemorrhage and/or brown discolored tissue which is due to hemosiderosis. Associated with these there may be prominent blood vessels visible on the pleural surface. Extensive areas of hemosiderosis may involve the entire lung parenchyma and prominent vascularization of the pleural surface can sometimes be present distally to an occluded vessel. This may occur particularly in the caudal lobes where the lung parenchyma underlying the pleural vessels is pale and comparatively free of hemosiderin.[16,21] Often there is no visible focal lesion associated with the thrombosed vessel.

In some cases there is distinct pleural scarring and contraction in association with atelectatic and consolidated areas.[21] Ventral, grayish or reddish consolidation consistent with a bronchopneumonia is not an infrequent finding.[16] This probably occurs due to interference to the normal host defense clearing mechanisms in the airways associated with the pulmonary damage caused by the disease.

Where interstitial changes have occurred the lung will have a firm consistency[17] with a fibrous-like texture. If the interstitial involvement is extensive and diffuse, the gross appearance of the lung is mottled and often reddish-brown due to hemosiderosis[16] (Figure 2).

In advanced cases the pulmonary artery may be conspicuously enlarged[3] (Figure 3) due

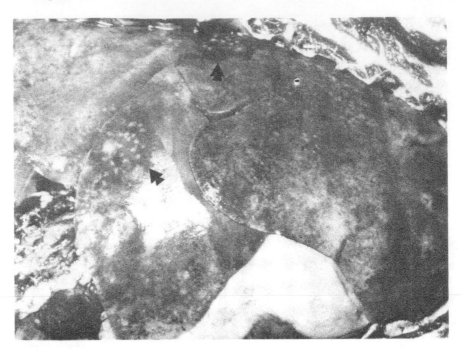

FIGURE 2.   Mottled lungs with scattered pale foci visible on the pleural surface. Some of these are surrounded by a zone of hemorrhage (arrows). The darker-appearing lung is due in part to hemosiderosis.

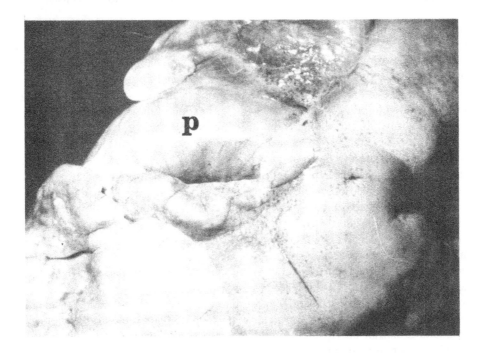

FIGURE 3.   Marked enlargement of the left pulmonary artery (P).

FIGURE 4.   Dilation of the proximal pulmonary trunk. Parasitic thrombi occlude two major branches (arrows). The intimal surface of the artery is roughened and the right ventricular free wall is enlarged.

to marked dilation of the vessel (Figure 4) and thickening of the vessel wall.[15,22] Apart from adult worms in the artery there is rugose and villose thickening of the intima[3,15] giving a fine, grainy, or pebbled appearance[11] (Figure 5). Occasionally, there may be organized or partially organized fibrin strands attached longitudinally in the arterial lumen[16] (Figure 6). The extent and degree of arterial dilation will vary, but it may extend caudally into the muscular branches.[16] Distally, however, the smaller vessels may be occluded or partially occluded by parasitic emboli and thromboemboli (Figure 7) and by an endovascular reaction.[1,11,13,15,22] In chronic cases, well-organized fibrous plugs occlude the distal vessels, but in early thromboembolism the degree of organization and gross degeneration of the worms within the thrombus will be variable.

A consequence of these arterial changes is the development of pulmonary hypertension. This is believed to be the main factor contributing to the dilation of the pulmonary artery in advanced disease.[6] As the pulmonary arterial pressure increases the arteries may also become tortuous.[5,6] The role of the intimal damage in the development of arterial dilation is not known, but the distension of the pulmonary arterial wall leads to a general stiffening and functional rigidity of the vessel. This loss of elasticity, which normally absorbs the increased pressure and volume of systole, produces an increase in systolic and diastolic pulmonary artery pressures.

It is apparent from the reports on the gross pathology that there is a predilection for the caudal lobes of the lung. Once the fifth-stage larvae enter the vascular system via the wall of a systemic vein they are carried to the pulmonary artery.[2,23] The right-sided pulmonary arteries, particularly the right caudal artery, are preferential sites for initial infection. Immature adults may be found there as early as 90 days following infection.[2,4,23] It is in these arteries that the most severe changes are likely to be found.[1,13,24-26] Even in chronic, advanced disease, where there may be total involvement of the pulmonary arterial system and the lung parenchyma, the more recent and most severe lesions occur in the right caudal lobe.[25]

FIGURE 5.   A dilated pulmonary artery with villose thickening and ridging of the intimal surface.

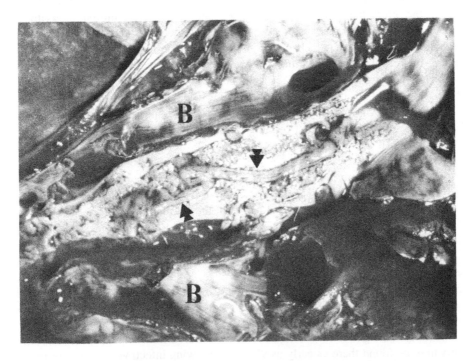

FIGURE 6.   Attached fibrin strands (arrows) along with intimal proliferation in the pulmonary artery. Bronchi (B).

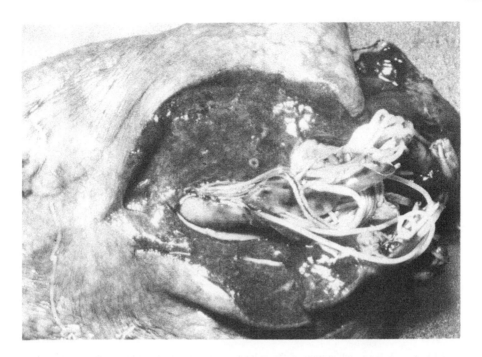

FIGURE 7.   Parasitic thrombus in a distal pulmonary artery.

The reasons for the right-sided preference have been studied. It has been shown that artificial adult filariae and larvae inserted into the venous system will be preferentially distributed in the pulmonary arteries on the right side.[27,28] The mean diameter of the right caudal lobar arteries are smaller than the left, but proximal to this, the right branch of the pulmonary artery has a diameter which is significantly greater than the left.[29] This, together with the blood flow following a curved pathway out of the right ventricle in association with the angles of deviation of the arteries as they branch from the pulmonary trunk, could contribute to this preferential distribution.[29]

B. Microscopic Pathology

Excluding the direct effects of thromboembolism, there are two types of pulmonary arterial lesions in heartworm disease:

1.   An intimal proliferative response which is associated with live worms and may in its early stages have an inflammatory response.[2,13,22] The media is frequently involved.
2.   A granulomatous response, often with widespread arterial damage and parenchymal involvement which occurs when dead parasites form emboli in the distal branches.[2,13]

1. Light Microscopic Changes in the Intimal Response

Experimentally, the initial lesions following introduction of infective larvae have been shown to be in the terminal arterial segments[9] particularly those in the caudal lobes.[10] With a subcutaneous injection of a large number of infective larvae (L_3), angiographic alterations have been observed as early as 91 days.[9] These initially appeared as saccular dilatations, but by 125 days filling defects caused by the worms were noted. With the elapse of time, the worms extended into and subsequently involved the proximal part of the pulmonary artery.[9] Within 6 to 9 months of inoculation the artery is markedly involved in that the initial focal lesions have become diffusely spread. Severe lesions have been observed as early as 5 months after inoculation.[5]

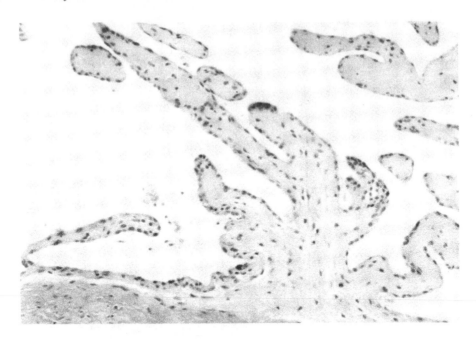

FIGURE 8.   Villose proliferation of the pulmonary arterial intimal surface lined by endothelium. (Hematoxylin-eosin; magnification × 200.)

Intimal proliferation develops rapidly in a villose or finger-like form (Figure 8) and may in the smaller branches of the artery envelop and mold around the filariae.[9,13] These villose protrusions have a central core of dense avascular connective tissue with loosely arranged fibroblasts radiating out from that core and are covered by endothelial cells which may show proliferation.[9,11,13,30] The connective tissue is often continuous with increased fibrous tissue in the intima and media. Disruption of the internal elastic lamellae is also apparent.[13,15] The form of the intimal proliferations may vary from being papillary,[31] to broad luxuriant growths, to masses of dense connective tissue resembling the gnarled roots of trees.[13] In the advanced lesion, there is the incorporation of smooth muscle elements within the fibrous tissue core,[9,15,32] which is considered to be a migration of smooth muscle cells from the media.[9] Endothelial cushions also develop under which there are accumulations of foamy macrophages,[30] this latter observation resembling atherosclerosis. The presence of inflammatory cells (particularly eosinophils, plasma cells, and, to a lesser extent, neutrophils and macrophages) is usually light to moderate in the advanced lesion.[11,13,15] Hemorrhage at the villi tips with hemosiderin containing macrophages in the core of the lesion is another finding.[11]

While the development of the intimal proliferation is rapid there is initially an intense inflammatory reaction. The earliest changes noted have been vacuolation and edema of the subendothelial connective tissue and the presence of small numbers of plasma cells and eosinophils.[13] With an increase in inflammatory cell numbers, the media and, sometimes, the adventitia surrounding the vessels become infiltrated. This endarteritis leads to a fibrous proliferation which produces the intimal and medial changes described earlier. The degree of proliferative change, particularly in the media, has been the subject of variation in some reports.[13,15,33] The smaller elastic and muscular arteries and arterioles have been considered by some to show medial hypertrophy,[13,33] while others have observed proliferative change in both the larger elastic and muscular arteries.[31] It has also been suggested that the nature of the reaction to viable worms may vary depending on location. In one of the early reports, two types of reactions were described: (1) an intimal proliferative lesion in the larger elastic branches of the pulmonary artery and (2) an intense cellular inflammatory response in the smaller arteries.[13] It is possible that the more intense cellular reaction is an early stage of the arterial reaction to the parasite.[2] The fact that a more advanced lesion was noted in the more proximal branches could suggest that a second wave of infection was occurring or that the particular branches concerned had only recently become infected.

2. Electron Microscopic Changes in the Intimal Response

Ultrastructural studies have in general tended to confirm light-microscope observations. Differences in fixation techniques between perfusion and immersion have led to different projections by the scanning electron microscope. With immersion fixation, it was shown that the endothelial cells have a relatively smooth surface with some small ridges.[6] In the presence of worms, the endothelial surface developed scattered, discrete rugose- or villose-like protrusions on the intimal surface which, with heavier worm burdens, became extensively covered with anastomosing ridges of variable height. These were variable in shape and were mostly oriented along the length of the vessel. The endothelium appeared to remain intact.[6] More detailed studies have been made both with scanning and electron microscopy following perfusion.[32,34,35] Examination at various stages of the disease has shown that 4 days after experimental transplantation of adult worms the endothelial cell junctions were disrupted and the cells themselves were rounded.[34] Beneath these cells were smooth muscle cell processes present in the tunica intima. In some areas where the endothelial cells had been lost, platelets and activated macrophages were adherent to the internal elastic laminae with the macrophages pseudopodia extending into the internal elastic laminae. In larger areas of endothelial loss, aggregations of platelets were adherent to the subendothelial structures. Many macrophages and some neutrophils were also present. These platelets were degranulated and other activated platelets were in the process of adhering. Adherence and activation of the platelets were believed to result from the exposure of subendothelial components to plasma proteins and blood cells.[4,34] It has been postulated that the platelets release a protein, platelet-derived growth factor which stimulates the migration and mitoses of smooth muscle cells and fibroblasts.[33] Intimal smooth muscle cells immediately beneath the internal elastic layer were shown to have extended pseudopodia into the internal elastic layer, and the cells themselves were oriented perpendicular to the luminal surface of the arteries. This had only occurred where there was platelet aggregation. In more advanced infection (30 days), the villose-like proliferations or protrusions were shown to be covered by endothelial cells which were either separated or showed widened intercellular junctions and pore formation.[10,32] The normal longitudinal orientation of the cells was lacking. Smooth muscle proliferation and connective tissue were present in the thickened intima.[35]

3. Light Microscopic Changes in the Granulomatous Response

Following death of the worms, which may occur either naturally, in prolonged chronic infection, or following adulticide therapy, parasitic emboli which are carried into the distal arterial branches provide the nidus for thrombus formation. The formation of such thrombi can become a serious complication causing obstruction to vessels which may already be partially occluded.[1] The presence of the dead parasites, apart from leading to thrombus formation, leads to a granulomatous reaction in the wall of the vessel which surrounds the parasitic segment. This reaction can also extend into the lung parenchyma.[13,36] The parasites will appear fragmented and collapsed and are usually surrounded by a zone of suppuration; the neutrophils may also invade the parasite remnants.[13] In association with the suppurative reaction there is fibrous organization of the thrombus and the suppurative zone may become surrounded by macrophages and other mononuclear cells which quite frequently are palisading epithelioid cells.[20] The wall of the vessel may show erosion[17] and be heavily invaded by inflammatory cells. In advanced cases, the walls of the vessels as well as the surrounding lung parenchyma have fibrous deposition as part of a chronic inflammatory response.[13]

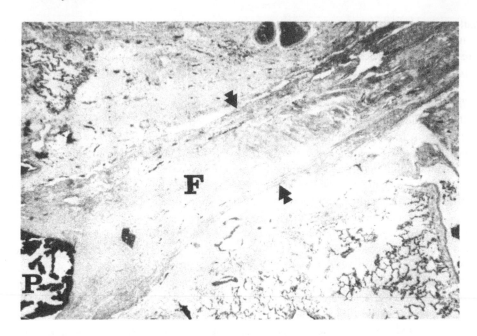

FIGURE 9.   Longitudinal section of a pulmonary artery showing a fibrosed thrombus (F). The proximal end of the thrombus is indicated (P) and the walls of the artery are arrowed. (Hematoxylin-eosin; magnification × 40.)

Parasitic remnants may be difficult to distinguish — they sometimes may be calcified and incorporated into the arterial wall by fibrous tissue.[13] The lumen of the vessel may eventually be plugged by mature fibrous tissue (Figure 9), although in some instances recanalization of the fibrous plug will occur (Figures 10 and 11).

Experimentally it has been shown that the reaction associated with the thrombi develops rapidly. Following the insertion of dead filariae into the pulmonary artery there is, within 24 hr, a severe acute endarteritis associated with the thrombus with infiltration of inflammatory cells, mainly neutrophils, through the arterial wall[37] (Figure 12). There is considerable disruption of the elastic fibers and the smooth muscle cells show degeneration. A suppurative reaction is present within both the thrombus which surrounds the dead worms and the lining the arterial intima. Extensive perivascular inflammatory edema and the suppurative reaction extend into the surrounding alveoli (Figure 13). By the 3rd day following insertion, fibrous organization is apparent with lymphocytes and plasma cells being part of the inflammatory reaction. By the 9th day, fibrous organization of the thrombus is well advanced and the arterial wall structure is indistinct (Figure 14). The zone of suppuration still surrounds the degenerate filariae and this, in turn, is surrounded by epithelioid cells in palisading formation (Figure 15). Within the parenchyma there is extensive consolidation in the area associated with the damaged artery, with interstitial fibrosis and granulomatous inflammation within the alveoli (Figure 11). One prominent finding has been hemorrhage which occurs within the first 24 hr. Grossly, it is apparent as a focal hemorrhage in the lung parenchyma, but, histologically, this has been shown to be an infiltrative process through the damaged arterial wall.[37]

The question of whether or not infarction occurs in dirofilariasis has not been fully elicited. Most evidence suggests that true infarction is not common. Initially, it was considered that arterial thrombosis resulted in infarction,[13] but, as indicated previously, it has been shown that an efficient collateral circulation does provide an adequate blood supply to the affected lung.[1,8,18] While the pulmonary artery supplies bronchioles, alveolar ducts, alveoli, and the

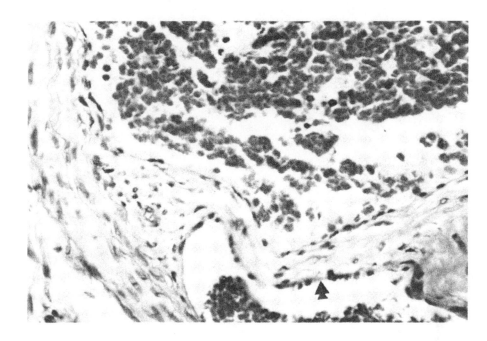

FIGURE 10. Part of a recanalized thrombus with the different channels separated by a fibrous trabecula (arrow). (Hematoxylin-eosin; magnification × 400.)[122]

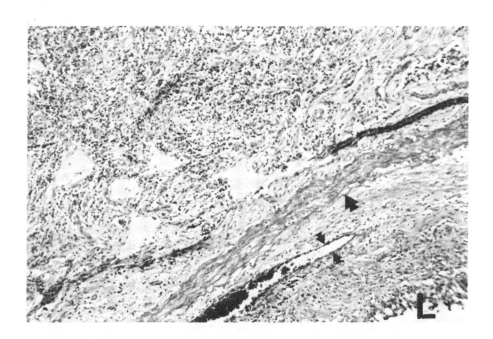

FIGURE 11. Granulomatous end- and periarteritis of a pulmonary artery. The arterial wall is indicated by the large arrow. Within the lumen (L) there is recanalization in the organized part of the arterial reaction (small arrows). (Hematoxylin-eosin; magnification × 100.)[122]

FIGURE 12.   Severe endarteritis with diffuse infiltration of neutrophils and erythrocytes through the arterial wall. The elastic fibers are disrupted and discontinuous. (Elastic stain; magnification × 200.)

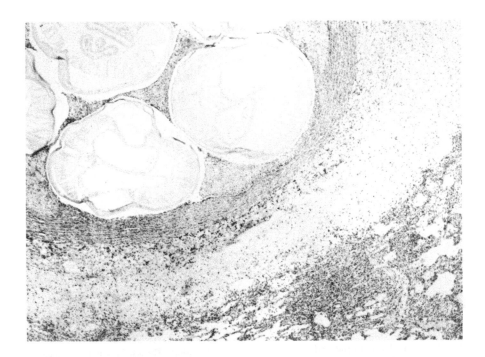

FIGURE 13.   Acute parasitic endarteritis with perivascular edema and inflammatory involvement of the surrounding alveoli. (Hematoxylin-eosin; magnification × 40.)

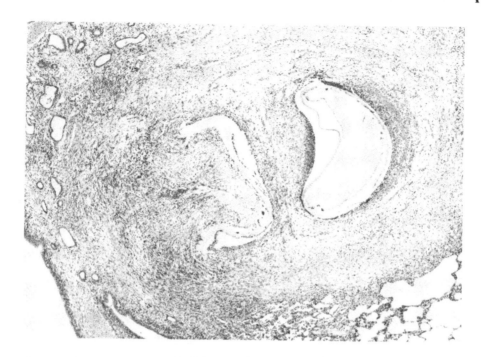

FIGURE 14. Advanced parasitic thromboendarteritis with extensive fibrous formation. The arterial wall is indistinct. A zone of suppurative inflammation surrounds the degenerating filariae. (Hematoxylin-eosin; magnification × 40.)[122]

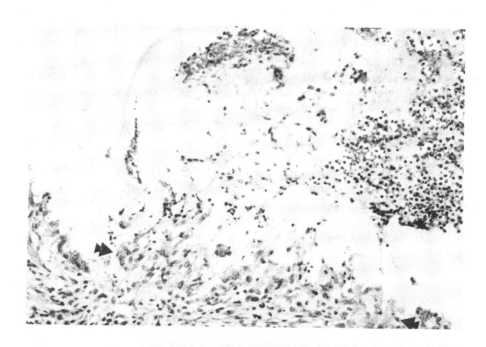

FIGURE 15. Palisading epithelioid cells (arrowed) associated with the suppurative reaction against a degenerate filaria. (Hematoxylin-eosin; magnification × 200.)[122]

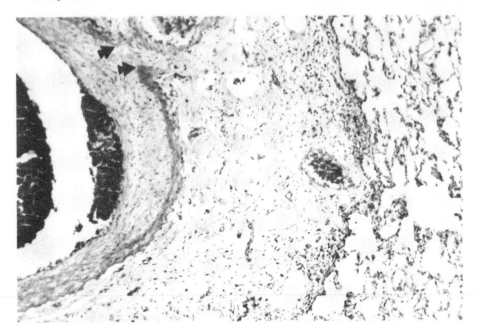

FIGURE 16.   Partially thrombosed artery showing old rupture site characterized by breakdown of the elastic layer (arrows). The artery is surrounded by alveolar fibrosis consistent with organization of hematoma. (Hematoxylin-eosin; magnification × 100.)[122]

pleura in the lungs of dogs,[38] the bronchial artery supplies the vasa vasorum of the pulmonary arteries and veins, the bronchi, and bronchioles.[18] With parasitic thromboembolism causing blockage in the pulmonary artery, there is anastomotic formation between the bronchial and pulmonary arteries. The vasa vasorum become very dilated, and it has been demonstrated histologically that communication between broken vaso vasorum and the pulmonary artery occurs distal to the obstruction.[18] Other collateral supply is derived from the esophageal branches of the left gastric artery, which penetrates the diaphragm through the esophageal hiatus and reaches the lung via the right pulmonary ligament.[8] Other possible sources of supply are the phrenic, pericardial, and intercostal arteries.[8] Vessels have also been observed penetrating the diaphragm from the anterior celiac artery and anastamosing with pleural vessels in the right caudal lobe in younger animals.[16] However, despite this development of systemic collateral supply, there is still a reduced ability of the animal to increase pulmonary blood flow during exercise.[5]

Focal hemorrhage in the pulmonary parenchyma is associated with arterial thrombosis. While it is probable in most instances that it is due to diffusion of red cells through a disrupted arterial wall into the surrounding alveolar spaces,[37] there are instances when hemorrhage into the larger airways can be a feature.[39-41] Where massive hemorrhage into the respiratory tract has occurred it has usually been in association with a severe pulmonary thrombo-arteritis.[40,42] While it is usual for rupture to occur either into the surrounding alveoli (Figure 16) or adjacent bronchus,[41] there is one report of rupture into the thoracic cavity.[40] Increase in blood pressure (pulmonary hypertension) has been postulated as a cause of the rupture, and it has been stated that circulatory interference with the vasa vasorum and subsequent hypoxia in the tunica media are thought to be the principal factors leading to rupture.[40]

C. Etiological Mechanisms of Pulmonary Arterial Pathology

Most of the evidence from the various observations supports the concept of villose proliferation occurring where there is direct contact between the parasite and the intimal sur-

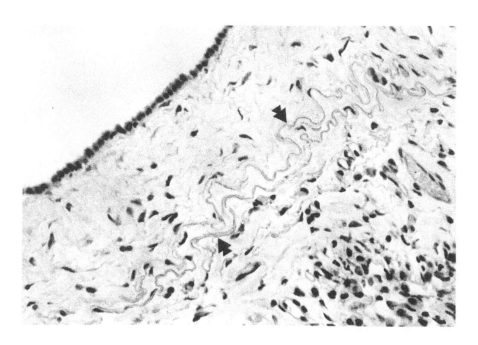

FIGURE 17.   Fibrous thickening of the lung pleura between the mesothelial cells which shows cuboidal change and the elastic fibers (arrowed). (Hematoxylin-eosin; magnification × 200.)

face.[2,13,32,35] This is supported by a case of systemic arterial involvement whereby intimal proliferation was present in the various arteries where worms were harbored.[42] Lesions within the pulmonary arteries were only present where the filariae were present. However, small focal lesions have been recorded in the pulmonary veins without parasitic contact[32] and it has been stated that intravascular lesions develop at downstream locations from the parasite.[2] While this, along with eosinophilic infiltration, suggests that the parasite is releasing a substance to which the host reacts, a search for immune complexes and IgE within actively proliferating lesions has failed to identify these components.[2] In addition, the insertion of flexible polyvinyl chloride threads of similar size and dimensions to adult worms causes a similar arterial intimal reaction to that caused by live worms.[43] It is likely, therefore, that the main arterial reaction is a result of direct contact, the mechanical irritation and damage of the endothelial surface provoking the intimal reaction. Despite this evidence more work on any possible immunological involvement is clearly required.

III. THE PATHOLOGY OF THE LUNG PARENCHYMA AND ITS DEVELOPMENT

The predominant lesions in dirofilariasis are centered on the pulmonary artery and its branches. The pulmonary parenchyma becomes secondarily involved as an inflammatory reaction spreads from the site of parasitic thrombosis which occurs following death of the adult worms.[11] Granulomatous inflammation and hemorrhage are early features. Alveoli become filled with inflammatory cells which are mainly eosinophils and macrophages.[11] Organization of the hemorrhage, along with the removal of red cells by macrophages, leads to areas of fibrosis, hemosiderosis, and consolidation which are grossly apparent. One other feature recently reported is the pleural changes associated with this[21] whereby there is fibrous thickening of the pleura between the mesothelium, which frequently shows cuboidal transformation, and the elastic fibers (Figure 17). Capillary proliferation within the thickened pleura often has the appearance of granulation tissue. With organization of the fibrous tissue

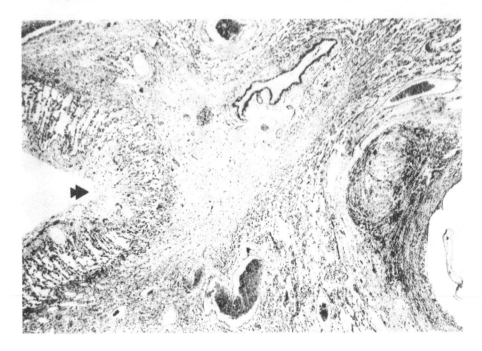

FIGURE 18. Invagination of the pleura (arrowed) associated with fibrosis surrounding an artery with parasitic thromboendarteritis and a periarterial granuloma. (Hematoxylin-eosin; magnification × 40.)[122]

distinct invagination of the pleura is present (Figure 18). The likely effect on pleural function is not known. Excess vascularization of the pleura, which is often noticeable at gross examination (Figure 19), could be due to both increased collateral pulmonary and systemic blood supply feeding an area distal to a branch of the thrombosed artery.[8,21,38] Later studies have revealed that anastomoses of both the pulmonary and systemic blood supply occur in these lesions.[16]

In advanced heartworm disease, widespread interstitial lesions which may be diffuse or focal can be a feature.[1,13-15,17,31,44-47] In the advanced lesions, there is increased interstitial fibrosis, with mixed cellular infiltrates of macrophages, lymphocytes, plasma cells, and neutrophils. Hemosiderosis is often a prominent feature both within the thickened interstitial tissue and the alveolar spaces.[1,17] Smooth muscle hypertrophy and hyperplasia were concentrated around the alveolar ducts[14] (Figure 20), and, in some cases, there is alveolar epithelialization.[15]

Although pulmonary hypertension in dirofilariasis is considered to be due mainly to arterial thromboembolism and other associated arterial changes, the importance of the interstitial pneumonitis as a contributor to pulmonary hypertension cannot be discounted, particularly when there is extensive capillary damage in the interstitial tissue.[47]

The etiology and pathogenesis of the interstitial changes have been the subject of some work, much of which has centered on the phenomenon of immune-mediated occult dirofilariasis whereby a microfilaremia is not detected despite infection with adult gravid worms.[45-50] Apart from persistent and often quite pronounced eosinophilias,[45,46] dogs with occult dirofilariasis have pronounced interstitial changes which grossly, histologically, and radiographically have some resemblance to tropical eosinophilia in man.[45,46,48] The cause of the occult dirofilariasis is believed to be due to a hypersensitivity reaction following exposure of dogs to microfilariae at an early age, possibly by intrauterine transplacental transmission.[46,51] Experimentally, to produce occult dirofilariasis, dogs are sensitized by intravenous inoculation of microfilariae and subsequently inoculated with third-stage larvae.[46] Although

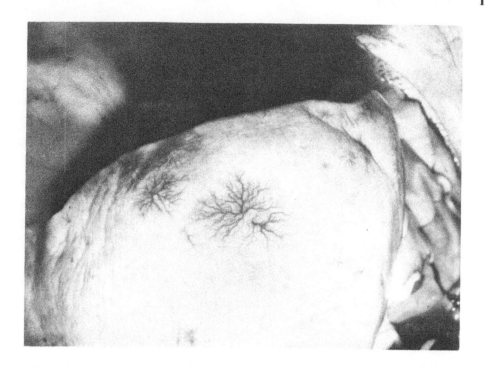

FIGURE 19.   Prominent vascularization of the pleura of a caudal lung lobe.

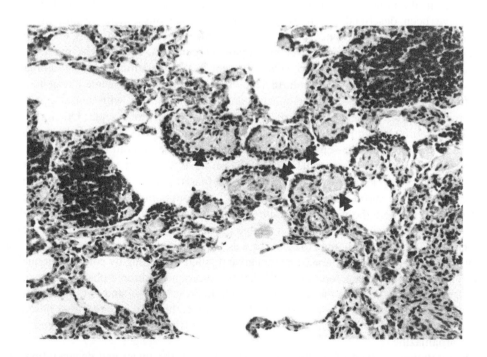

FIGURE 20.   Interstitial thickening in the lung parenchyma with prominent, smooth, muscle hypertrophy (arrowed) of the alveolar ducts. (Hematoxylin-eosin; magnification × 100.)

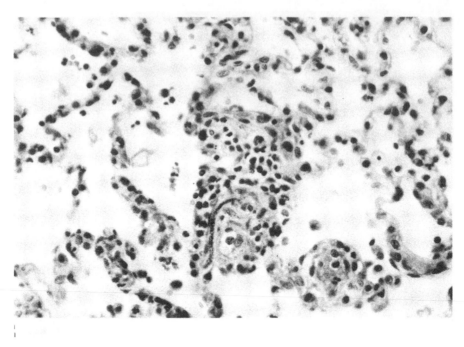

FIGURE 21.   Focal inflammatory lesion associated with a microfilaria in an alveolar septa. Inflammatory cells are present in the alveoli. (Hematoxylin-eosin; magnification × 200.)

the adult worms which develop are gravid, the microfilariae have been shown to be trapped, particularly in the lungs, rather than circulate in the bloodstream. Two types of pulmonary lesion have been described.[46,47] In what is considered the more acute lesion, there are numerous scattered foci of nondegenerate microfilariae associated with many eosinophils, lymphocytes, macrophages, and fewer neutrophils (Figure 21). Most of the microfilariae are present in the capillaries with some also present in the interstitium. The inflammatory reaction can also extend into the alveoli. In the more chronic phase, fragmented, degenerating microfilariae are surrounded by macrophages and epithelioid cells, with dense aggregates of plasma cells often present in the perivascular areas of the alveolar septa. Other changes noted in the more chronic phase have been increased numbers of Type II alveolar epithelial cells lining the interalveolar septa and increased collagen and smooth muscle cells.[45] Hyperplasia and hypertrophy of capillary endothelium have also been noted. Ultrastructural studies of the chronic phase have shown that the microfilarial cells are swollen and fragmented and that the vacuole surrounding the microfilariae is in communication with the phagocytic cell lysosomes. Where the microfilariae are within the capillary lumina and surrounded by hypertrophied endothelial cells, the macrophage processes which contain lysosomes project between the capillary endothelium to be in close association with the microfilariae.[47] It would appear, therefore, that microfilariae are retained and killed in the lung by polymorphonuclear cells, lymphocytes, and macrophages, and that the subsequent chronic phase is a response to persistent antigen.[47] While the lung has been the main subject of study in this reaction it is feasible that other organs may be involved. For example, the focal lung reactions described are similar to those present in the livers of microfilaremic dogs challenged with diethylcarbamazine.[52]

The morphological type of reaction in association with the intact and degenerating microfilariae, along with antibody titers to the microfilariae,[49] is considered to be consistent with a cell-mediated hypersensitivity reaction. The reason for the retention of the microfilariae is thought to be due to cellular adherence to the microfilariae which, along with a cell-mediated response, results in embolization of the microfilariae within the pulmonary vasculature.[47] Other reasons could include structural and physiological damage to the micro-filariae by immune factors which makes them incapable of transversing the pulmonary microvasculature, and the obstructive effect of hypertrophy or proliferation of capillary endothelial cells in response to microfilarial products or mediators released during the in-flammatory reaction. The reaction, therefore, precludes the later development of a patent infection, but does not appear to interrupt the maturation of infective larvae into adult worms.[2]

The experimental work on immune-mediated occult dirofilariasis, therefore, does show that an interstitial pneumonitis can occur as a result of microfilarial retention and that this may become diffuse and extensive[48] with the damage to the pulmonary microvasculature leading to hemosiderosis. The presence of the reaction in nonoccult dogs[13,31] has been explained by the fact that the response in occult dirofilariasis may be more exaggerated, where all microfilariae are retained and destroyed in the lung.[47]

Most cases of interstitial pneumonitis show no morphological evidence of retained mi-crofilariae.[16] While the advanced nature of the lesion with previous removal of all micro-filarial debris is a possible answer, other factors have to be considered. There are viral and bacterial conditions which can cause similar reactions in the alveolar septa, and such a finding is not uncommon in old dogs.[46,53] The dirofilarial antigen which may provoke the interstitial reaction could be present in another form. It has been shown that pups given a subcutaneous, crude, aqueous extraction of adult filariae developed marked alveolar septal thickening with a reduction in the alveolar space.[54] There was infiltration of the alveolar walls with macrophages, neutrophils, and some lymphocytes. Proliferation of the alveolar septal interstitial cells has also been noted. This reaction appears to be more pronounced in dogs which have had the antigen as well as filarial segments inserted into the pulmonary artery than in dogs with only filarial segments and no antigen.[20,37] Because no microfilariae have been involved in these septal reactions, it can be assumed that the dirofilarial antigen which has provoked the reaction is not necessarily derived from microfilariae.

IV. FACTORS MODIFYING THE PATHOLOGY OF THE LUNGS IN DIROFILARIASIS AND THE SEQUELAE OF TREATMENT

Experimentally, it has been shown that the inflammatory reaction in the pulmonary arteries with parasitic thromboembolism can be enhanced by prior subcutaneous inoculation of the animal with a crude antigenic extract.[20,37] This enhanced reaction is in addition to the interstitial reaction just described. The granulomatous reaction in the artery is characterized by a more marked epithelioid and multinucleated giant cell response against the dead filariae[20,37] (Figure 22). The significance of this in terms of the host survival is not known. However, it is possible that the hypersensitive reaction, if severe enough, could be detrimental.

One of the major problems in heartworm disease is the consequences of adulticide therapy. There appears to be a difference in the way filariae are killed comparing levamisole to thiacetarsamide, the former being associated with reduced clinical pulmonary signs and with a different arterial wall reaction.[16] Following such therapy, the parasites are washed into the distal tributaries of the pulmonary artery inducing thrombosis and a granulomatous inflammatory response. Other complications have included hemoptysis and disseminated intravascular coagulation.[55] Detailed study on the effects of adulticide treatment has shown that there is considerable resolution of the villous proliferations in the larger branches of the pulmonary artery 6 weeks following treatment with an arsenical.[56] Initially, however, the endothelial covering of the villi is damaged and platelets become adherent.[57] This "complicated" villous lesion may be due to the toxic effect of the arsenical therapy, as areas of endothelial cell loss have also been noted in areas where there is no villous protrusion.

In the more distal and smaller branches of the pulmonary artery where the worms accu-mulate following treatment, the granulomatous reaction is more severe. The degree of damage

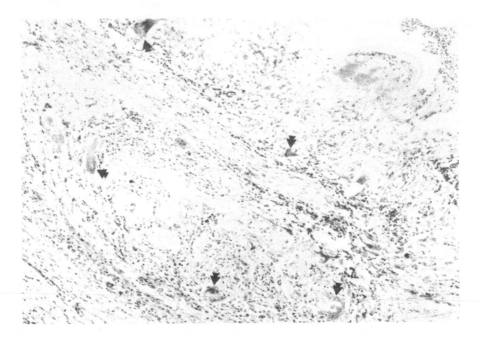

FIGURE 22.   A severe chronic granulomatous reaction with prominent, multinucleated, giant cells (arrowed) as part of the endarteritis associated with parasitic thrombosis. (Hematoxylin-eosin; magnification × 100.)

and obstruction in these smaller arteries is generally worse following treatment than in untreated natural disease.[56] Increased permeability of the damaged vessels also occurs. However, at about 6 weeks post-treatment there is evidence that resolution of these lesions is occurring. The onset of resolution, however, is dependent both on the results of the death of filariae and on the persistence of any infection. The degree of resolution is probably governed by the duration and severity of the pathologic process and upon the general health of the animal.[6]

Both aspirin and prednisolone have been shown to alter the severity of the arterial pathology when used simultaneously with an adulticide.[58-62] The effect of aspirin treatment appears to be a reduction in the formation of thromboemboli thus leaving many of the small caudal pulmonary arteries comparatively unobstructed. The degree of villous proliferation in these arteries is also reduced. Aspirin is also believed to reduce the degree of platelet aggregation and the release of the proposed platelet-derived growth factor. As a consequence, there is a reduction in the amount of myointimal proliferation.[61] A reduction in the severity of the reaction has also been noted when dead filarial segments were inserted into the pulmonary vasculature while the dog was on a course of aspirin therapy.[63]

The main effect of prednisolone on arterial compromise appears to be deleterious.[60-62] This is due, mainly, to suppression of the normal protective nature of the inflammatory response. Suppression of inflammatory cell function leads to persistence of the thromboemboli. In addition, some of the filariae appear to be protected against the effect of arsenicals in that they survive and remain viable. Prednisolone does, however, appear to reduce the degree of reaction in the lung parenchyma. The nature of this drug and its beneficial effects in the treatment of immune-mediated occult cases[2] could support an immune basis for some of the reaction that occurs in the interstitial tissue of the lung in dirofilariasis.

V. THE EFFECT OF HEARTWORM DISEASE ON THE HEART

The progressive development of heartworm disease leads to obstruction of blood flow through the small pulmonary vessels. As indicated earlier, the larger and medium-sized vessel dilate and become tortuous.[5,6] The intimal reaction, the presence of the parasites, and interstitial and pneumonic complications all lead to increased tension in the walls of these larger vessels which produces a rigidity leading to an increase in systolic and overall pulse pressure[5] contributing to increased pulmonary vascular resistance.[5,64] As the vascular resistance increases, the pulmonary pressure also increases to maintain normal pulmonary flow.[5] With the onset of pulmonary hypertension there is cardiac enlargement which is initially due to ventricular dilation.[5,15,30] This increases the work capacity of the right ventricle, and with progression of the disease hypertrophy will occur.[22]

Accordingly, in severe disease, the heart will appear grossly enlarged (cor pulmonale) due to both hypertrophy and dilatation of the right ventricle and right atrium and enlargement of the pulmonary artery[3,15] (Figures 3 and 4). The development of congestive heart failure is usually a sequela to the prolonged hypertrophy in that the hypertrophied cell has decreased electrical and mechanical function[5] and reduced nutritional capacity. Because of the reduced cardiac output as the ventricle fails, venous pressure rises and the typical signs of congestive heart failure, namely ascites, hydrothorax, hydropericardium, and edema, will develop.[5] The effect is compounded by an expanded blood volume because of decreased renal perfusion leading to retention of salt and water through the renin-angiotensin pathway.[4,5] Dilation of the right ventricle may also lead to regurgitation of blood through the tricuspid valve.[16] Left ventricular atrophy associated with severe cor pulmonale has also been reported.[1]

While the onset of right-sided congestive heart failure is generally gradual, it may sometimes be acute. This can occur with extensive thrombosis and thromboembolism following adulticide treatment or by the inexplicable natural death of a large number of worms.[2] Sudden entanglement of worms around the cusps and/or chordae tendineae of the tricuspid valve can also lead to acute right-sided failure.[16] In some of these cases the chordae tendinae are thickened[36] and shortened. Another cause of acute congestive heart failure can be the venae cavae or caval syndrome.[65,66]

Valvular lesions have not been extensively reported in association with dirofilariasis. Slight, chronic, fibrous thickening of the atrioventricular valves has been noted, but was considered not to be due to dirofilariasis.[17] Interpretation of such changes is difficult because of the prevalence of endocardiosis of these valves in dogs, although the right valve is less affected than the left valve. One case of marked fibrosis and calcification of the tricuspid valve which was reported was shown microscopically to be fragments of necrotic worms surrounded by partially calcified fibrous tissue.[36]

Mural endocardial lesions do not appear to be common, which suggests that heartworms do not usually reside there or, if present in heavy infections,[2,4] are not in contact for long enough periods of time to cause any endocardial reaction. Some endocardial "stippling" has been noted in the right atrium and in the right ventricular outflow tract of dogs with caval syndrome.[2] This is probably the result of tricuspid regurgitation whereby the jetting of blood causes a reaction on the endocardium. The outflow tract lesions could be associated directly with presence of worms confined to one area or with the turbulence produced by nonlaminar blood flow in the area.[16]

Histological studies of the myocardium are generally unrewarding. Myocardial hypertrophy can be seen, and a few nonspecific, mainly lymphocytic, nodules have been observed.[16] A mild focal interstitial myocarditis has also been described.[31] However, the significance of these changes in relation to heartworm disease is not known.

VI. THE EFFECT OF HEARTWORM DISEASE ON THE LIVER

A. Right-Sided Congestive Heart Failure

As the predominant clinical feature of dirofilariasis related to interference in the pulmonary

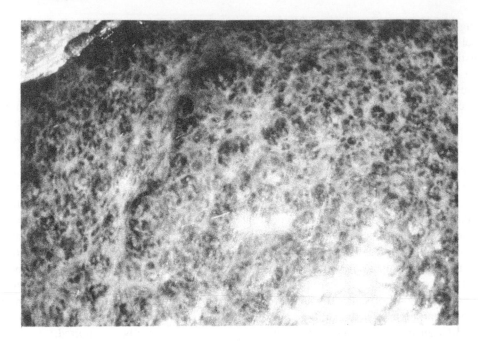

FIGURE 23. The effect of chronic, passive, venous congestion in dirofilariasis showing a distinct lobular or ''nutmeg'' pattern of the liver due to periacinar fibrosis.

circulation and the consequential ''cor pulmonale'', it is to be expected that, with the progression of heartworm disease, secondary changes will occur in the liver due to right-sided congestive heart failure. Because of the progressive and prolonged nature of the disease, the liver most frequently shows evidence of chronic, passive, venous congestion. Most commonly, it is enlarged to a variable degree, generally darker because of the congestion, and there is a lobular and sometimes granular pattern on the capsular surface.[1,15,17,31] The granular or ''dimpled'' pattern which is a feature of prolonged congestion is often referred to as a ''nutmeg'' liver[3] (Figure 23). Some livers may have fibrous thickening of the capsule and interlobular adhesions as a result of an acute episode of postsinusoidal venous congestion whereby the venous back pressure results in lymphatic dilatation and exudation of high protein-content lymph through the liver capsule (Figure 24). In more prolonged cases, the liver may become diffusely fibrotic[40] (Figure 23). Following fixation, the lobules have been shown to have a yellow-tan center and a brownish-red periphery with large conglomerations of yellow-tan-colored lobules around the medium-sized vessels and bile ducts.[15] These discolorations reflect variations in the vascularity and pigmentation as a result of venous stasis.

Microscopically, the findings may include periacinar (centrilobular) necrosis associated with the congestion,[3] and dilation of the central veins and sinusoids. This congestion may be more marked in peripheral areas of the liver, and, as the disease progresses, there is mild fibrosis in both periacinar and centriacinar (portal) areas.[15] The lymphatics, including sub-capsular vessels, may be dilated and pigment containing macrophages are found, particularly in areas around the distended central veins. This pigment is usually a mixture of both hemosiderin and bile pigment, the latter accumulating because of biliary stasis associated with the congestion.

More severe liver pathology is seen in the complicating disorder of the caval syndrome and in the reaction of microfilaremic dogs to the administration of diethylcarbamazine.

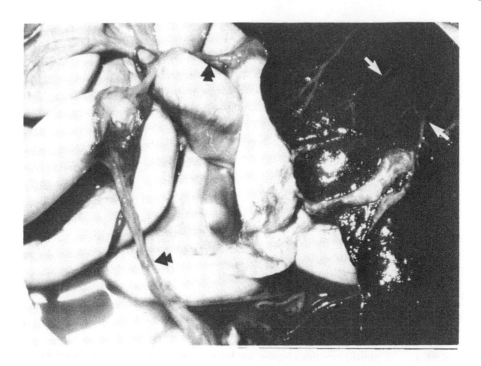

FIGURE 24.   Acute venous congestion of the liver in dirofilariasis. The capsular lymphatics are dilated (white arrows) and there is interlobular fibrin deposits. Fibrin strands (black arrows) in the peritoneal cavity indicate the exudation of high-protein lymph through the liver capsule.

B. Caval Syndrome

Caval syndrome is characterized by a very acute hemodynamic change[65,66] as reflected in an acute clinical syndrome, often resulting in death. Because this syndrome is so acute, all reported necropsy findings are not necessarily associated with the acute clinical signs.[16] The liver is reported swollen and congested and, as a result of lymph exudation, there may be fibrin deposition on the capsular surface and between the liver lobes leading to adhesions.[67] Fibrin strands may also extend from the liver to the surrounding viscera. The ascitic fluid which develops in some cases may be copious and usually has a high protein content which clots on exposure to air (Figure 24). The acute circulatory failure which this represents could be associated with acute intravascular hemolysis, the possible interference of tricuspid valve function, and reduced venous return associated with the presence of filariae in the right atrium and venae cavae. It is more likely associated with the latter.[16] The filariae may also be present in the hepatic and iliac[16] veins and, as a consequence, parasitic thrombi in these vessels can be a feature of this syndrome.[65,66] Generally, these thrombi are recent and nonocclusive. Often, the walls of the posterior vena cava and hepatic veins are thickened, which would reflect chronic, congestive heart failure prior to the onset of the clinical signs of caval syndrome. A detailed histological study has shown that one of the features in the liver of dogs with caval syndrome was a cavernomatous change of the hepatic veins with a large number of dilated vessels appearing to replace the centrilobular venule.[66] However, it would appear from more recent observations that many, if not all, of these dilated vessels are distended lymphatics.[16,68] This is supported by the frequent exudation of fibrin via the hepatic capsule and by dilation of the subcapsular lymphatics (Figure 24). There was also a variable degree of bile stasis, microfilariae present in the sinusoids and veins, and increased numbers of polymorphs and macrophages in the sinusoids which occasionally formed small nodular aggregates.[66] It is probable that this latter observation is not associated with the

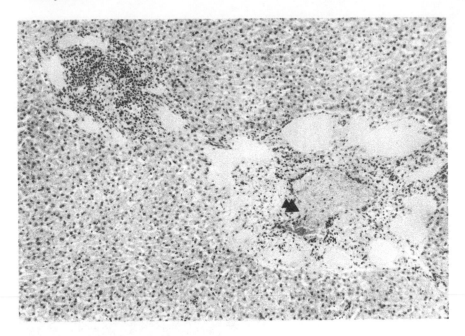

FIGURE 25. A partially constricted hepatic vein (arrowed) surrounded by distended lymphatics as part of the hepatic reaction to the administration of diethylcarbamazine (DEC) to a microfilaremic dog. (Hematoxylin-eosin; magnification × 100.)

caval syndrome because, although not a frequent finding in dogs, it has been seen in experimental infection.[11] The cells, which included histiocytes (macrophages), eosinophils, and plasma cells, were aggregated into granulomata. No microfilariae were seen associated with this although observations on other tissues such as the lungs would suggest that these granulomata are a response to lodgement of microfilariae in the tissues.[47] Granulomata associated with microfilariae have also been noted in many organs, including the liver, following treatment with ivermectin.[69]

It would seem that the extent of hepatic involvement in caval syndrome cases was related to the extent of cor pulmonale prior to clinical signs developing and to the extent of venous congestion associated with the filarial burden in the heart and venae cavae.[16]

C. Reaction to Diethylcarbamazine (DEC)

The main feature of the diethylcarbamazine reaction is the vascular change which occurs in the liver. The etiology and pathogenesis of this phenomenon are discussed in Chapter 9, but the changes which occur result from constriction of the hepatic veins.[52] Grossly, the liver may appear swollen and dark from severe congestion and will ooze blood on sectioning. There is serosanguineous ascitic fluid, edema and hemorrhage of the gall bladder and other perihepatic tissues, edematous, hepatic lymph nodes, and, to a less extent, hydroperitoneum and congestion of the intestines.[52,70]

Microscopically, the liver is congested, hemorrhagic, and there is occasional thrombosis formation in the veins of the periacinar region. The main feature is constriction of the larger hepatic veins with marked dilatation of the associated lymphatics and marked perivascular edema, congestion, and hemorrhage (Figure 25). In general, the portal areas do not appear to be affected. The severity of the vascular change appears to be correlated with the number of microfilariae present in the liver. The mast cells associated with this vascular reaction are not degranulated, suggesting that histamine-mediated anaphylaxis is not the mechanism by which these changes occur.[52]

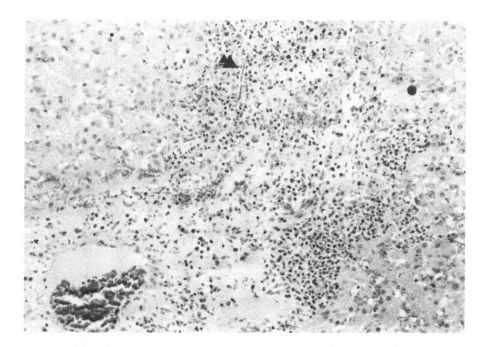

FIGURE 26. Scattered inflammatory foci in the liver of a dog treated with diethylcarbamazine (DEC). One focus is associated with a microfilaria. (Hematoxylin-eosin; magnification × 200.)

Granulomata formation also appeared to be a feature in the livers of both reactive and nonreactive dogs following diethylcarbamazine administration to microfilaremic dogs.[52] These dogs had scattered foci of inflammation which varied from being extensive to small clusters of five to ten cells. Not all of these foci had visible microfilariae associated with them, but, when they were present, the predominant inflammatory cell was the eosinophil. However, lymphocytic, neutrophilic, and mixed inflammatory cell infiltrates were frequent. In the dogs which reacted to diethylcarbamazine, many of the larger inflammatory foci were associated with vascular changes (Figure 26). While many of the foci were associated with microfilariae, there were numerous microfilariae present in the sinusoids and veins which had no associated cellular reaction.[52] In an earlier report on diethylcarbamazine-reactive dogs, a moderate infiltration of inflammatory cells consisting of macrophages, leukocytes (not defined), and plasma cells was described. Discrete foci were not reported.[70]

VII. THE EFFECT OF HEARTWORM DISEASE ON THE KIDNEY

As part of the multifaceted effects of dirofilariasis, renal disease has been shown to be of considerable importance. As previously mentioned, the functional role of the kidney can be important in conserving salt and water when renal perfusion is reduced in congestive heart failure.[4,5] There is increased granularity of the juxtaglomerular apparatus and thickening of the zona glomerulosa in the adrenal, these being indicators of increased renin and aldosterone formation.[71] With the reduction in renal blood flow, there is reduced glomerular filtration which may be associated with a mild to moderate azotemia.[72] Other prerenal disorders may be found in association with the shock-like reaction in the caval syndrome and with possible concomitant, disseminated intravascular coagulation.[55,72] In case reports of disseminated intravascular coagulation in heartworm-affected dogs, two out of the three were affected with caval syndrome and had elevated plasma urea levels. While this elevation could have been due to reduced renal perfusion, obstruction of the renal glomeruli by fibrinous thrombi was another possibility.[65]

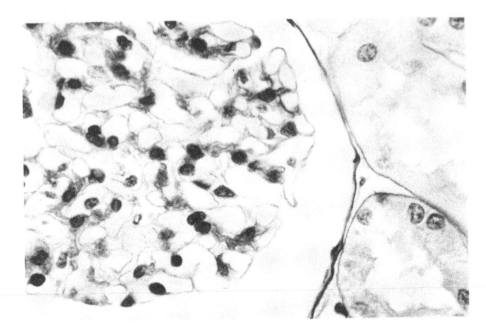

FIGURE 27.   Normal renal glomerulus showing thin capillary loops (2 μm section). (Periodic Acid Schiff stain; magnification × 1000.)

With hemoglobinuria, the gross appearance of the kidney is a dark red color, this becoming more brown with the uptake of hemoglobin by the tubular epithelium with eventual conversion to hemosiderin. Although high hemoglobin concentration can damage renal tubular epithelium, there have been no reports of severe renal tubular damage in caval syndrome.[73-75] Terminal renal failure, however, is not uncommon in caval cases in which treatment has been delayed.[16] However, in noncaval dogs, there is deposition of hemosiderin in the renal tubules.[76] The degree of hemosiderin present appears to be related to the actual filarial burden rather than the severity of the disease. This observation, along with the presence of hemosideruria,[77] indicates that intravascular hemolysis of a subclinical nature occurs in dirofilariasis without worms being present in the right atrium or venae cavae. The cause of this hemolysis is not known, but, in view of the relationship with the numbers of worms, mechanical damage of the erythrocytes seems likely.

The most important aspect of the renal pathology in heartworm disease is that centered on the glomeruli. However, the study of the glomerulus requires careful specimen preparation. It is only with light microscopic examination of fine 1 to 2 μm-thick sections (Figure 27) and with ultrastructural examination that any certain interpretation can be made. Examination of 4 to 6 μm-thick paraffin-embedded blocks is generally unrewarding.

It was known in the 1940s that albuminuria with thickened glomerular basement membranes (Figure 28) and protein casts in the kidneys were present in association with heartworm disease.[78] The pathogenesis of this phenomenon has been open to conjecture but it is probably multifactorial.[72,79-82] Lesions in association with microfilariae in the glomeruli have been noted.[22,80-82] These lesions included swelling and fragmentation of the basement membranes with loss (denudation) of endothelial cells and fusion and atrophy of foot processes. The severity of these lesions was correlated with the degree of entrapment of microfilariae, some of which were in direct contact with exposed basement membrane. However, deposition of immune complexes was not able to be demonstrated.

As a follow up to this work, the effect of levamisole treatment on the microfilariae was studied and thickening of glomerular capillary basement membranes was noted prior to

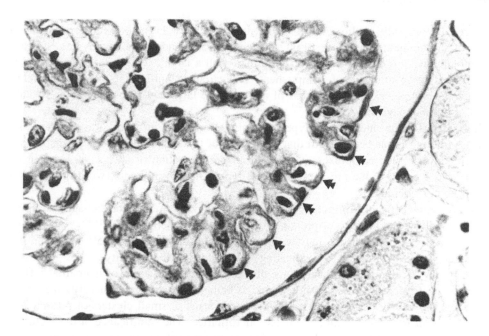

FIGURE 28.   Renal glomerulus from a dog with dirofilariasis showing thickened glomerular basement membranes (arrows) within the capillary loops (2 μm section). (Periodic Acid Schiff stain; magnification × 1000.)

treatment.[79] However, 30 hr post-treatment there was noticeable mesangial cell proliferation and partial occlusion of capillary lumina by phagocytosed microfilariae, the latter showing degeneration. The glomerular basement membrane appeared more thickened. At 5 days post-treatment there was a slight segmented or lobulated appearance to the glomeruli and the mesangium was prominent, but, in general, the cellularity of the glomeruli was normal. However, there were some spikes and electron-dense deposits on the epithelial side of the basement membrane. Dead microflariae in both cortex and medulla were surrounded by macrophages, plasma cells, and lymphocytes. Similar focal lesions were also noted in the liver. It was postulated that the severity of these glomerular lesions was related to the microfilarial numbers in both treated and nontreated cases and, therefore, considered to result from mechanical damage. Failure to demonstrate both immunoglobulin and ultrastructural changes, characteristic of an immune-complex problem, would support mechanical damage by the microfilariae, although it has been suggested that in untreated cases there is usually enough *D. immitis* antigen present to form biologically active immune complexes in moderate antigen excess.[83]

Most of the evidence, however, supports a glomerulopathy which is immune based.[72,79,83] The findings to support this include electron-dense deposits on the epithelial side of the basement membrane as well as demonstration of the immunoglobulin IgG and complement component C_3 in the capillary wall and mesangium of the glomeruli.[72,79] In addition, subendothelial deposits which could be immune complexes have been noted. However, these deposits were different, both in distribution and morphologically, from the electron-dense deposits usually associated with immune-complex glomerulopathy.[72] Such changes appear to be a common finding in dogs infected with *D. immitis*,[72,84,85] but the conclusive evidence that the antibody involved is specific for the parasitic antigen or that the suspect antigens are located in conjunction with immunoglobulin and complement has not been produced.[72]

Immune complexes are formed by deposition of circulating or of locally formed antigen-antibody-complement complexes in glomerular capillary walls and mesangium.[72] These complexes are usually soluble and of low molecular weight.[86] The reason for their lodgement in the renal glomeruli is believed to be due to their formation in the presence of a moderate excess of antigen being, therefore, of intermediate size and binding a relatively large quantity of complement.[72] In contrast, complexes formed with a large excess of antigen are small, do not bind large amounts of complement, and circulate until catabolized or excreted. Conversely, complexes with an excess of antibody are large and insoluble and are usually removed by phagocytosis.[72]

The reason for the intermediate type of complex formation in heartworm disease is not known, although it has been shown that depression of the antibody response predisposes to the formation of such complexes. The possibility of formation of such complexes in the glomeruli as a result of the presence of microfilariae and possibly adult worms would also have to be considered.[87,88]

The consequences of the immune-complex glomerulopathy are that the damaged basement membranes allow the passage of protein with a resultant proteinuria.[72,78,86,89,90] The degree and incidence of proteinuria in dogs with heartworm disease are variable[89] and when it is present most of it is albumin,[78,90] indicating some selective permeability.

Other renal lesions noted in heartworm disease include focal areas of interstitial inflammation.[11,91,92] While inflammation is reported as being associated with microfilariae, this is not always the case.[16] In most instances, the foci consist of small clusters of plasma cells (often only four to five) without any visible microfilariae. The significance of these clusters is unknown, but it is likely that there is some antigen provoking an antibody response.

Aberrant adults have been reported in most body locations including the kidneys.[93] These were in a subcapsular cystic cavity and were believed to have escaped from the arteries. Amyloidosis of the glomeruli has also been reported in association with *D. immitis*.[94] This is usually regarded as being related to an immunologic disturbance, however, the prevalence and significance in relation to canine dirofilariasis are not known.

Renal disease plays an important role in canine dirofilariasis. The nature of the changes and, more importantly, the pathogenesis of the changes that occur are complex and require further elaboration.

VIII. THE PATHOLOGY OF OTHER ORGANS AND UNUSUAL LOCATIONS OF *DIROFILARIA IMMITIS*

In dogs with pulmonary hypertension and congestive heart failure the zona glomerulosa of the adrenal gland may be enlarged. As mentioned previously, this indicates increased aldosterone formation consequential to reduced renal perfusion.[71] However, the main effect of heartworm disease on the other organs of the body relates to the reaction associated with mislocated or ectopic worms. Apart from this, the changes can be secondary to treatment whereby adverse reactions to the drugs used may provoke a reaction in tissues remote from the target worms.

The use of levamisole hydrochloride treatment in dogs has resulted in hemolytic anemia,[95] thrombocytopenia,[96] and nervous disorders, the most frequent of which have been associated with distemper and granulomatous meningoencephalomyelitis.[97] The reason for this reaction is not known, but levamisole does play a role in stimulating immunity by enhancing the activity of the cell-mediated immune response.[98,99] Microfilaremic dogs treated with levamisole and ivermectin may have focal inflammation with phagocytosis of the microfilariae in organs such as spleen, skeletal and cardiac muscles, diaphragm, pancreas, and the lymph nodes.[16,68]

The lymph nodes, particularly those in the thoracic cavity, and the spleen have shown variable degrees of hemosiderosis which in the lymph nodes involves mainly the medulla.[15,17] Histologically, the nodes have erythrocytes in the sinuses, both free and phagocytosed, and macrophages containing hemosiderin. The reticular lining cells of the sinuses also contain hemosiderin granules. Because the bronchial nodes are the ones most likely to show this change, it is probable that these changes result from the hemorrhage which commonly occurs both in the interstitium and alveoli of the lung.

There are many reports of *D. immitis* being sited in unusual locations.[100] These can be classified into two categories. Firstly, there are those worms which are located within blood vessels as an extension from the focal point of infection in the pulmonary arteries. The filariae in the venae cavae in association with the caval syndrome would be in this category.[65,66] There are also references to worms in the aorta, femoral arteries, and abdominal arteries.[100-105] A saddle aortic parasitic thromboembolism noted in one report[93] resulted in ischemic necrosis of the rear limb musculature which histologically showed diffuse muscle-fiber swelling, fragmentation, vacuolation, and nuclear dissolution, with irregular areas of hyalinization, sarcolemma-cell proliferation, mononuclear-cell infiltration, and fatty replacement. Severe inflammatory change was present in the arterial walls, which included arteries in the pancreas and a popliteal node. *Adult worms were also present in cysts within the right kidney, thigh muscle, and popliteal node.*[93]

A further report describes adult worms not only in the pulmonary arteries and right ventricle, but also in the left atrium, the right atrium, the venae cavae, aorta, left and right external iliac arteries, left and right femoral arteries, left and right popliteal arteries, and the testicular arteries.[42] There was thickening and roughening of the intima of all the arteries where the worms were found which, histologically, was either villose-like intimal proliferation or granulomatous inflammation. In addition to this, there was intimal thickening and hyalinization of the coronary arteries, some of which were occluded by microfilariae.

The second category refers to worms located in sites other than blood vessels. These sites have included the bronchioles,[100] an interdigital cyst,[104] the epidural space of the spinal cord,[105,106] the abdominal cavity,[107] and the brain.[108-112] The clinical signs are dependent on the exact location and degree of pathological change that occur. Cerebral infarction has been one consequence of aberrant filaria in the brain,[111] whereas in another case, there was a minimal inflammatory reaction associated with the worm.[112] The most common site for aberrant *D. immitis* is the eye.[104,113-121] In most instances the immature adults have been found in the anterior chamber. There is usually an associated conjunctivitis, keratitis, and uveitis. The cornea may often be cloudy. It has been postulated that the immature worms migrate directly to the eye,[109,114] but the dimensions of the intraocular vascular system would tend not to support this.[117] A more plausible possibility is the migration of the L_3 infective-stage larvae from the subconjunctival tissues into the intraocular chamber where they then mature to the adult stage.[117] In many of these cases, successful surgical removal of the worm has been implemented.

The siting of aberrant worms is not surprising even though they have a predilection for the pulmonary-arterial system. The reasons for this predilection are not fully known. Although aberrant location is relatively rare, when it does occur it constitutes a diagnostic problem.[100] Most cases have burdens within the pulmonary arteries, and, therefore, any animal with heartworm showing atypical clinical signs should be considered as having a possible aberrant parasitic infection.

REFERENCES

1. **Knight, D. H.,** Heartworm heart disease, *Adv. Vet. Sci. Comp. Med.,* 21, 107, 1977.
2. **Knight, D. H.,** Heartworm disease, in *Textbook of Veterinary Internal Medicine: Diseases of the Dog and Cat,* 2nd ed., Ettinger, S. J., Ed., W. B. Saunders, Philadelphia, 1983, chap. 54.

3. **Otto, G. F. and Jackson, R. F.,** Heartworm disease, in *Textbook of Veterinary Internal Medicine,* 1st ed., Ettinger, S. J., Ed., W. B. Saunders, Philadelphia, 1975, chap. 36.

4. **Calvert, C. A. and Rawlings, C. A.,** Diagnosis and management of canine heartworm disease, in *Current Veterinary Therapy, Eight: Small Animal Practice,* Kirk, R. W., Ed., W. B. Saunders, Philadelphia, 1983, 348.

5. **Rawlings, C. A., McCall, J. W., and Lewis, R. E.,** The response of the canine's heart and lungs to *Dirofilaria immitis, J. Am. Anim. Hosp. Assoc.,* 14, 17, 1980.

6. **Rawlings, C. A., Keith, J. C., and Schaub, R. G.,** Development and resolution of pulmonary disease in heartworm infection: illustrated review, *J. Am. Anim. Hosp. Assoc.,* 17, 71, 1981.

7. **Munnell, J. F., Weldon, J. S., Lewis, R. E., Thrall, D. E., and McCall, J. W.,** Intimal lesions of the pulmonary artery in dogs with experimental dirofilariasis, *Am. J. Vet. Res.,* 41, 1108, 1980.

8. **Thrall, D. E., Badertscher, R. R., Lewis, R. E., and McCall, J. W.,** Collateral pulmonary circulation in dogs experimentally infected with *Dirofilaria immitis:* its angiographic evaluation, *Vet. Radiol.,* 21, 131, 1980.

9. **Knight, D. H.,** Evolution of pulmonary artery disease in canine dirofilariasis: evaluation of blood pressure measurements and angiography, in *Proceedings of the Heartworm Symposium '80,* Otto, G. F., Ed., Veterinary Medicine Publ., Edwardsville, Kan., 1981, 55.

10. **Rawlings, C. A., Schaub, R. G., Lewis, R. E., and McCall, J. W.,** Heartworm-induced pulmonary vascular lesions: development and resolution, in *Proceedings of the Heartworm Symposium '80,* Otto, G. F., Ed., Veterinary Medicine Publ., Edwardsville, Kan., 1981, 63.

11. **Tulloch, G. S., Pacheco, G., Casey, H. W., Bills, W. E., Davis, I., and Anderson, R. A.,** Prepatent clinical, pathologic and serologic changes in dogs infected with *Dirofilaria immitis* and treated with diethylcarbamazine, *Am. J. Vet. Res.,* 31, 437, 1970.

12. **Jackson, R. F., Otto, G. F., Bauman, P. M., Peacock, F., Hinrichs, W. L., and Maltby, J. H.,** Distribution of heartworms in the right side of the heart and adjacent vessels of the dog, *J. Am. Vet. Med. Assoc.,* 149, 515, 1966.

13. **Adcock, J. L.,** Pulmonary arterial lesions in canine dirofilariasis, *Am. J. Vet. Res.,* 22, 655, 1961.

14. **Gross, D. R., Williams, G. D., Hobson, H. P., and Humphries, J. P.,** Heartworm disease with pulmonary hypertension, *Southwest. Vet.,* 28, 233, 1975.

15. **Patterson, D. F. and Luginbuhl, H.,** Clinico-pathologic conference, *J. Am. Vet. Med. Assoc.,* 143, 619, 1963.

16. **Sutton, R. H. and Atwell, R. B.,** unpublished data, 1985.

17. **Winter, H.,** The pathology of canine dirofilariasis, *Am. J. Vet. Res.,* 20, 366, 1959.

18. **Liu, S.-K., Yarns, D. A., Carmichael, J. A., and Tashjian, R. J.,** Pulmonary collateral circulation in canine dirofilariasis, *Am. J. Vet. Res.,* 30, 1723, 1969.

19. **Bloomer, W. E., Harrison, W., Lindskog, G. E., and Liebow, A. A.,** Respiratory function and blood flow in the bronchial arteries after ligation of the pulmonary artery, *Am. J. Physiol.,* 157, 317, 1949.

20. **Atwell, R. B., Sutton, R. H., and Moodie, E. W.,** Pulmonary changes associated with dead filariae *(Dirofilaria immitis)* and concurrent antigenic exposure in dogs, *J. Comp. Pathol.,* in press.

21. **Sutton, R. H. and Atwell, R. B.,** Lesions of pulmonary pleura associated with canine heartworm disease, *Vet. Pathol.,* 22, 637, 1985.

22. **Simpson, C. F. and Jackson, R. F.,** Pathophysiology of heartworm disease, in *Proceedings of the Heartworm Symposium '74,* Morgan, H. C., Ed., Veterinary Medicine Publ., Bonner Springs, Kan., 1975, 38.

23. **Kume, S. and Itagaki, S.,** On the life cycle of *Dirofilaria immitis* in the dog as the final host, *Br. Vet. J.,* 111, 16, 1955.

24. **Carlisle, C. H.,** Canine dirofilariasis: its radiographic appearance, *Vet. Radiol.,* 21, 123, 1980.

25. **Atwell, R. B. and Carlisle, C. H.,** The distribution of filariae, superficial lung lesions and pulmonary arterial lesions following chemotherapy in canine dirofilariasis, *J. Small Anim. Pract.,* 23, 667, 1982.

26. **Atwell, R. B.,** Early stages of disease of the peripheral pulmonary arteries in canine dirofilariasis, *Aust. Vet. J.,* 56, 157, 1980.

27. **Atwell, R. B. and Carlisle, C. H.,** Distribution of infused artificial filariae within the pulmonary arteries of dogs, *J. Small Anim. Pract.,* 23, 725, 1982.

28. **Atwell, R. B. and Razekahni, A.,** Inoculation of dogs with artificial larvae similar to those of *Dirofilaria immitis:* distribution within the pulmonary arteries, *Am. J. Vet. Res.,* 47, 1044, 1986.

29. **Buoro, I. B. J., Atwell, R. B., and Heath, T.,** Angles of branching and the diameters of pulmonary arteries in relation to the distribution of pulmonary lesions in canine dirofilariasis, *Res. Vet. Sci.,* 35, 353, 1983.

30. **Hennigar, G. R. and Ferguson, R. W.,** Pulmonary vascular sclerosis as a result of *Dirofilaria immitis* infection in dogs, *J. Am. Vet. Med. Assoc.,* 131, 336, 1957.

31. **Hirth, R. S., Huizinga, H. W., and Nielsen, S. W.,** Dirofilariasis in Connecticut dogs, *J. Am. Vet. Med. Assoc.,* 148, 1508, 1966.

32. **Schaub, R. G. and Rawlings, C. A.,** Pulmonary vascular response during phases of canine heartworm disease: scanning electron microscopic study, *Am. J. Vet. Res.,* 41, 1082, 1980.

33. **Porter, W. B.,** Chronic cor pulmonale in dogs with *Dirofilaria immitis* (heartworms) infestation, *Trans. Assoc. Am. Physicians,* 64, 328, 1951.

34. **Keith, J. C., Schaub, R. G., and Rawlings, C.,** Early arterial injury-induced myointimal proliferation in canine pulmonary arteries, *Am. J. Vet. Res.,* 44, 181, 1983.

35. **Schaub, R. G., Rawlings, C. A., and Keith, J. C.,** Platelet adhesion and myointimal proliferation in canine pulmonary arteries, *Am. J. Pathol.,* 104, 13, 1981.

36. **Balch, R. K., Fonseca, J., Rice, W. G., and Leach, B. F.,** Canine filariasis in the Far East, *J. Am. Vet. Med. Assoc.,* 131, 298, 1957.

37. **Atwell, R. B., Sutton, R. H., and Buoro, I. B. J.,** Early pulmonary lesions caused by dead *Dirofilaria immitis* in dogs exposed to homologous antigens, *Br. J. Exp. Pathol.,* 67, 395, 1986.

38. **McLaughlin, R. F., Tyler, W. S., and Canada, R. O.,** A study of the subgross pulmonary anatomy in various mammals, *Am. J. Anat.,* 108, 149, 1961.

39. **Hobbs, W. R.,** Canine filariasis, *Cornell Vet.,* 30, 383, 1940.

40. **Giles, R. C. and Hildebrandt, P. K.,** Ruptured pulmonary artery in a dog with dirofilariasis, *J. Am. Vet. Med. Assoc.,* 163, 236, 1973.

41. **Kotani, T.,** Pathological studies on canine dirofilariasis, *Bull. Univ. Osaka Prefect. Ser. B,* 35, 79, 1983.

42. **Liu, S.-K., Das, K. M., and Tashjian, R. J.,** Adult *Dirofilaria immitis* in the arterial system of a dog, *J. Am. Vet. Med. Assoc.,* 148, 1501, 1966.

43. **Atwell, R. B., Buoro, I. B. J., and Sutton, R. H.,** Experimental production of canine pulmonary endarterial pathology similar to that produced by *Dirofilaria immitis, Vet. Rec.,* 116, 539, 1985.

44. **Adams, E. W.,** Obstruction of hepatic veins, *North Am. Vet.,* 37, 299, 1956.

45. **Lord, P. F., Schaer, M., and Tilley, L.,** Pulmonary infiltrates with eosinophilia in the dog, *J. Am. Vet. Radiol. Soc.,* 16, 115, 1975.

46. **Wong, M. M., Suter, P. F., Rhode, E. A., and Guest, M. F.,** Dirofilariasis without circulating microfilariae: a problem in diagnosis, *J. Am. Vet. Med. Assoc.,* 163, 133, 1973.

47. **Castleman, W. L. and Wong, M. M.,** Light and electron microscopic pulmonary lesions associated with retained microfilariae in canine occult dirofilariasis, *Vet. Pathol.,* 19, 355, 1982.

48. **Wong, M. M.,** Experimental occult dirofilariasis in dogs with reference to immunological responses and its relationship to tropical eosinophilia in man, *Southeast Asian J. Trop. Med. Public Health,* 5, 480, 1974.

49. **Wong, M. M. and Suter, P. F.,** Indirect fluorescent antibody test in occult dirofilariasis, *Am. J. Vet. Res.,* 40, 414, 1979.

50. **Wilkins, R. J., Hurvitz, A. I., Gula, I., and Tilley, L. P.,** Clinical experience with a fluorescent antibody test in dogs with occult dirofilariasis, in *Proceedings of the Heartworm Symposium '77,* Otto, G. F., Ed., Veterinary Medicine Publ., Bonner Springs, Kan., 1978, 45.

51. **Mantovani, A. and Jackson, R. F.,** Transplacental transmission of microfilaria of *Dirofilaria immitis* in the dog, *J. Parasitol.,* 52, 116, 1974.

52. **Sutton, R. H., Atwell, R. B., and Boreham, P. F. L.,** Liver changes, following diethylcarbamazine administration, in microfilaremic dogs infected with *Dirofilaria immitis, Vet. Pathol.,* 22, 177, 1985.

53. **Reis, J. S. and Rhodes, W. H.,** Linear opacities in canine thoracic radiographs, *J. Am. Vet. Radiol. Soc.,* 9, 57, 1968.

54. **Atwell, R. B., Sutton, R. H., and Moodie, E. W.,** Preliminary report of the pulmonary pathology associated with subcutaneous injections of *Dirofilaria immitis* antigen, *Vet. Res. Commun.,* 6, 59, 1983.

55. **Kociba, G. J. and Hathaway, J. E.,** Disseminated intravascular coagulation associated with heartworm disease in the dog, *J. Am. Anim. Hosp. Assoc.,* 10, 373, 1974.

56. **Rawlings, C. A., Losonsky, J. M., Schaub, R. G., Greene, C. E., Keith, J. C., and McCall, J. W.,** Post adulticide changes in *Dirofilaria immitis*-infected beagles, *Am. J. Vet. Res.,* 44, 8, 1983.

57. **Keith, J. C., Rawlings, C. A., and Schaub, R. G.,** Treatment of canine dirofilariasis: pulmonary thromboembolism caused by thiacetarsamide: microscopic changes, *Am. J. Vet. Res.,* 44, 1272, 1983.

58. **Rawlings, C. A., Keith, J. C., Lewis, R. E., Losonsky, J. M., and McCall, J. W.,** Aspirin and prednisolone modification of radiographic changes caused by adulticide treatment in dogs with heartworm infection, *Am. J. Vet. Res.,* 182, 131, 1983.

59. **Schaub, R. G., Keith, J. C., and Rawlings, C. A.,** Effect of acetylsalicylic acid on vascular damage and myointimal proliferation in canine pulmonary arteries subjected to chronic injury by *Dirofilaria immitis, Am. J. Vet. Res.,* 44, 449, 1983.

60. **Keith, J. C., Rawlings, C. A., and Schaub, R. G.,** Pulmonary thromboembolism during therapy of dirofilariasis with thiacetarsamide: modification with aspirin or prednisolone, *Am. J. Vet. Res.,* 44, 1278, 1983.

61. **Rawlings, C. A., Keith, J. C., Losonsky, J. M., Lewis, R. E., and McCall, J. W.,** Aspirin and prednisolone modification of postadulticide pulmonary arterial disease in heartworm-infected dogs: arteriographic study, *Am. J. Vet. Res.,* 44, 821, 1983.

62. **Rawlings, C. A., Keith, J. C., Losonsky, J. M., and McCall, J. M.,** An aspirin-prednisolone combination to modify postadulticide lung disease in heartworm-infected dogs, *Am. J. Vet. Res.,* 45, 2371, 1984.

63. **Atwell, R. B., Sutton, R. H., and Carlisle, C. H.,** The reduction of pulmonary arterial thromboembolic disease *(D. immitis)* in the dog associated with aspirin therapy, in *Proceedings of the Heartworm Symposium '83,* Otto, G. F., Ed., Veterinary Medicine Publ., Edwardsville, Kan., 1983, 115.

64. **Wallace, C. R. and Hamilton, W. F.,** Study of spontaneous congestive heart failure in the dog, *Circ. Res.,* 11, 301, 1962.

65. **Jackson, R. F., von Lichtenberg, F., and Otto, G. F.,** Occurrence of adult heartworms in the venae cavae of dogs, *J. Am. Vet. Med. Assoc.,* 141, 117, 1962.

66. **von Lichtenberg, F., Jackson, R. F., and Otto, G. F.,** Hepatic lesions in dogs with dirofilariasis, *J. Am. Vet. Med. Assoc.,* 141, 121, 1962.

67. **Sawyer, T. K. and Weinstein, P. P.,** Experimentally induced canine dirofilariasis, *J. Am. Vet. Med. Assoc.,* 143, 975, 1963.

68. **Saidla, J. E.,** Dilation of the hepatic lymphatics, in *Proceedings of the Heartworm Symposium '77,* Otto, G. F., Ed., Veterinary Medicine Publ., Bonner Springs, Kan., 1978, 63.

69. **McManus, E. C. and Pulliam, J. D.,** Histopathologic features of canine heartworm microfilarial infection after treatment with ivermectin, *Am. J. Vet. Res.,* 45, 91, 1984.

70. **Furrow, R. D., Powers, K. G., and Parbuoni, E. L.,** *Dirofilaria immitis.* II. Gross and microscopic hepatic changes associated with diethylcarbamazine citrate (DEC) therapy in dogs, in *Proceedings of the Heartworm Symposium '80,* Otto, G. F., Ed., Veterinary Medicine Publ., Edwardsville, Kan., 1981, 117.

71. **Nichols, J. and Hennigar, G.,** Effects of pulmonary hypertension on adrenal and kidneys of dogs infected with heart-worms *(Dirofilaria immitis),* Lab. Invest., 13, 800, 1964.

72. **Osborne, C. A., Hammer, R. F., O'Leary, T. P., Pomeroy, K. A., Jeraj, K., Barlough, J. E., and Vernier, R. L.,** Renal manifestations of canine dirofilariasis, in *Proceedings of the Heartworm Symposium '80,* Otto, G. F., Ed., Veterinary Medicine Publ., Edwardsville, Kan., 1981, 67.

73. **Sawyer, T. K.,** The venae cavae syndrome in dogs experimentally infected with *Dirofilaria immitis,* in *Proceedings of the Heartworm Symposium '74,* Morgan, H. C., Ed., Veterinary Medicine Publ., Bonner Springs, Kan., 1975, 45.

74. **Jackson, R. F.,** The venae cavae or liver-failure syndrome of heartworm disease, *J. Am. Vet. Med. Assoc.,* 154, 384, 1969.

75. **Jackson, R. F.,** The venae cavae syndrome, in *Proceedings of the Heartworm Symposium '74,* Morgan, H. C., Ed., Veterinary Medicine Publ., Bonner Springs, Kan., 1975, 48.

76. **Sutton, R. H. and Atwell, R. B.,** Renal haemosiderosis in association with canine heartworm disease, *J. Small Anim. Pract.,* 23, 773, 1982.

77. **Ishihara, K., Kitagawa, H., Ojima, M., Yagata, Y., and Suganuma, Y.,** Clinico-pathological studies on canine dirofilarial hemoglobinuria, *Jpn. J. Vet. Sci.,* 40, 525, 1978.

78. **Pollock, S.,** Canine filariasis complicated by albuminuria, *North Am. Vet.,* 29, 429, 1948.

79. **Casey, H. W. and Splitter, G. A.,** Membranous glomerulonephritis in dogs infected with *Dirofilaria immitis,* Vet. Pathol., 12, 111, 1975.

80. **Simpson, C. F., Gebhardt, B. M., Bradley, R. E., and Jackson, R. F.,** Glomerulosclerosis in canine heartworm infection, *Vet. Pathol.,* 11, 506, 1974.

81. **Simpson, C. F. and Jackson, R. F.,** Liver and kidney lesions in heartworm disease as a result of destruction of microfilariae, in *Proceedings of the Heartworm Symposium '83,* Otto, G. F., Ed., Veterinary Medicine Publ., Edwardsville, Kan., 1983, 15.

82. **Osborne, C. A.,** The glomerulus in health and disease: a comparative review of domestic animals and man, *Adv. Vet. Sci. Comp. Med.,* 21, 207, 1977.

83. **Klei, T. R., Crowell, W. A., and Thompson, P. E.,** Ultrastructural glomerular changes associated with filariasis, *Am. J. Trop. Med. Hyg.,* 23, 608, 1974.

84. **Aikawa, M., Abramowsky, C., Powers, K. G., and Furrow, R.,** Dirofilariasis. IV. Glomerulonephropathy induced by *Dirofilaria immitis* infection, *Am. J. Trop. Med. Hyg.,* 30, 84, 1981.

85. **Abramowsky, C. R., Aikawa, M., Powers, K. G., and Swinehart, G.,** Immunopathology of filaria nephropathy in dogs, *Lab. Invest.,* 44, 1A, 1981.

86. **Winter, H. and Majid, N. H.,** Glomerulonephritis — an emerging disease?, *Vet. Bull. London,* 54, 327, 1984.

87. **Couser, W. G., Steinmuller, D. R., Stilmant, M. M., Salant, D. J., and Lowenstein, L. M.,** Experimental glomerulonephritis in the isolated perfused rat kidney, *J. Clin. Invest.,* 62, 1275, 1975.

88. **Damme, B. J. C., Fleuren, G. J., Bakker, W. W., Vernier, R. L., and Hoedemaeker, Ph. J.,** Experimental glomerulonephritis in the rat induced by antibodies directed against tubular antigens. V. Fixed glomerular antigens in the pathogenesis of heterologous immune-complex glomerulonephritis, *Lab. Invest.,* 38, 502, 1978.

89. **Barsanti, J. A.,** Serum and urine proteins in dogs infected with *Dirofilaria immitis,* in *Proceedings of the Heartworm Symposium '77,* Otto, G. F., Ed., Veterinary Medicine Publ., Bonner Springs, Kan., 1978, 53.

90. **Buoro, I, B. J. and Atwell, R. B.,** Urinalysis in canine dirofilariasis with emphasis on proteinuria, *Vet. Rec.,* 112, 252, 1983.

91. **Bailey, R. W.,** A comparison study of various arsenical preparations as filaricides of *Dirofilaria immitis, J. Am. Vet. Med. Assoc.,* 133, 52, 1958.

92. **Jones, T. C. and Hunt, R. D.,** *Veterinary Pathology,* 5th ed., Lea & Febiger, Philadelphia, 1983, 799.

93. **Stuart, B. P., Hoss, H. E., Root, C. R., and Short, T.,** Ischemic myopathy associated with systemic dirofilariasis, *J. Am. Anim. Hosp. Assoc.,* 14, 36, 1978.

94. **Drazmer, F. H.,** Renal amyloidosis and glomerulonephritis secondary to dirofilariasis, *Canine Pract.,* 5, 66, 1978.

95. **Atwell, R. B., Johnstone, I., Read, R., Reilly, J., and Wilkins, S.,** Haemolytic anaemia in two dogs suspected to have been induced by levamisole, *Aust. Vet. J.,* 55, 292, 1979.

96. **Atwell, R. B., Thornton, J. R., and Odlum, J.,** Suspected drug-induced thrombocytopenia associated with levamisole therapy in a dog, *Aust. Vet. J.,* 57, 91, 1981.

97. **Sutton, R. H. and Atwell, R. B.,** Nervous disorders in dogs associated with levamisole therapy, *J. Small Anim. Pract.,* 23, 391, 1982.

98. **Symoens, J. and Rosenthal, M.,** Levamisole in the modulation of the immune response: the current experimental and clinical state, *J. Retic. Soc.,* 21, 175, 1977.

99. **Guerro, J.,** Immunological modulating influence of levamisole, in *Proceedings of the Heartworm Symposium '77,* Otto, G. F., Ed., Veterinary Medicine Publ., Bonner Springs, Kan., 1978, 104.

100. **Otto, G. F.,** Occurrence of the heartworm in unusual locations and in unusual hosts, in *Proceedings of the Heartworm Symposium '74,* Morgan, H. C., Ed., Veterinary Medical Publ., Bonner Springs, Kan., 1975, 6.

101. **Turk, R. D., Gaafar, S. M., and Lynd, F. T.,** A note on the occurrence of nematodes, *Dirofilaria immitis* and *Ancylostoma braziliense* in unusual locations, *J. Am. Vet. Med. Assoc.,* 129, 425, 1956.

102. **Mihashi, C.,** Parasitic infection of dog hindquarters. Three cases of arterial embolism, *Jpn. J. Vet. Med.,* 565, 21, 1972.

103. **Hoerlein, B. F., Horne, R. D., Wood, J. P., Bartels, J. E., and Hoff, E. J.,** An aortic occlusion in a dog caused by heartworms, *Auburn Vet.,* Winter, 71, 1972.

104. **Schnelle, G. B. and Jones, T. C.,** *Dirofilaria immitis* in the eye and in an interdigital cyst, *J. Am. Vet. Med. Assoc.,* 107, 14, 1945.

105. **Luttgen, P. J. and Crawley, R. R.,** Posterior paralysis caused by epidural dirofilariasis in a dog, *J. Am. Anim. Hosp. Assoc.,* 17, 57, 1981.

106. **Shires, P. K., Turnwald, G. H., Qualls, C. W., and King, G. K.,** Epidural dirofilariasis causing paraparesis in a dog, *J. Am. Vet. Med. Assoc.,* 180, 1340, 1982.

107. **Dibbell, C. B.,** *Dirofilaria immitis* in abdominal cavity of dog, *J. Am. Vet. Med. Assoc.,* 118, 298, 1951.

108. **Mandeleker, L. and Brutus, R. L.,** Feline and canine dirofilarial encephalitis, *J. Am. Vet. Med. Assoc.,* 159, 776, 1971.

109. **Donahoe, J. M. R. and Holzinger, E. A.,** *Dirofilaria immitis* in the brains of a dog and a cat, *J. Am. Vet. Med. Assoc.,* 164, 518, 1974.

110. **Kotani, T., Tomimura, T., Ogura, M., Mochizuki, H., and Horie, M.,** Pathological studies on the ectopic migration of *Dirofilaria immitis* in the brain of dogs, *Jpn. J. Vet. Sci.,* 37, 141, 1975.

111. **Kotani, T., Tomimura, T., Ogura, M., Yoshida, H., and Mochizuki, H.,** Cerebral infarction caused by *Dirofilaria immitis* in three dogs, *Jpn. J. Vet. Sci.,* 37, 379, 1975.

112. **Carlisle, M. S., Webb, S. M., Sutton, R. H., Hampson, E. C., and Blum, A. J.,** Case report — adult *Dirofilaria immitis* in the brain of a dog, *Aust. Vet. Pract.,* 14, 10, 1984.

113. **Beller, J.,** Parasite in the eye of a dog, *Small Anim. Clin.,* 2, 681, 1962.

114. **Bellhorn, R. W.,** Removal of intraocular *Dirofilaria immitis* with subsequent corneal scarring, *J. Am. Anim. Hosp. Assoc.,* 9, 262, 1973.

115. **Brightman, A. H., Helper, L. C., and Todd, K. S.,** Heartworm in the anterior chamber of a dog's eye, *Vet. Med. Small Anim. Clin.,* 72, 1021, 1977.

116. **Eberhard, M. L., Daly, J. J., Weinstein, S., and Farris, H. E.,** *Dirofilaria immitis* from the eye of a dog in Arkansas, *J. Parasitol.,* 63, 978, 1977.

117. **Blanchard, G. L. and Thayer, G.,** Intravitreal *Dirofilaria immitis* in a dog, *J. Am. Anim. Hosp. Assoc.,* 14, 33, 1978.

118. **Thornton, J. G.,** Heartworm invasion of the canine eye, *Mod. Vet. Pract.,* 59, 373, 1978.

119. **Smith, C. R., Foreman, M. M., Heil, J., and Hammond, C.,** Heartworm removal from the eye, *Canine Pract.,* 6, 48, 1979.

120. **Guterbock, W. M., Vestre, W. A., and Todd, K. S.,** Ocular dirofilariasis in the dog, *Mod. Vet. Pract.,* 62, 45, 1981.

121. **Forney, M. M.,** *Dirofilaria immitis* in the anterior chamber of a dog's eye, *Vet. Med.*, 85, 49, 1985.
122. **Atwell, R. B.,** Aspects of Natural and Experimental Dirofilariasis, Ph.D. thesis, University of Queensland, Brisbane, Australia, 1983.

Chapter 7

THERAPY IN RELATION TO ASPECTS OF THE PATHOPHYSIOLOGY OF DIROFILARIASIS

Richard B. Atwell

TABLE OF CONTENTS

I. INTRODUCTION

While a reasonably standard approach to the use of adulticides, microfilaricides, and prophylactic drugs has been achieved, there is still confusion and debate over the use of support and symptomatic therapy, in particular that associated with the amelioration of pathology in the lungs. The views expressed in this chapter are based on the literature, on our own experiments and observations, and on a personal view regarding such therapy which needs to be selected based on a clear understanding of the true nature and variable extent of the underlying pathology. It is, therefore, incorrect to have one standard regimen of support therapy for every case of dirofilariasis. All dogs will have subclinical pathology (do we need to treat that?), and those with clinical disease will have obvious arterial pathology, but to a variable extent in different dogs. Thus, each clinically affected dog needs to be individually assessed as to what support therapy is indicated (to ameliorate the disease) before, during, and after the specific adulticidal therapy program. It is only on a firm understanding of the basic pathophysiology of this disease that rational decisions can be made regarding the type of support therapy indicated, taking due consideration of possible side effects and drug interactions.

II. HOW DOES THIS DISEASE DEVELOP AND PROGRESS?

Table 1 outlines the various pathological aspects of dirofilariasis. Some heartworm infections (in contrast to the immunologically over-responsive, immune-mediated occult cases) can be very mild and not develop to the status of a case of heartworm disease in which obvious clinical signs of the underlying pathology exist. Why does the disease develop and what is the basic pathophysiology? There is no doubt that any infection will cause some "degree of disease", but the reserve capacity of the pulmonary artery tree, plus the mild insults of a low infection level over a prolonged period of time, does not allow enough disease to develop to produce severe clinical signs.

In fact, acute insertion of worms and microfilariae,[1,2] or of 30 polyvinylchloride threads resembling adult worms, inserted for a 6-week period[3] did not induce disease in the recipient animals. Thus, it is not just worm numbers alone (or potentially obstructive but inert artificial worms) that cause clinical signs of disease. Obviously, large numbers of filariae will cause caval syndrome both experimentally[4] and clinically,[5] but this syndrome does not necessarily have the associated signs of moderate to severe heartworm disease. As always, some cases are unexplainable, e.g., a 2-year-old Boxer with 497 worms (many small and immature) without signs of caval syndrome or of severe right-sided congestive heart failure (RSCHF).[6]

Some animals which are never treated with diethylcarbamazine (DEC) in infected areas develop little clinical disease over a decade. Usually, these animals have few worms at necropsy and it is proposed that self-induced immunity has developed to reduce the expected increase in the adult population. This adult burden caused damage well within the reserve capacity of the dog both in terms of pulmonary artery reserve and in terms of any exercise expectations of the domestic pet vs., e.g., the racing greyhound.

However, the development of disease is in part related to worm numbers. More accurately, it is related to their rate of insult (i.e., worm numbers over a certain period of time) and to the secondary effects of worms *in situ*. So it seems that live adult filariae, following their earlier, intensely inflammatory reaction with the distal pulmonary artery tree,[8-10] cause little local disease as adults in the larger proximal pulmonary arteries except local intimal proliferation due to direct contact. When worm diameters and arterial diameters are compared, it is obvious that less direct worm contact is possible as the arterial diameter increases helping to explain why little direct local intimal disease occurs with filariae in the large pulmonary arteries. Similar proliferation has been seen with artificial worms (polyvinyl threads) which

Table 1
CLINICAL AND PATHOLOGICAL ASPECTS OF DIROFILARIASIS

Condition	Associated clinical data
Pulmonary arterial disease	Various forms of intimal and adventitial disease; inflammation due to live adult, live immature, dead (granulomatous)
Thromboembolism	Distal pulmonary arterial occlusive disease
Pneumonitis	Particularly periarterial and in immune-mediated occult cases, interstitial disease as well — immune-complex deposition?
Bronchitis	Associated with periarterial pneumonitis
Pulmonary parenchyma	Hemorrhage (focal, alveolar, and gross), hemosiderosis, fibrosis, collapse, pleural retraction
Pleural lesions	Noninflammatory effusions, adhesions, pneumothorax, hemothorax, collateral pulmonary vasculature from thoracic and abdominal systemic arteries
DIC	Rarely reported
Thrombocytopenia	Particularly in caval syndrome and thromboembolism
Platelet activation	Associated with parasites and vascular turbulence, intravascular hemolysis — subclinical in routine cases and grossly in caval syndrome
Anemia	Normochromic, normocytic, metabolic and increased red blood cell mechanical fragility, accentuated shear force with direct hemolysis and induced microangiopathic (fibrin) hemolysis in caval cases?, usually responsive in occult and caval cases
Caval syndrome	Shock, hemolysis and a varying extent of heartworm disease
RSCHF and tricuspid regurgitation	Secondary to pulmonary hypertension and annular dilation, secondary also to filarial entanglement
Right ventricular hypertrophy	Eventually leading to ischemic myocardial failure
Syncope	Due to increased exercise/excitement demand and reduced ability to maintain cerebral and systemic perfusion
Left ventricular atrophy	Seen in severe cases
Liver	Engorgement, hepatic effusion, and ascites; splenomegaly infrequent
Kidney	Glomerular and tubular disease, protein-losing nephropathy
Ectopic infections	Central nervous system, eye, systemic circulation, myopathy, infarction, abdominal and pleural spaces
General catabolism	Muscle wastage, skin changes
Bronchial/pulmonary arterial fistula	Bronchial and pulmonary hemorrhage
Hemosiderosis	Lungs, kidney, spleen — hemolysis, hemorrhage, worm by-products?
Microfilariae	Amicrofilaremic occult states, systemic microgranulomas (subclinical?) with microfilaricides, direct glomerular involvement? DEC reaction
Postadulticide reactions	Associated with filarial death; thrombosis and thromboembolism, pulmonary hemorrhage and inflammation, granulomatous arteritis

From Atwell, R. B., *Vet. Annu.*, 26, 255, 1986. With permission.

were *in situ* in the pulmonary arteries without the development of signs of heartworm disease, suggesting that it is most likely local, direct contact rather than worm-related factors that induce such macroscopic intimal disease.[11]

Obviously, large numbers of immature filariae arriving at the terminal pulmonary arteries over a short period of time (as occurs with experimentally induced disease) will cause severe local arterial disease and so progressive disease with continued exposure to immature adults, once again reinforcing worm burden in relation to a time period as being the criterion as to whether clinical disease develops.[8]

So worm numbers over the associated period of insult (all insult over a short period, e.g., experimentally induced disease, or over longer, perhaps interrupted periods, e.g., low infection rate in natural disease) will determine the extent of and rate of development of pulmonary disease. The reserve of the pulmonary artery tree will then determine what clinically apparent disease will occur in the animal so infected. Obviously, the working dog will be more affected due to its higher exercise demands and so will develop more obvious disease earlier than would the inactive, nonworking, domestic pet.

There is no doubt that extensive arterial disease occurs with immature filariae in the distal arterial bed, but once the worms have matured and moved proximally, a healing process leads to some degree of chronic fibrotic disease. However, once healed and with the worms more proximally located, the distal disease should not progress provided further immature filariae are not introduced.

In spite of this, there is, however, evidence of continuing distal arterial disease not directly related to the proximity of either adult or immature worms.[12] The fact that resolution of microscopic arterial lesions can occur after the death of filariae[13] seems to support the existence of an on-going "insulting" process associated with the presence of live filariae, but not necessarily with their physical proximity.

In mild experimental infections, pulmonary arterial pressures rise with infection and fall following adulticide therapy.[14] However, vascular resistance is still elevated suggesting that residual damage is present which is supported by the accentuated response to hypoxia and isoprenaline stimulation. Knight,[8] using more heavily infected dogs, showed that the hypertension eventually reaches a plateau, whereas the extent of individual arterial disease, as seen with angiograms, tended to increase with more arterial sacculation and dilation as worms become more centrally located. From these two studies, it appears that a critical number of filariae over a certain period of time is needed to instigate hypertension. With large numbers of filariae in the pulmonary arteries, hypertension increases to a plateau, whereas the luminal pathology of those affected arteries progresses producing sacculation and even tortuosity. This would support the belief that filariae, after causing local, distal arterial lesions and moving centrally, continue to secrete "products" or are associated with embolic phenomena that cause more distal arterial disease further increasing the resistance in those arteries and, thus, the progressive angiographic findings, whereas, overall, the pulmonary arterial pressure is dissipated by passive recruitment of normally expansive, unaffected arteries. Obviously, if all branches of the tree contain filariae then no such recruitment is possible and the pressure will continue to rise as this resistance-inducing distal arterial pathology continues to develop.

Once the source of the arterial insult associated with adult worms is removed then, with time, the overall pressure will fall.[1] However, vascular resistance will remain elevated due to residual, nonreversible arterial disease producing an inelastic, nonexpandable status in affected vessels, particularly the distal arterial tree.

Sudden alterations to the extent of any residual hypertension are usually due to thromboembolism and can induce RSCHF. However, such hypertension is somewhat reversible as recanalization can occur, particularly if the animals are given time at rest and appropriate therapy.

Thus, the severity of disease relates to worm numbers, duration of insult, the number of filariae arriving in the pulmonary arteries over a period of time (worm/time interaction), the extent of inflammatory and thrombotic disease, and, most importantly, the host's response (at the intimal level) to the presence of worms and their by-products and to the intimal pathology itself. In fact if this host response was reduced then less arterial disease would occur. So, once the distal arterial pathology due to early forms of the parasite has resolved and the filariae are in the larger arteries and the right ventricle, how does further disease occur?

Obviously, it is distal arterial and arteriolar, microscopic, hypertensive pathology that leads to further development of the disease based on cor pulmonale pathophysiology. This consists of an embolic process associated with the existing, proximal arterial (intimal) disease-inducing thromboembolism and with a reaction to worm by-products loosely called "antigen". It is hypothesized that worm by-products lead to inflammatory intimal disease, i.e.,

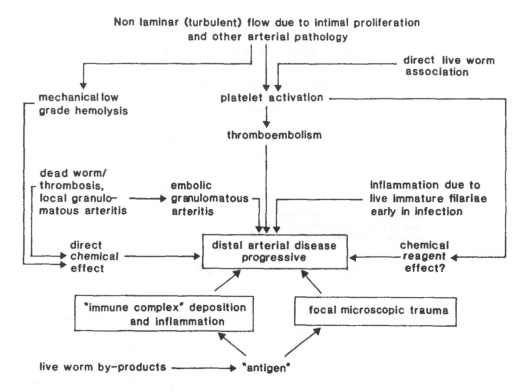

FIGURE 1. Hypothesis of the cause of distal pulmonary arterial disease. Immune-complex deposition (by-products of filariae) inducing a "vasculitis"; thromboembolism secondary to worm or intimal disease-induced *thrombosis; platelet aggregation due to platelet activation associated with turbulence, arterial damage, or worm* activating products (nonimmunological); biochemical damage to the distal arterial bed inducing cell damage or permeability changes associated with platelet or clot release factors; secondary, local arterial, periarterial, and parenchymal involvement due to intimal permeability alterations and tissue deposition of inflammatory "factors".

"antigen" release into a host that is immunologically responsive to a variable individual degree. This is lodged at the distal arteriolar level or below and presumably produces a hypersensitive (Type 3) arteritis and/or pneumonitis. The extreme example of this, of course, is the vasculitis and pneumonitis due to death of large antigenic components, namely microfilariae, as occurs in immune-mediated occult cases. Thromboembolism also occurs and is probably associated with platelet activation possibly due to the presence of parasites themselves,[15] but probably also due to turbulence-induced platelet activation associated with the abnormal intima. Both mechanisms of activation could lead to active release of platelet contents, which, apart from any direct chemical/intima interactions, can lead to fibrin deposition and thus to the possibility of thromboembolic lodgement in the distal vasculature.

Thus, the distal arterial bed is probably progressively confronted by immune-complex inflammation (antigenic by-products) and thromboembolic deposition (secondary to platelet activation) even while the adult infection remains proximally located. In addition to this, of course, is the gross and microscopic embolism, thromboembolism, and secondary granuloma formation from dead worm fragmentation. Figures 1 and 2 outline this hypothesis.

It would seem that the disease the infection produces is related originally to the direct and later to the secondary effects of the presence of a critical number of filariae over a certain period of time. It, thus, goes without saying that the age of the host in relation to the infection status is not of direct importance except with regard to altered immunological response or to other geriatric diseases that may alter drug choice, etc. Rather, it is the possible age of and degree of infection related to the degree of worm insult over a period of time which is the more important consideration in the assessment of the disease status.

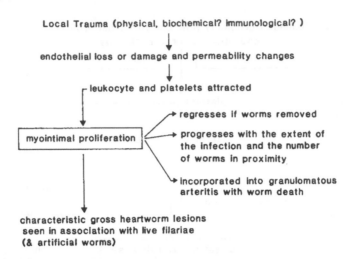

FIGURE 2.   Adult filarial contact pathology.

Thus, in summary, the disease progression is as follows: immature filariae are distributed within the pulmonary artery tree according to blood distribution patterns. These have been simulated with adult[16] and immature infections[17] and have confirmed the right caudal lobe to be most affected. The early arterial pathology is related to the number of filariae and the number impacted into each branch of the distal arterial tree. Extensive focal inflammation occurs in association with these filariae. Some filariae will impact a vessel and, with secondary thrombosis and myointimal proliferation, cause lumen obstruction. If a critical number of arteries are so affected (obstructive, obliterative arteritis, thrombosis, fibrosis, and possibly vasoconstriction due to clot-breakdown-products) then vascular resistance will rise and so signs of cor pulmonale will develop due to pulmonary hypertension.[12]

Once a critical percentage of the distal arterial tree has been so affected (obstructed, inflamed, thrombosed, fibrosed, obliterated) then arterial changes (dilation, sacculation, tortuosity) due to the direct effects of hypertension will occur leading to developing signs of RSCHF. The right ventricle will initially dilate and then hypertrophy and may eventually fail. This failure is due to myocardial ischemia, excessive exercise and/or stress, obstructive thrombosis/granuloma formation in the remaining viable pulmonary arteries, or combinations of these factors.

Rapid development of signs or sudden deterioration of the disease status is usually associated with worm death (geriatric, postadulticide, self cure?) and/or with thrombosis and secondary thromboembolism. The response to therapy will depend upon how permanent are the changes in the pulmonary arterial tree. Once extensive arterial luminal fibrosis is formed (obstructive obliterative disease) and there is significant periarterial fibrosis, it is likely that little resolution will be seen with selected drug therapy. If, however, the extent of thrombotic disease can be limited or reduced and the extent and severity of the inflammatory arterial and periarterial disease can be reduced, overall therapy may be more successful. Recanalization of arteries can occur allowing some blood flow, but the vessel wall may have lost its elastic capacity and so its ability to dramatically increase pulmonary flow rates without a rise in pressure in response to demand.

III. THE USE OF ANTITHROMBOTIC AND ANTI-INFLAMMATORY THERAPY

The chief cause of the signs associated with heartworm disease is the development of severe arteritis of the pulmonary arteries leading to increased vascular resistance. In general,

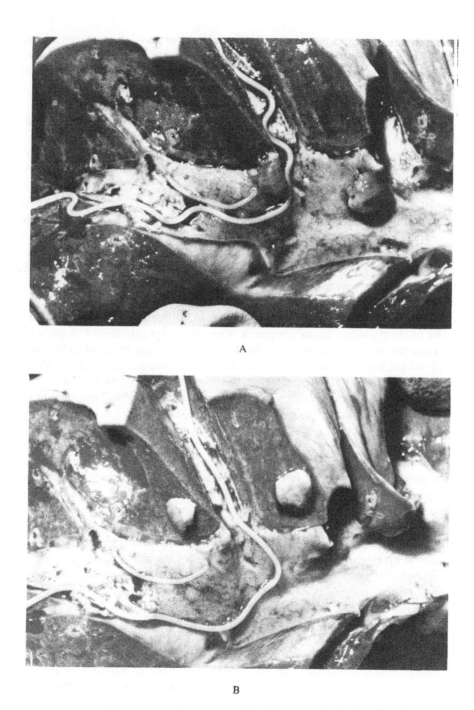

A

B

FIGURE 3. (A) The development of thrombin accumulations on the intima of a mildly damaged artery. (B) These accumulations were easily dislodged by gentle finger pressure.

the disease process is such that most forms of pulmonary arterial pathology occur associated with inflammation and fibrosis of the intima and the arterial wall and with the perivascular areas as well.

In a model of the disease, Atwell et al.[6,18] showed that the pathology of the lung could be induced by the use of dead filariae introduced into the venous system and, hence, to the pulmonary arteries in dogs which had had subcutaneous exposure to antigenic extracts of adult worms. In this model, severe arterial inflammation, thrombosis and thromboembolism, filarial embolism, hemorrhage (both perivascular and parenchymal), and secondary fibrosis were induced. In fact, all of the pulmonary pathologies of the disease, including septal thickening, were produced.

It would, therefore, seem obvious that, wherever possible, the beneficial effects of anti-inflammatory and antithrombotic therapy should be assessed in an attempt to reduce the effects of inflammation on the pulmonary arteries and surrounding parenchyma and also of the effects of thromboembolism (Figure 3) so as to reduce the extent of resultant obstructive arterial disease leading to secondary fibrosis. However, such therapy is not without side effects and these need to be carefully assessed. In dogs with severe disease, significant arterial compromise has already occurred, and it would, therefore, be even more indicated to attempt to prevent further compromise because of controllable inflammation and thrombosis.

Researchers at the Department of Small Animal Medicine and Surgery at the University of Georgia, Athens, have produced numerous papers on the development of arterial disease and on the effects of arsenical, prednisolone, and aspirin therapy.[19] Although some of the experimental groups of dogs are difficult to follow through the various publications, it seems that the chief effect of aspirin (7 mg/kg daily for 6 to 12 months) is antiplatelet and it has been shown to clearly reduce the development of myointimal proliferation.[19] Aspirin at these low dose rates has no anti-inflammatory effect, so the effects are believed to be solely due to its antiplatelet effects. Histologically, aspirin therapy has also been associated with reduced thromboembolic disease in the experimental, dead worm model of the disease.[20]

The major concern is the secondary effects of bleeding which has been seen in clinical cases with doses as low as 5 mg/kg given on alternate days.[6] There is a report that relates to the use of aspirin at 3 mg/kg every 6th day as being antiplatelet and it would appear that this lower dose could be used and would possibly have the same effect of reducing the development of intimal disease without the potential side effects of bleeding, particularly seen in the gastrointestinal tract.[21] Even though the experiments to show that this lower dose is effective in the experimental disease have not been performed, it is my opinion that this lower dose is effective, and its use certainly has been associated with negligible side effects in contrast to the higher doses where bleeding has been a concern in clinical cases. However, the experiments need to be performed to establish the lowest effective dose.

Prednisolone given daily at 1 mg/kg for 4 weeks after arsenical therapy did significantly reduce the degree of periarterial inflammation (assessed radiographically) when compared to an aspirin-treated group of dogs (10 mg/kg daily for 4 weeks) and to the control group, all of which had nine filariae (presumably all female adults of a similar age) inserted intravenously 4 weeks earlier.[22] However, filariae "survived" the arsenical therapy in the prednisolone-treated dogs; three of six dogs had live filariae with overall worm mortality of 70%. Filarial viability was not assessed on motility characteristics at necropsy but by the use of staining quality and other guidelines used to compare the mortality of the different groups after fixation. Scoring the most severely affected lobe of each group radiographically, the aspirin group scored 70 and the prednisolone group 36 when compared to the control group (100). While the aspirin group had the same level of overall pulmonary disease as the control group 3 weeks after arsenical therapy, there was a slower rate of development of parenchymal disease in this group. Thus, prednisolone had a significant effect on the radiographic extent of postadulticide pulmonary involvement.

It must be remembered, however, that the fixed standard insult of experimental disease needs to be related to real clinical cases in which greater than nine filariae are usually present and in which disease development (both arterial and periarterial parenchymal) is continual and is sometimes severe and chronic. While it is suggested that prednisolone (1 mg/kg daily for 4 weeks) used immediately after aresenical therapy reduces worm mortality, its use (on

Table 2
EFFECTS OF THE USE OF ASPIRIN AND
PREDNISOLONE ON LUNG PATHOLOGY
ASSESSED RADIOGRAPHICALLY[22,26]

Group dogs	Live filariae/total	Lung disease	Luminal disease
Prednisolone only	16/54	1	3 +
Aspirin only	1/54	2 +	1
Aspirin and prednisolone	16/54	1	1
Control	0/54	3	3

Note: 1 = mild; 3 + = severe.

an alternate day basis), both before arsenical therapy (to help reduce the extent of pulmonary inflammation and so stabilize the patient) and after arsenical therapy, needs to be considered when it is known that severe periarterial inflammatory disease, both experimentally[23] and clinically,[24] is a major component to this disease.

Speaking specifically of luminal effects based on arteriographic assessment, aspirin (10 mg/kg daily for 4 weeks) reduced and prednisolone (1 mg/kg daily for 4 weeks) increased arterial occlusion. There was a resultant arterial dilation due to thrombosis, filarial presence (live and dead), and intimal disease when compared to the control group of dogs. Thus, when used alone, prednisolone will markedly reduce periarterial inflammatory disease, but will be associated with thrombotic lumen obstruction, whereas aspirin-treated dogs will have pulmonary disease as severe as control dogs 3 weeks after arsenical therapy, but markedly reduced luminal disease. Similar results with increased intimal reaction, thrombosis, and residual fibrosis have been seen histologically in prednisolone-treated (1 mg/kg) experimental dogs in which dead, adult, female filariae had been inserted.[25]

Combination therapy showed that aspirin added to prednisolone therapy limited luminal disease to that of the aspirin-only group (assessed arteriographically) with few obstructive lesions and that the pulmonary disease (assessed radiographically) was similar to the results of the prednisolone-only group mentioned above.[26] No proximal arterial enlargements were noted as with the previous prednisolone-only group, however, 16 of 54 worms appeared to be alive in 4 of the 6 dogs, mortality once again being assessed on fixed-worm morphology without control-worm comparison. Even so, there is no doubt that prednisolone given at 1 mg/kg as arsenical therapy commences, and for 30 days thereafter has been associated with, reduced worm mortality.[22,26] Table 2 outlines the results of experimental aspirin and prednisolone therapy.[22,26]

In animals in which only partial clearance of filariae occurred, residual pulmonary parenchymal lesions persisted.[27] These had occurred following the development of obvious arteriographic abnormalities (dilation/obstruction), but had been reduced, along with the arterial lesions, following the death of filariae, suggesting that the periarterial parenchymal lesions were associated with live filariae and with arterial disease. As this periarterial disease can be severe, it is obvious that using luminal arteriographic measurements alone may not give a representative view of the extent of actual arterial wall involvement and, therefore, of the real significance of the periarterial extent of the disease.

However, the author believes that treatment and modulation of the periarterial and parenchymal inflammatory disease of the active, natural infection is essential in the moderate to severely affected case before adulticide use and especially after filarial death. Alternate-day prednisolone (0.5 to 1.0 mg/kg) is used to avoid the side effects to the drug, and therapy is not used for 1 week following arsenical therapy; this is believed to be the period when worm death is occurring. Aspirin used at 3 mg/kg twice weekly (or, in the future, the use of specific, thromboxane blockers or thromboxane release/receptor-antagonistic drugs) at the same time is also considered essential to reduce the extent of luminal disease, but care is required in cases where bleeding is evident, e.g., hemoptysis.

The use of aspirin may also have another bearing on pathophysiology. It has been shown that platelet activation is present in dogs with heartworm disease. Whether this is due to turbulence and trauma associated with arterial lesions or with secretory products from, or directly with, the parasite itself[15] is not known, but modulation of this associated activation by aspirin therapy could also help to reduce downstream arterial disease. Most of the work done in the Department of Small Animal Medicine and Surgery, University of Georgia, Athens, relates to worm-associated damage with endothelial contact.[13] However, pulmonary vein lesions have also been seen, and it is possible that platelet activation, secondary aggregation and mediator release, and thrombin deposition could be instrumental in the production of some of the downstream pathology that is seen.

In the future, it will be most likely that the use of antiserotonin therapy (e.g., S_2 blocking drugs) or similar antimediator therapy will be used to aid with the treatment of thromboembolism, helping to overcome secondary vasoconstriction, bronchoconstriction, and local vascular permeability effects. Similarly, the use of more specific antiplatelet (function and/or effect) drugs will be used to modulate the intimal disease.

IV. THE IMPORTANCE OF CAGE REST IN DISEASED DOGS

Exercise is contraindicated in severe cases of disease. This is associated with several factors. Pulmonary artery pressure rises with stress and exercise[12] and so increases the pressure load on the right ventricle which, if hypertrophic, could already be partially ischemic. Any further reduction in the output of blood through the damaged pulmonary arteries (due to rises in pulmonary-artery resistance and pressure) to the left ventricle could reduce cardiac output and so possibly reduce myocardial perfusion, particularly subendocardial perfusion in the pressure-overloaded right ventricle, already needing additional perfusion due to the accentuated oxygen requirements of exercise.

Exercise in the decompensated patient will also reduce renal blood flow, induce fluid and salt retention, and so produce a volume overload in a patient that already has a pressure overload and may be in RSCHF, or is close to decompensation. In fact, such additional loading could lead to decompensation.

Exercise may also induce coughing bouts which will, along with the direct hypertensive effects of exercise, further increase pulmonary-arterial pressure and possibly lead to arterial rupture in an area where the arterial wall is weakened.

Exercise increases blood-flow rates and, as blood turbulence is a neglected but, nevertheless, important factor in this disease, could induce turbulence-associated pathology. Turbulence-associated platelet activation is known to occur with caval cases,[4] and the chronic, low-grade hemolysis in dirofilariasis is believed associated with mechanical effects due to nonlaminar flow. Increased platelet activation could lead to further fibrin deposition and, so, to showers of thromboembolic material further occluding the distal microscopic bed. It must also be remembered that the capillary bed of the lung is the significant area of resistance when compared to the systemic circulation and, so, occlusion due to such showers becomes very important in terms of vascular resistance.

After adulticide therapy, thrombosis and thromboembolism can lead to increased hypertension, and so any other factors that also increases pressure, e.g., exercise, should be avoided. Also, it is more likely that rupture is possible in the first few days after induced arterial disease associated with dead worm deposition.[23] Diapedesis occurs into the periarterial and alveolar tissue and, before the inflammatory process is organized, sudden pressure changes could lead to separation of tissues in the inflamed wall and thus rupture.

Additionally, clot dislodgement and dispersal is probably more likely with the associated tachycardia and increased blood-flow rates seen with exercise. Thus, there are numerous reasons to avoid exercise particularly following adulticide therapy in the moderate to severely affected case.

V. POSSIBLE MECHANISMS OF COUGHING IN INFECTED DOGS

The cause of coughing is most likely multifactoral. The extreme is the presence of filariae in the airways causing local irritation as would any foreign body.[28] Similarly, arterial-bronchial fistulas and the associated inflammation must be stimulatory to cough receptors. Exercise commonly induces coughing in severe cases of dirofilariasis.

The extent to which periarterial inflammation involves the adjacent airway may also exert some local stimulatory effects. In fact, a zone of inflammation usually is seen completely surrounding the airway. Whether this is simply an extension of the inflammatory process from the artery or actually represents local antigen deposition and a local secondary response is not known. Presumably, local airway inflammation of this nature would be sufficient to activate cough receptors.

The secondary effects of thromboembolism can also induce bronchospasm and, in such cases, coughing could be related to the release of clot-breakdown mediators inducing coughing secondarily to this bronchospasm.[29]

Additionally, there is the extensive pneumonitis seen both in immune-mediated occult cases and in clinical and experimental cases of heartworm disease. There could be cough-receptor activation associated with this process, although it probably would not be a direct effect but rather via mediators to the more proximately located receptors. Severe cases of the disease, usually in the middle-aged to older dog, will often be occult (80%), have few live adults, increased eosinophil levels, and extensive pulmonary disease and thrombosis.[6] It is almost as if a self-cure phenomenon has been enacted and the secondary effects of that is extensive disease, particularly in the lung, being manifested clinically as excessive coughing and exercise intolerance. Coughing, in such cases, is probably, mainly inflammatory in nature, but could be due to a combination of some of the above causes.

It would, therefore, seem that the use of low-dose prednisolone could have beneficial effects with regard to lessening the degree of coughing. Bronchodilator therapy would be indicated in cases where thromboembolism is significant, in fact, the use of antiserotonin therapy in such cases would directly reduce the secondary effects of thromboembolism, bronchoconstriction, vasoconstriction, etc. (see Appendix Figures). However, such therapy is still experimental.

As the cause of coughing is believed to be mainly inflammatory in nature the use of antimicrobial therapy is not necessary. Based on the response in clinical cases, cage rest and low-dose prednisolone will suffice as antitussive therapy in most situations. However, specific antitussive therapy may be necessary in cases that do not response, particularly, if there is the danger of a bronchial-arterial fistula developing.

VI. CONGESTIVE HEART FAILURE

Chronic RSCHF, seen with heartworm disease, is directly related to the extent of pulmonary arterial disease and right ventricular afterload leading to ventricular dilatation and, eventually, to hypertrophy (with the added possibility of anatomical hypoxia due to the altered capillary/muscle fiber ratio). This process will maintain competency until there is a sudden change in the pressure load (e.g., further hypertension due to thrombosis) or an accentuated volume load, this being due to reduced renal perfusion because of progressively limited pulmonary perfusion or from reduced renal blood flow following the peripheral redistribution of blood as occurs with exercise. Both factors then lead to fluid/salt retention and so to a volume overload. Induction of tricuspid regurgitation resulting from the effects of cor pulmonale on the valvular annulus or directly associated with worm entanglement can induce an acute volume overload and in some cases acute RSCHF. Aldosterone levels are accentuated in cases of RSCHF supporting the concept that renal perfusion is reduced in such cases.[6]

Hypoxia of the right ventricular myocardium (particularly subendocardial areas) associated with reduced cardiac output, reduced coronary perfusion, and/or increased cardiac workload and oxygen demands can lead directly to failure as the existing pressure load progressively exceeds right ventricular work capacity. Such a situation is probably accentuated by the potentially ischemic effects of severe hypertrophy.

Pleural effusion, sometimes evident in cases without RSCHF and ascites, is not always seen associated with severe RSCHF cases.[6] Presumably, it is therefore associated with something directly related to reduced venous/lymphatic return from the parietal pleura rather than due to a general increase in central venous pressure. Using rubber threads in the anterior vena cava, pleural effusion has been experimentally induced in association with surgical manipulation in that area without the simultaneous induction of ascites or RSCHF. Such effusion seen in natural cases without RSCHF could be associated with a similar etiology.[3]

In summary, RSCHF in this disease is a combination of a progressive pressure load and associated volume load on the right ventricle, which can also be directly affected by myocardial hypoxia and, in some cases, by valve entanglement.

VII. THERAPY FOR SEVERE CASES AND CASES OF RIGHT-SIDED CONGESTIVE HEART FAILURE

Nichols and Hennigar[30] in 1964 showed that severe RSCHF cases had increased zona granulosa mass and increased radial tissue percentage of the sectioned adrenal gland. Using antialdosterone therapy the extent of ascites was reduced in these dogs when compared to mildly infected and noninfected animals. Thus, the volume overload of the RSCHF was presumably reduced by antagonizing the angiotensin pathway activated by reduced renal perfusion due to cor pulmonale.

In treating the cause of RSCHF, correction of the primary defects must be investigated. The RSCHF is due to a pressure overload caused by arterial disease (a combination of inflammation, thrombosis, and nonreversible chronic fibrotic arterial and periarterial disease), secondary tricuspid regurgitation (effects of cor pulmonale, sometimes valve entanglement), and due to a volume overload induced by poor renal perfusion.

Except for worm entanglement, of which severe cases usually die within 24 hours as a result of acute and extreme RSCHF, the other causes relate directly to reduced pulmonary perfusion. Thus, primary treatment of such cases (and of moderately affected dogs) needs to be based on reducing the degree of inflammation, particularly reducing periarterial (dilation-limiting) inflammation, and reducing the progression of thrombosis.[31] Thus, the use of prednisolone and aspirin needs to be considered in these cases. Usually, they are administered for 2 to 6 weeks to stabilize the patient prior to assessment for adulticide and subsequent therapy. Prednisolone is given at a dose of 0.5 to 1 mg/kg on alternate mornings to avoid the possible side effects (polydipsia, polyuria, polyphagia, hepatopathy) to this drug. Obese animals are dosed at lean body weight. This drug is also indicated in cases of immune-mediated occult disease usually at higher initial doses with a reducing dose rate for 1 to 2 weeks. The other drug of choice is aspirin. It is used at varying dose rates, but in the author's opinion, based on the experience of the side effects of higher dose rates, the drug should be used at 3 mg/kg twice weekly to reduce platelet function.

Moderate to severely affected dogs should be thus treated, have restricted exercise (preferably caged), and have restricted salt and water (70 mℓ/kg/day) intake.

Diuretics may also be used, particularly if ascites is severe and diaphragmatic pressure is causing respiratory distress. Antialdosterone diuretics may be more effective and may be used in association with other diuretics. Digoxin is used only if a significant digoxin-responsive arrhythmia is present.[31,32] Although debatable, the author[31] feels that the inotropic use of digoxin in RSCHF cases associated with increased afterload is not indicated and, because of reduced efficiency in such cases, may easily lead to toxicity.[31]

Vasodilation, while not proven to be very useful clinically, may have benefit, particularly if some of the pulmonary hypertension is due to reversible vasoconstriction associated with thromboembolism.[29] Additional vasomotor pulmonary hypertension could also be induced if animals are hypoxemic, and vasodilatory therapy could therefore be beneficial in such situations. Hydralazine and other vasodilatory agents have been used,[33] but further work in this area is required as such drug therapy is not without risk and its use needs to be carefully monitored. Perhaps a response to such a drug could be of prognostic value, indicating the extent of reversible pulmonary hypertension, but, as with some other prognostic procedures, costs and the risks of an invasive procedure need to be carefully assessed. As mentioned above, the future use of antiserotonin therapy could avoid the need for vasodilatory therapy by preventing any vasoconstriction.

However, anti-inflammatory and antithrombotic therapy associated with cage rest is based on treating the primary causes of disease and on avoiding exercise-induced complications and as such would seem justified. The response to such therapy can also be used to assess prognosis in that a poor response is indicative of permanent nonreversible hypertension, whereas a reduction in the degree of RSCHF would suggest that pulmonary perfusion capacity may have been improved and a better prognosis could therefore be given.

Once stabilized the case can then be assessed for adulticide therapy remembering that severe cases of the disease in older dogs often have very few live adult worms, most of the RSCHF being due to chronic arterial pathology. The use of antigen tests in such cases could be very helpful to ascertain the actual live adult burden.

This basic approach to therapy should be continued after adulticide therapy corresponding to the individual response in each case. Prednisolone is used at a quarter to half the dose rate of that of the experimental work at Georgia where daily administration for 30 days showed reduced adulticide effectiveness.[22] Because the periarterial inflammatory disease, seen associated even with noninfected, antigen *(Dirofilaria immitis)* challenged dogs, is so severe in the first week following dead worm deposition,[23] it is felt that prednisolone therapy used prior to adulticide therapy should be continued using the alternate-day regime, perhaps stopping therapy for a week once adulticide therapy is started. Alternate-day prednisolone therapy has not been tested to see if it reduces adulticide effectiveness, but clinical experience with this regime would suggest any such effect to be minimal.[6] As most clinical side effects to adulticide therapy occur from the second week onwards, it would seem an ideal time to have these important anti-inflammatory effects reintroduced.

It is stressed that dead worms induce severe arterial and periarterial inflammation and thrombosis, and anti-inflammatory (alternate day) and antithrombotic (twice weekly) therapy to reduce these side effects would seem indicated for 2 to 4 weeks post adulticide depending on the case at hand and on the extent of pulmonary disease associated with the death of filariae. More work needs to be performed in this area to attempt better dosage regimes and, most importantly, durations for therapy. Our work with the early pathology associated with dead filariae would suggest that such therapy would be most indicated.[23] However, recognition of the apparent antiadulticide effect of daily prednisolone and concurrent thiacetarsamide therapy needs to be considered.

VIII. PERSISTENT INFECTION AFTER THERAPY

Post-treatment (adulticide and microfilaricide) microfilaremias are most likely associated with viable, adult, fertile filariae.[6] As reduced effectiveness of adulticide is seen against female filariae,[34] it is more likely that female filariae are present still producing microfilariae with or without male worms. If the microfilarial levels are rising this is even more likely to be the case. Persistent and drug-resistant microfilarial populations are reported, but with the use of ivermectin, which is reportedly 99.9% effective, this will possibly not occur.[35] Thus, consideration of a second course of adulticide is necessary in such situations possibly incorporating the use of an antigen test to give an indication of the extent of the remaining infection.

The other problem also exists of residual adult filariae as an occult infection after therapy. For example, male filariae are all killed as they are more sensitive to chemotherapy, immature filariae may have survived but are sexually immature, and the adult females may have become sterile due to the drug effects and/or immune mechanisms.[12] In such cases, most clinicians, not knowing the true adult heartworm status, would proceed with prophylactic therapy having, presumably, effectively cleared the infection. While it is hard to accept that over the years many of the dogs we have treated could well have harbored a small number of residual filariae (supposedly without significant clinical effect), this is most likely the situation. The dogs did not develop further problems due to the low number of worms involved, due to the extent of pulmonary reserve after resolution of the effects of dead filariae, and, more importantly, due to effective continuing prophylaxis. Antigen tests could well help resolve this type of situation being used 8 to 12 weeks post therapy in the assessment of adulticide effectiveness (see Chapter 4).

The other alternative is to give a second course of adulticide. However, one has to ask whether a second course of potentially (and unpredictably) toxic adulticide to kill a low worm residual in the moderate to severely affected, older case is warranted. Would it not be better to let the few remaining worms die of old age and concentrate on prophylaxis? The chance of filarial entanglement of the tricuspid valve is very low with low worm numbers, and not a lot of direct pathology would be produced by the few remaining filariae. This has probably been the real situation with many cases over the last 2 decades with adulticide efficacy as we now know it to be. As most of these cases (presumably with a few remaining live filariae without a microfilaremia) clinically improved after adulticide therapy, it would seem justifiable to accept that maybe the one to three live worms left would not be a major problem. The experiments to support such a statement have not been performed, but, based on the use of artificial worms, it would seem to be worth investigating.[3]

IX. FUTURE PROGNOSTIC ASSESSMENT OF THE MODERATE TO SEVERELY AFFECTED CASE

Costs will always be a factor in what workup procedures are possible. Cost effectiveness will have to be increasingly considered, in spite of the concerns for litigation, if veterinary medicine is to stay below what is seen as the affordable level for the majority of the public. However, this section gives a guide to what present (experimental) and futuristic, prognostic assessment procedures could be available to further assess pulmonary artery-perfusion reserve and the capacity for absorbance of emboli following therapy.

A. Prediction of Arsenical Toxicity

Prediction of arsenical toxicity is not possible at present, but the work of Holmes et al.[36] with the use of indocyanine green to predict hepatic arsenical metabolism and thus predict worm mortality could possibly also suggest toxic potential. The procedure is marginally invasive and does give a prediction of expected worm mortality based on predicted arsenic levels allowing for the hepatic metabolic rate.

B. Thromboembolic Prediction

This should be partially related to worm-number (mass) assessment, now available with antigen detection tests, and is being evaluated. However, there are other thrombogenic factors, namely the extent of arterial pathology and associated turbulence, and thus a test like the D-dimer test being developed to assess the degree of thrombosis will hopefully be of use in this area. Assessment of the degree of existing thrombosis and of its potential, in association with worm-number estimation, could be predicted in each case. This D-dimer test could also aid with planning the dose and duration of antithrombotic therapy indicated for each case (e.g., aspirin) and be of specific, diagnostic and prognostic use in cases of thromboembolism.

C. Pulmonary Perfusion Capacity

Although angiography can show the extent of obstruction to flow it does not give quantitative assessment. Pressure recordings and resistance calculations do give more objective data but are technically invasive, costly, place the patient at risk, and need to be performed under stress to assess the full extent of the hypertensive disease. The once-used circulation time could be reworked to give a more accurate estimation of the ability of dye (blood) to pass the pulmonary bed, allowing for body surface area, weight, heart rate — the results perhaps being then related to the results of other noninvasive assessments, e.g., echocardiography. If we could assess the actual perfusion deficit we would have a more accurate prognostic assessment for each case. Most other procedures give the infection or the gross or secondary disease status and are not able to give a functional status (the extent of the loss of function) in each case.

If the reliable electrocardiogram (EKG) criteria of right ventricular hypertrophy (RVH) could be related to pulmonary artery pressure (or to perhaps dP/dT or pulmonary vascular resistance) in a series of cases, perhaps a correlation could be produced that would give a guide to the extent of hypertension. This is probably less likely due to the lag phase in the hypertrophic response. However, the echocardiogram could be incorporated in this proposed correlation between noninvasive, easily obtained factors and invasively derived hemodynamic determinants. Obviously, this research needs to be performed, but the emphasis must be based on easily used, readily available, noninvasive technology rather than invasive and less practical procedures.

The availability of a relatively noninvasive, not excessively technical test to assess the perfusion deficit (and/or the extent of thrombosis, particularly the potential for thrombosis) would have great benefit, enabling more objective case assessments to be performed.

D. Standardized Exercise Test and Clinical Assessment

If such a test were existent in veterinary medicine for cardiac cases (as exists for people) perhaps more objective information could be obtained from our basic clinical assessment. If we could equate the pulmonary reserve with a heart and respiratory rate response to a known workload setting allowing for dog weight, resting heart rate, and other factors, perhaps predictive data could be obtained comparatively easily. Obviously, maximal testing would be contraindicated and the potential dangers of such a test in some cases would have to be considered. Individual dog variation, e.g., psychological factors, would also be a problem. Incorporated into such a procedure could be the vasodilation assessment predicting reserve dilatory capacity mentioned previously. While less objective, the response to anti-inflammatory and antithrombotic therapy could also be used to assess the degree of reversible "obstructive" disease.

E. Antigen Tests

There is no doubt that the antigen tests will probably be the best, noninvasive, prognostic support in the near future allowing more accurate diagnosis, worm-number assessment, and an assessment of adulticide therapy, as well as being a possible guide to the thrombotic, and, therefore, to the thromboembolic potential of selected cases.

REFERENCES

1. **Rawlings, C. A.**, Acute response of pulmonary blood flow and right ventricular function to *D. immitis* adults and microfilariae, *J. Am. Vet. Med. Assoc.*, 41, 244, 1980.
2. **Carlisle, C. H.**, The experimental production of heartworm disease in the dog, *Aust. Vet. J.*, 46, 190, 1970.
3. **Buoro, I. B. J. and Atwell, R. B.**, unpublished data, 1982.
4. **Buoro, I. B. J. and Atwell, R. B.**, Development of a model of caval syndrome in dogs infected with *D. immitis*, *Aust. Vet. J.*, 61, 267, 1984.
5. **Fujii, I.**, A clinical study of venae cavae embolism by heartworms of dogs, *Bull. Azubu Vet. Coll.*, 30, 105, 1975.
6. **Atwell, R. B.**, unpublished data, 1982.
7. **Kume, S.**, Epizootiology of *D. immitis* in Tokyo, in *Canine Heartworm Disease: A Discussion of the Current Knowledge*, Bradley, R. E., Ed., University of Florida, Gainesville, Fla., 1970, 39.
8. **Knight, D. H.**, Evaluation of pulmonary artery disease in canine dirofilariasis: evaluation by blood pressure measurement and angiography, in *Proceedings of the Heartworm Symposium '80*, Otto, G. F., Ed., Veterinary Medicine Publ., Edwardsville, Kan., 1981, 55.
9. **Atwell, R. B.**, Early stages of disease of the peripheral pulmonary arteries in canine dirofilariasis, *Aust. Vet. J.*, 56, 157, 1980.
10. **Honaga, S., Kongure, K., Hamasaki, H., Tagawa, M., and Kuradawa, K.**, Clinical pathological observations in prepatent period of canine dirofilariasis, *Jpn. J. Vet. Sci.*, 40, 603, 1978.
11. **Atwell, R. B., Buoro, I. B. J., and Sutton, R. H.**, Experimental production of lesions in canine pulmonary arteries similar to those produced by *D. immitis* infection, *Vet. Rec.*, 116, 539, 1985.
12. **Knight, D. H.**, Heartworm disease, in *Internal Veterinary Medicine*, 2nd ed., Ettinger, S. J., Ed., W. B. Saunders, Philadelphia, 1983, 1097.
13. **Schaub, R. G. and Rawlings, C. A.**, Pulmonary vascular response during phases of canine heartworm disease. Scanning electron microscopic study, *Am. J. Vet. Res.*, 41, 1082, 1980.
14. **Rawlings, C. A.**, Cardiopulmonary function in the dog with *D. immitis* infection: during infection and after treatment, *Am. J. Vet. Res.*, 41, 319, 1980.
15. **Clemmons, R. M., Yamaguchi, R. A., Schaub, R. G., Fleming, J., Dorsey Lee, M. R., and McDonald, T. L.**, Interaction between canine platelets and adult heartworms: platelet recognition of heartworm surfaces, *Am. J. Vet. Res.*, 47, 322, 1986.
16. **Atwell, R. B. and Carlisle, C. H.**, Distribution of infused artificial filariae within the pulmonary arteries of dogs, *J. Small Anim. Pract.*, 23, 725, 1982.
17. **Atwell, R. B. and Rezakhani, A.**, Inoculation of dogs with artificial larvae similar to those of *D. immitis*: distribution within the pulmonary arteries, *Am. J. Vet. Res.*, 47, 1044, 1986.
18. **Atwell, R. B., Sutton, R. H., and Moodie, E. W.**, Preliminary report on the pulmonary pathology associated with subcutaneous injection of *D. immitis* antigen, *Vet. Res. Commun.*, 6, 59, 1983.
19. **Rawlings, C. A., Keith, J. C., and Schaub, R. G.**, Effects of acetylsalicylic acid on pulmonary arteriosclerosis induced by a one-year *D. immitis* infection, *Arteriosclerosis*, 5, 355, 1985.
20. **Atwell, R. B., Sutton, R. H., and Carlisle, C. H.**, The reduction of pulmonary arterial thromboembolic disease *(D. immitis)* in the dog associated with aspirin therapy, in *Proceedings of the Heartworm Symposium '83*, Otto, G. F., Ed., Veterinary Medicine, Publ., Edwardsville, Kan., 1983, 115.
21. **Booth, N. H.**, Non-narcotic analgesics, in *Veterinary Pharmacology and Therapeutics*, 5th ed., Booth, N. H. and McDonald, L. E., Eds., Iowa State University Press, Ames, 1982, 302.
22. **Rawlings, C. A., Keith, J. C., Lewis, R. E., Losonsky, J. M., and McCall, J. W.**, Aspirin and prednisolone modification of radiographic changes caused by adulticide treatment in dogs with heartworm infection, *J. Am. Vet. Med. Assoc.*, 182, 131, 1983.
23. **Atwell, R. B., Sutton, R. H., and Buoro, I. B. J.**, Early pulmonary lesions caused by dead *D. immitis* in dogs exposed to hemologous antigens, *Br. J. Exp. Pathol.*, 67, 395, 1986.

24. **Adcock, J. L.,** Pulmonary arterial lesions in canine dirofilariasis, *Am. J. Vet. Res.,* 22, 655, 1961.
25. **Atwell, R. B. and Sutton, R. H.,** unpublished data, 1984.
26. **Rawlings, C. A., Keith, J. C., Losonsky, J. M., and McCall, J. M.,** An aspirin-prednisolone combination to modify post adulticide lung disease in heartworm-infected dogs, *J. Am. Vet. Med. Assoc.,* 45, 2371, 1984.
27. **Rawlings, C. A., Losonsky, J. M., Lewis, R. E., and McCall, J. W.,** Development and resolution of radiographic lesions in canine heartworm disease, *J. Am. Vet. Med. Assoc.,* 178, 1172, 1981.
28. **Boring, J. G.,** Radiographic diagnosis of heartworm disease, in *Proceedings of the Heartworm Symposium '74,* Morgan, H. C., Ed., Veterinary Medicine Publ., Bonner Springs, Kan., 1975, 32.
29. **Thomas, D., Stein, M., Tanake, G., Rege, V., and Wessler, S.,** Mechanism of bronchoconstriction produced by thromboemboli in dogs, *Am. J. Physiol.,* 206, 1207, 1964.
30. **Nichols, J. and Hennigar, G.,** Effects of pulmonary hypertension on adrenal and kidneys of dogs infected with heart worms *(D. immitis), Lab. Invest.,* 13, 600, 1964.
31. **Atwell, R. B.,** Treatment of severe canine dirofilariasis and associated cardiac decompensation, *Aust. Vet. Pract.,* 12, 132, 1982.
32. **Calvert, C. A. and Thrall, D. E.,** Treatment of canine heartworm disease coexisting with right-sided heart failure, *J. Am. Vet. Med. Assoc.,* 180, 1201, 1982.
33. **Lombard, C. W.,** Pulmonary vasodilation in heartworm dogs with cor pulmonale, in *Proceedings of the Heartworm Symposium '83,* Otto, G. F., Ed., Veterinary Medicine Publ., Edwardsville, Kan., 1983, 110.
34. **Blair, L. S., Malatesta, P. F., Gerckens, L. S., and Ewanciw, V.,** Efficacy of thiacetarsamide in experimentally infected dogs at 2, 4, 6, 12 or 24 months post-infection with *Dirofilaria immitis,* in *Proceedings of the Heartworm Symposium '83,* Otto, G. F., Ed., Veterinary Medicine Publ., Edwardsville, Kan., 1983, 130.
35. **Jackson, R. F., Seymour, W. G., and Beckett, R.,** Routine use of 0.05 mg/kg of Ivermectin as a microfilaricide, in *Proceedings of the Heartworm Symposium '86,* Otto, G. F., Ed., American Heartworm Society, Washington, D.C., 1986, 37.
36. **Holmes, R. A., McCall, J. W., Prasse, K. W., and Wilson, R. C.,** Thiacetarsamide sodium: pharmacokinetics and the effects of decreased liver function on efficacy against *Dirofilaria immitis* in dogs, in *Proceedings of the Heartworm Symposium '86,* Otto, G. F., Ed., American Heartworm Society, Washington, D.C., 1986, 57.

Chapter 8

CHEMOTHERAPY AND CHEMOPROPHYLAXIS

Charles H. Courtney

TABLE OF CONTENTS

I. INTRODUCTION

Treatment of canine dirofilariasis has evolved over the last 35 years into the reasonably safe and effective procedure used today. Unless precluded by an animal's poor condition, adulticidal treatment is always indicated once a diagnosis of dirofilariasis is made. An untreated dog is a reservoir of infection for others, and a substantial number of infected dogs will eventually develop severe disease. In a highly endemic area like Florida, approximately one third of untreated dogs will develop severe clinical signs within 1 to 2 years of diagnosis. Another third will show mild to moderate chronic changes best described to the pet owner as "seeming to age faster than normal", and the remaining third will usually remain asymptomatic for life. Unfortunately, there is, at present, no means to predict the course of disease in any individual patient.

In the past, surgical removal of adult heartworms has been advocated,[1] but the 10% mortality from surgery far exceeds the less than 1% mortality from medically treated cases.[2] Surgery is no longer an acceptable method for routine treatment except in special cases such as caval syndrome[3] or intraocular and other ectopic infections.[4]

Treatment of canine dirofilariasis should be viewed as a five-step process: (1) the animal is thoroughly examined and evaluated prior to treatment, (2) supportive or symptomatic therapy is implemented as necessary before, during, and after adulticidal therapy (see Chapter 7 for further discussion), (3) adulticidal treatment and (4) microfilaricidal treatment are administered, and (5) the animal is started on prophylactic medication. Minor modifications to this sequence may occur (i.e., microfilaricides need not be given in cases of occult disease), but, in most instances, treatment should follow this five-step approach.

II. EVALUATION OF THE PATIENT

Prior to treatment, the patient should be evaluated as thoroughly as feasible to identify those individuals that should not be treated, have their treatment delayed, or be treated with great care. Specifically, the clinician is seeking any sign that the patient has an increased risk of developing one of the two major untoward effects of adulticide treatment: (1) toxic reactions (hepatotoxicity and/or nephrotoxicity) and (2) post-treatment pulmonary embolism and thromboembolism associated with the death of worms.

There is considerable variation from one veterinarian to the next as to the degree of pretreatment clinical evaluation considered necessary. An important cause of this variation is that published protocols for pretreatment clinical evaluation have often been designed through trial and error or clinical "impression" rather than through the rigors of controlled experimentation, and, in many instances, no hard evidence exists for the predictive value of specific tests. Costs are also an important factor in determining an assessment protocol.

Clinicians are in general agreement that the evaluation should first begin with a thorough physical examination of all patients with an emphasis on the cardiopulmonary system; beyond that step, opinions begin to diverge as to what is necessary. The American Heartworm Society recommends that laboratory tests should include a hemogram, urinalysis, estimations of plasma urea concentrations, and selected liver tests, with thoracic radiographs being listed as "helpful".[5] However, Australian workers have shown that prior assessment of clinical condition by urinalysis or plasma chemistries did not predict toxic reactions to thiacetarsamide in the asymptomatic or mildly affected dog.[6] Similarly, a large scale retrospective study of 270 dogs treated for dirofilariasis at Louisiana State University[7] concluded that "pretreatment clinical and laboratory findings could not be used to identify dogs that would adversely react to the administration of thiacetarsamide". However, it is presumably beneficial for the clnician to know if an animal, particularly if aged, has other concurrent diseases or subclinical organ dysfunction and to establish baseline values for that animal regardless of the value of these tests for predicting reactions to thiacetarsamide.

Most deaths after adulticidal treatment occur from embolism of dying worms and associated thromboembolism rather than toxic reactions. Thoracic radiographs may be invaluable in predicting the likelihood of severe post-treatment embolic reactions. Asymptomatic dogs with little or no radiographic signs of pulmonary vascular disease are rarely so affected, but the more evidence there is of pulmonary disease, the more likely the occurrence of clinically evident complications.[8] Sometimes, asymptomatic dogs are seen that have advanced lung pathology[9] presumably associated with the slow onset of disease and the lack of clinical disease due to the capacity for adaptive cardiopulmonary compensation.

With the symptomatic dog, pretreatment evaluation should be more rigorous since mortality rates are greater in symptomatic (3 to 5%)[10] than in asymptomatic dogs (1%).[2] After a thorough, physical examination, thoracic radiographs and the use of antigen titers (see Chapter 4) are perhaps the best predictors of potential complications.[10] Urine and liver function tests are likely to be of greater value here because of the increased likelihood of finding damage to these organs in the symptomatic dog. The value of a complete blood count is still open to question.

III. SUPPORTIVE THERAPY

Supportive therapy should be given as required and any concurrent diseases should be treated to the extent possible prior to adulticidal treatment (see Chapter 7 with regard to preadulticide therapy).

IV. ADULTICIDAL THERAPY

A number of compounds have been used to kill adult heartworms, but the "ideal" adulticide does not exist. Adulticides include the phenylarsenoxides (one of which, thiacetarsamide, is the most widely used adulticide), antimonials, and levamisole.

A. Thiacetarsamide

Otto and Maren[11] first proposed that thiacetarsamide be used in the treatment of dirofilariasis after having discovered it in a search for drugs active against human filariae during World War II; the work being performed used *Dirofilaria immitis* in dogs as the model. A phenylarsenoxide, *p*-arsenosobenzamide, was found to regularly kill all adult heartworms present in dogs at well-tolerated dosages. Since the compound was only slightly soluble in water, thiacetarsamide sodium, a more soluble analog, was synthesized. Fifteen daily doses of thiacetarsamide at 2.2 mg/kg was found to be highly effective and was used for a number of years until Kume and Ohishi[12] found that 4.4 mg/kg daily for 3 days was just as effective, although worms died at a somewhat slower rate. One year later, Baily[13] found that only two daily doses were required at this higher dose rate. Jackson[14] subsequently modified this schedule further by dividing the single daily dose into two doses of 2.2 mg/kg given morning and afternoon. This virtually eliminated shock-like reactions occasionally seen following the higher single daily dosage.

Since its introduction in 1963, this twice-daily, 2-day treatment regimen has been generally accepted as the standard treatment for adult heartworms and was thought to have an efficacy in excess of 90%. However, by the mid 1970s, some clinicians had begun to question its efficacy.[15] At first, it appeared that reduced efficacy was associated with an improperly assayed batch of thiacetarsamide,[16,17] but three studies sponsored in the U.S. by the American Heartworm Society[18-20] refuted this. Using different batches and different brands of thiacetarsamide, the three studies gave similar results: an overall, average worm kill of 63% with only 40% of treated dogs being completely cleared of heartworm. Nothing was gained by extending the treatment schedule to 3 days (i.e., six doses); mean worm kill was 70% and only 39% of treated dogs were completely cleared of heartworm. More recently, Atwell improved efficacy to near 90% by giving the entire 2-day dose of thiacetarsamide over a single, 4-hr period as a slow intravenous drip, but problems with hepatotoxocity and shock-like reactions were encountered.[21] Studies at the University of Florida suggested that slight increases in the dose rate of thiacetarsamide resulted in a dramatic improvement in worm kill.[22] Efficacy was increased from 78% in 16 conventionally treated dogs (2.2 mg/kg b.i.d. for 2 days) to 94% in 19 dogs given a 20% increased dose (2.64 mg/kg b.i.d. for 2 days), and the percentage of dogs completely cleared of heartworms increased from 38 to 74%, respectively. In 11 dogs given a 33% increased dose (2.93 mg/kg b.i.d. for 2 days) efficacy of worm kill and percent of dogs completely cured was 95 and 91%, respectively. No adverse reactions were noted except for some evidence of increased thromboembolism in dogs given the two higher dosages. This was attributable to the greater (and presumably more rapid) worm kill in these dogs. At a 50% increase in dose rate (3.3 mg/kg b.i.d. for 2 days), hepatotoxic reactions and death occurred in 2 of 11 dogs, although efficacy was 100% in all surviving dogs.

Surprisingly, little is known about the pharmacodynamics of thiacetarsamide. In dogs given 0.45 mg/kg of arsenic (as thiacetarsamide) daily for 15 days, 17% of the total dose of arsenic given on the first day was excreted in the urine in the first 24 hr, with two thirds of the urinary arsenic being recovered during the first 2 hr after treatment.[23] On subsequent days, arsenic appeared in feces and the level of arsenic in urine increased. Over 85% of the total arsenic injected during the 15 days of treatment was recovered in urine and feces by the time the dogs were necropsied 48 hr after the last dose. Nearly twice as much total arsenic was recovered from feces as from urine during this time. The bulk of the unexcreted arsenic was found to remain in the liver. In a University of Georgia study, the elimination half life of thiacetarsamide (as arsenic) was estimated to be 43 min after a single injection.[24] Actual blood levels of arsenic seem to fall very quickly after a single dose of thiacetarsamide.[21,24] Similarly, studies at the University of Florida[25] found that blood arsenic levels fell below the limits of detection within 30 min after a single injection. However, after the remaining three injections of the usual treatment schedule were given (2.2 mg/kg b.i.d. for 2 days), it became apparent that this rapid decrease represented distribution to a deeper tissue level rather than true elimination. Commencing with the second injection, this deeper tissue became "filled" and detectable blood levels of arsenic were then present for about 7 days. These studies should be interpreted with a degree of caution since arsenic levels rather than thiacetarsamide levels were measured and it is not known to what extent the drug had been metabolized.

Thiacetarsamide should only be given intravenously since injection by any other route (or perivascular leakage) leads to severe tissue necrosis at the injection site, and thiacetarsamide is not effective by the oral route.[23] Because of this danger of necrosis with perivascular injection of the drug, a variety of injection techniques have been developed to ensure proper administration. The most popular are to flush the needle (or butterfly needle and tubing) with saline (0.9%) to insure its accurate placement prior to injection of thiacetarsamide or to use indwelling intravenous (i.v.) catheters similarly flushed with saline prior to use. Saline can also be used after adulticide to ensure complete injection of all thiacetarsamide. In general, the best method for preventing perivascular injection is careful attention to proper injection technique coupled, more importantly, with proper restraint of the dog, including sedation if necessary. In the author's experience, more perivascular leaks occur as a result of an inadequately restrained dog than any other cause. If perivascular injection does occur, local application of dimethylsulfoxide (DMSO) and steroids has been recommended to reduce the severity of any potential necrosis.[26] Instillation of saline and of steroids has also been used to dilute the effect of chemical irritation, but no controlled studies have been done to confirm the effectiveness of any of these regimens. If sloughing occurs the area needs to be surgically debrided and treated as a devitalized wound with appropriate antibiotic cover.

It is often recommended that various substances be given with (and often mixed with) the thiacetarsamide, including vitamins, steroids, or lipotrophic agents. Even disregarding the unknown effect of these mixtures on the efficacy of thiacetarsamide, there has been no proof that they in any way benefit the dog. On the contrary, there is evidence that corticosteroids (given daily for 30 days at 1 mg/kg from the first dose of thiacetarsamide) may reduce the effectiveness of thiacetarsamide.[27]

The most commonly recommended treatment protocol is that of Jackson[14] in which i.v. injections are given twice daily for 2 days at a dose rate of 2.2 mg/kg. Care about potential toxicity is needed with a per kilogram dose rate with very large or very small breeds of dogs and, in those that are obese, when actual body weights are considered. Perhaps in the future, body surface-area calculations should be used in such extreme situations.

Several steps can be taken to monitor the dog's health during the course of thiacetarsamide treatment, remembering that the most important adverse reactions are (1) drug toxicities during the course of or very shortly after treatment and (2) pulmonary embolism and associated pulmonary complications commencing a week or more after the conclusion of treatment. During the treatment period, it is advisable to feed the dog prior to each injection. This has been presumed, but never proven, to protect the liver from the toxic effects of the drugs, but its real value is that of an indicator of toxic reactions; suspicion of toxicity would be high in a dog suddenly "off feed" after one or more injections of thiacetarsamide. The dog should be carefully monitored for signs of depression, persistent vomiting, anorexia, or icterus since all are valid reasons to suspend treatment due to presumed toxicity. The value of repeated urinalyses or liver tests during treatment of the asymptomatic dog remains to be proven.

After the last injection, the dog is returned to the owner with instructions to sharply (or preferably completely) restrict exercise for several weeks, usually by strict confinement to a small area. For the first few days the owner should observe the dog daily for the above-mentioned signs of drug toxicity. After the first few days, the chances of drug toxicity are lessened and the dog should be monitored for signs of thromboembolism — anorexia, listlessness, depression, rapid respiration, and coughing. If any of these signs are present, the dog should be promptly presented to the veterinarian for further evaluation.

Toxic reactions requiring the suspension of thiacetarsamide treatment may occur in up to 4% of dogs treated.[7] Persistent vomiting, anorexia, and icterus are signs of hepatotoxicity. Treatment involves the use of parenteral fluids, dietary alteration (increased carbohydrate, reduced fat, and lower protein), and vitamin supplementation. Steroids and lipotrophic agents are not known to be of use in treating toxicity. If treatment was suspended because of hepatotoxicity, it can often be safely resumed within a month or so, provided the dog is otherwise healthy. There is some evidence that increased tolerance of the dog's liver to thiacetarsamide is apparently induced in response to the earlier thiacetarsamide exposure.[30] The exact mechanism and, therefore, any interpretation of that response is not known as of yet. Does the liver that has had prior experience with thiacetarsamide eliminate the drug from the bloodstream more rapidly or is it simply better able to tolerate therapeutic blood levels? The effect of that response on efficacy of the drug is likewise unknown. However, dogs which rapidly eliminated thiacetarsamide (measured as arsenic) from the blood were found to be more likely to harbor live heartworms following therapy than were dogs that eliminated the drug more slowly.[25] Less commonly, nephrotoxic reactions may occur.[31] These are likewise treated with fluid therapy, although in severe cases peritoneal dialysis may be indicated.[26]

Pulmonary embolism, commencing a week or more after treatment, is by far the most common complication of thiacetarsamide treatment. A fever accompanied by dyspnea, tachycardia, pale mucosa, and a cough are the main clinical signs. Since more dogs die of post-treatment embolism than any other cause, such cases should be treated aggressively by strict rest, aspirin, dipyridamole, and corticosteroids, with careful assessment necessary using coagulograms and platelet counts (see Chapters 4 and 7). Routine use of prednisolone after thiacetarsamide therapy in all dogs, regardless of the presence or absence of clinical signs of embolism, does not appear to be justified, although aspirin therapy during this time should help all but those patients with hemoptysis.[32] It may be necessary to pretreat dogs with prednisolone (2 mg/kg s.i.d.) if they have immune-mediated occult disease and are showing clinical signs of immune-mediated pneumonitis. One should then give thiacetarsamide 1 to 2 days after the prednisolone treatment to avoid the reduced mortality of adult worms reported with concurrent daily prednisolone therapy and thiacetarsamide.[33]

B. Levamisole

Levamisole also has a degree of adulticidal activity, although its use for this purpose is more popular in some parts of Australia than it is in the U.S. Kelly[34] reported that the two dose schedules widely used were (1) 2.5 mg/kg daily for 14 days, then 5 mg/kg daily for 14 days, and, finally, 10 mg/kg daily for 14 days; (2) 2.5 mg/kg daily for 7 days, then 5 mg/kg daily for 21 days, and, finally, 10 mg/kg daily for 21 days. Many different schedules have been used, but most trials have used small groups of animals.[35]

The main adverse reaction with these schedules is reported to be ataxia and is usually reversible if levamisole is immediately withdrawn. Commencing with a very low dose and gradually increasing the dosage over time will minimize toxic signs.[34] The main toxic signs encountered with levamisole use in other studies included vomition and neurologic disturbances, but many other toxic side effects are possible, including hemolytic anemia, thrombocytopenia, bone-marrow suppression, toxic skin problems (e.g., toxic epithelionecrolysis), and alopecia.[36] Vomition is usually not severe and can be controlled with atropine or antiemetic agents. Neurologic signs can include panting, apprehension, nervousness, paresis, ataxia, apparent hallucinations, and seizures.[36] However, the above dose schedules may not be sufficient to kill all female worms, and subsequent treatment with thiacetarsamide may be required.[34] These doses are considerably less than those reported as 100% effective against adult heartworms in a group of 28 experimental dogs (10 mg/kg or more twice daily in various schedules).[37] In the U.S., no schedule has been found in which levamisole consistently killed all heartworms,[35] although the doses used were lower than those reported above.[36] The most consistent finding was that levamisole killed all male heartworms and a varying proportion of female worms.

C. Other Compounds

Several other drugs have been used or proposed as adulticides. Jackson and Otto[28] showed that *p*-arsenosobenzamide was highly (97%) effective against adult worms, but its water-soluble derivative, thiacetarsamide, is the only organic arsenical marketed in much of the world. Malpharsan, an organic arsenical that can be given by intramuscular injection, has been used in Japan but is not available elsewhere. Stibophen is an antimonial whose adulticide activity was never very good and is no longer used.

V. MICROFILARICIDAL THERAPY

Six or more weeks after adulticidal treatment, depending upon the seriousness of any post-treatment reactions, the dog should be given microfilaricidal treatment. (This may be unnecessary if levamisole was used as an adulticide or if an occult infection is being treated.) This step is necessary for several reasons: (1) although most signs of heartworm disease are caused by adult worms, studies have shown that microfilariae are not completely harmless and may be associated with specific pathology, e.g., a membranous glomerulonephritis;[38,39] (2) if left untreated, microfilariae may survive for up to 2 years[40] with the dog serving as a reservoir of infection during this time; (3) if no attempt is made to eliminate microfilariae, incomplete kill of fertile adult worms or reinfection at a later date may go undetected; and (4) microfilariae must be eliminated before a dog can safely be given diethylcarbamazine prophylaxis.

When to treat with a microfilaricide has been a matter of some controversy. Microfilaricides have been given, "traditionally", some 3 to 6 weeks after thiacetarsamide, the most appropriate interval never having been determined experimentally. Otto and Jackson have discouraged treatment earlier than 6 weeks because of occasional reactions "which resemble those caused by emboli of dead worms following adulticide therapy".[26] Regardless of the exact timing, the vast majority of dogs are treated with a microfilaricide after, rather than before, thiacetarsamide is given.

Until recently, no scientific evidence supported this choice of treatment sequence. It was based on two assumptions: (1) living microfilariae were much less harmful than adult worms and their continued presence would not exacerbate reactions to thiacetarsamide or the severity of post-treatment thromboembolism; and (2) if pretreated with a microfilaricide, the adult worms would continue to release microfilariae so that a second course of microfilaricide would be needed after adulticidal treatment.[41] Tulloch[42] first suggested that microfilaricidal treatment should precede the use of an adulticide, but this proposal was based on the observation that three of seven dogs treated with "an analogue of tetramisole" showed no evidence of living or dead worms at necropsy. He hypothesized that a proportion of microfilaremic dogs did not harbor adult heartworms, thus, a microfilaricide would be the only treatment required. He did not, however, suggest any means to diagnose such a condition prior to treatment, although with the immunodiagnostic kits now available, such a diagnosis may be possible.

Garlick et al.[43,44] proposed that adverse reactions to thiacetarsamide could be eliminated if microfilariae were reduced in number to below 2000/mℓ prior to adulticidal treatment. Although his study involved over 500 cases, he did not compare his proposed regimen with the standard treatment (i.e., adulticide first). In contrast, Courtney and Boring[45] were unable to demonstrate any difference in the occurrence of adverse reactions to thiacetarsamide in a comparison of dogs receiving thiacetarsamide either before or after a microfilaricide (dithiazanine iodide), nor were they able to find any association between the microfilarial count at the time of thiacetarsamide treatment and toxic reactions to thiacetarsamide. The death of microfilariae is known to cause hepatic microgranulomas in animals with high microfilaremias,[46] and microfilaricidal therapy, with either levamisole or dithiazanine iodide, may cause elevations in serum enzymes suggestive of hepatic damage.[45,47] Thus, pretreatment with a microfilaricide may, in fact, compromise rather than enhance liver function.

Dithiazanine iodide, a cyanine dye, is the only microfilariacide licensed for use in the U.S. It is generally given at a rate of 4.4 mg/kg once daily for 7 days. If the dog is still microfilaremic after the initial 7-day course of dithiazanine, treatment is continued at an increased dosage of 11 mg/kg once daily until the dog is free of microfilariae or 10 days of treatment at this higher dose has elapsed.[26] Vomiting and diarrhea are common, but are seldom serious complications of dithiazanine therapy. The pet owner should be warned that vomitus and feces containing dithiazanine will permanently stain carpets, etc. Dithiazanine may not be a completely benign drug, however. Its package insert states that it is sometimes nephrotoxic and should not be used in dogs with renal disease, although Osborne and co-workers feel that the danger of renal disease is minimal.[31] Dithiazanine therapy has also caused transient increases in alanine aminotransferace (ALT) levels.[45] Some similar changes have occurred after levamisole therapy,[47] suggesting that such rises could either be a reaction to dead and dying microfilariae or could be due to direct drug effects.

Levamisole, although not licensed for use in dogs in the U.S., is, nevertheless, widely used as a microfilaricide. When given at the rate of 11 mg/kg once daily, microfilariae will usually be eliminated in 7 to 11 days.[47] The partial adulticidal effect of levamisole at this dosage gives it an added advantage over other microfilaricides since it may kill adult worms that survived thiacetarsamide therapy. Nuisance vomiting is a common complication that is usually controlled with atropine or antiemetic agents. Occasionally, more severe neurologic reactions such as nervousness, ataxia, apparent hallucinations, and seizures occur. These signs will usually cease upon withdrawal of levamisole. Its continued use is absolutely contraindicated in dogs showing neurological toxicity.

Ivermectin is highly effective as a microfilaricide at a dose of 0.05 mg/kg and will suppress microfilarial counts at doses as low as 0.003 mg/kg.[48] There is also some tenuous evidence that ivermectin may potentiate the adulticidal effect of thiacetarsamide.[48] Although a rapid initial kill of microfilariae occurs, with nearly all being cleared from the peripheral blood during the first 24 hr after treatment, up to 3 weeks may be required for complete elimination of microfilariae.[49] At microfilaricidal doses, some shock-like reactions have been attributed to rapid destruction of large numbers of microfilariae.[50] Some dogs become inappetent and depressed for 1 to 3 days after therapy. Additionally, rough-coated collies may be more sensitive to ivermectin than other breeds, showing depression, ataxia, coma, and death.[51,52] Ivermectin is also effective against microfilariae of *Dipetalonema reconditum*.[53]

Several other drugs are no longer in widespread use as microfilaricides. Fenthion is a microfilaricide of variable efficacy in topical, injectable, and oral formulations.[54-57] Diethyl-carbamazine is microfilaricidal at high doses, but severe reactions may occur.[9] Faudin® was a relatively toxic antimonial compound, so its use was discontinued when safer microfilaricides became available.[9]

VI. PROPHYLACTIC THERAPY

Currently, two methods of heartworm prophylaxis are recommended: (1) daily diethyl-carbamazine citrate (DEC) administration or (2) twice-yearly courses of thiacetarsamide at the adulticidal dose schedule.[5] Daily DEC prophylaxis is the preferred method. Japanese researchers showed that daily doses of DEC at 11 mg/kg commencing prior to infection and continued for at least 1 month after infection were highly effective.[58] At a dose rate of 5.5 mg/kg given either daily or every other day, protection was complete provided treatment was continued for at least 60 days after infection. Toxic reactions are seen in microfilaremic dogs and these are discussed in detail in Chapter 9. Because these toxic reactions occur in microfilaremic dogs, the presence of microfilariae should be excluded by recommended blood examination before DEC prophylaxis is begun. If a dog on DEC prophylaxis develops a microfilaremia, DEC prophylaxis can be continued provided it has not been interrupted.[5] However, missing as few as 3 days of DEC therapy in a microfilaremic dog may result in a severe reaction when DEC prophylaxis is re-commenced.[59,60] Although male sterility has been associated with DEC prophylaxis, controlled studies suggest that this phenomenon is a rare idiosyncrasy that is reversed by withholding DEC.[61]

It is currently recommended that daily DEC prophylaxis commences at the onset of the local mosquito season and continues for at least 60 days after completion of the mosquito season.[5] Dogs should be examined for the presence of microfilariae prior to beginning each annual course of DEC prophylaxis, usually in early Spring in most temperate climates. Where mosquitoes are present for 10 or more months of the year DEC prophylaxis must be given year round and dogs probably should be examined for the presence of microfilariae at six-month intervals, particularly if dosage has been irregular and/or unreliable. Although DEC is effective when given every other day,[58] daily dosing is recommended because alternate-day doses may confuse some clients. Puppies in endemic areas should be given DEC during the mosquito season at as young an age as possible (at least from 4 to 6 weeks of age). Using liquid DEC preparations, puppies as young as 2 weeks of age can be dosed effectively. Alternatively, they must be kept in mosquito-proof quarters until DEC is commenced.

Low doses of *dl*-tetramisole (5 mg/kg/day) given for a period of at least 30 days after infection have been shown to be an effective heartworm prophylactic.[62] This is approximately the equivalent of 2.5 mg/kg/day of levamisole. This dose schedule has never been widely adopted, probably because the alternative, DEC, is so very safe, effective, and inexpensive.

Ivermectin holds great promise for prophylaxis and is likely to be licensed for heartworm prophylaxis in the U.S.* and Australia in the near future. In a series of three field trials involving 143 dogs in Florida and Georgia, oral doses of ivermectin as low as 0.001 mg/kg once monthly were 100% effective in preventing natural infection by *Dirofilaria immitis*.[63] In studies using experimental infections, it was found that larval stages were most sensitive to ivermectin during the first month after infection when as little as 0.003 mg/kg was 100% effective.[64] In contrast, 3-month-old larvae were not completely controlled (95 to 98% effective) by oral doses of 0.05 to 0.2 mg/kg.

VII. TREATMENT OF HEARTWORM INFECTIONS IN UNUSUAL SITES AND UNUSUAL HOSTS

Treatment of caval syndrome in the dog and feline dirofilariasis is discussed in Chapters 11 and 12, respectively. Occasionally, *D. immitis* infections occur at unusual sites in dogs, particularly in the eye.[65] Intraocular worms can usually be removed surgically by snaring the worm with a small hook introduced through a corneal stab incision made near the limbus.[4]

Wild canids[65] and domestic ferrets[66] are commonly infected with *D. immitis* in endemic areas, so those kept as zoo specimens or pets should be given regular DEC prophylaxis (and probably in the future, ivermectin therapy). In contrast, wild felids are rarely infected,[65] so regular prophylaxis of zoo specimens is probably not justified. The California sea lion, a popular aquarium performer, is commonly infected with heartworms in endemic areas. Infected sea lions are treated with adulticides (thiacetarsamide or levamisole), microfilaricides (levamisole or dithiazanine), and given daily DEC prophylaxis as for dogs.[67] However, with these unusual hosts little experimental data are available to substantiate dose rates and effectiveness.

REFERENCES

1. **Jackson, W. F.,** Surgical treatment of heartworm disease, *J. Am. Vet. Med. Assoc.*, 154, 383, 1969.
2. **Mitchell, W. C.,** Panel report — heartworms in dogs: 1% treatment mortality, *Mod. Vet. Pract.*, 60, 418, 1979.
3. **Jackson, R. F., Seymour, W. G., Growney, P. J., and Otto, G. F.,** Surgical treatment of the caval syndrome of canine heartworm disease, *J. Am. Vet. Med. Assoc.*, 171, 1065, 1977.
4. **Thornton, J. G.,** Heartworm invasion of the canine eye, *Mod. Vet. Pract.*, 59, 373, 1978.
5. **American Heartworm Society,** Recommended procedures for the management of canine heartworm disease, in *Proceedings of the Heartworm Symposium '83,* Otto, G. F., Ed., Veterinary Medicine Publ., Edwardsville, Kan., 1983, 181.
6. **Carlisle, C. H., Prescott, C. W., McCosker, P. J., and Seawright, A. A.,** The toxic effects of thiacetarsamide sodium in normal dogs and in dogs infested with *Dirofilaria immitis,* Aust. Vet. J., 50, 204, 1974.

---

* Ivermectin was licensed in the U.S. in March 1987 for use at a dose rate of 0.006 to 0.012 mg/kg once monthly.

7. **Hoskins, J. D., Hagstad, H. V., and Hribernik, T. N.,** Effects of thiacetarsamide sodium in Louisiana dogs with naturally-occurring canine heartworm disease, in *Proceedings of the Heartworm Symposium '83,* Otto, G. F., Ed., Veterinary Medicine Publ., Edwardsville, Kan., 1983, 134.

8. **Knight, D. H.,** Heartworm heart disease, *Adv. Vet. Sci. Comp. Med.,* 21, 107, 1977.

9. **Jackson, R. F.,** Treatment of the asymptomatic dog, in *Proceedings of the Heartworm Symposium '74,* Morgan, H. C., Ed., Veterinary Medicine Publ., Bonner Springs, Kan., 1975, 51.

10. **Jackson, W. F.,** Management of the symptomatic patient, in *Proceedings of the Heartworm Symposium '74,* Morgan, H. C., Ed., Veterinary Medicine Publ., Bonner Springs, Kan., 1975, 56.

11. **Otto, G. F. and Maren, T. H.,** Possible use of an arsenical compound in the treatment of heartworm in dogs, *Vet. Med. Kansas City Mo.,* 42, 48, 1947.

12. **Kume, S. and Ohishi, I.,** Observations of the chemotherapy of canine heartworm infection with arsenicals, *J. Am. Vet. Med. Assoc.,* 131, 475, 1957.

13. **Baily, R. W.,** A comparison study of various arsenical preparations as filaricides of *Dirofilaria immitis, J. Am. Vet. Med. Assoc.,* 133, 152, 1958.

14. **Jackson, R. F.,** Two-day treatment with thiacetarsamide for canine heartworm disease, *J. Am. Vet. Med. Assoc.,* 142, 23, 1963.

15. Second discussion period, in *Proceedings of the Heartworm Symposium '74,* Morgan, H. C., Ed., Veterinary Medicine Publ., Bonner Springs, Kan., 1975, 103.

16. **Jackson, R. F., Seymour, W. G., and Growney, P. J.,** An evaluation of thiacetarsamide as an adulticide against *Dirofilaria immitis,* in *Proceedings of the Heartworm Symposium '77,* Otto, G. F., Ed., Veterinary Medicine Publ., Bonner Springs, Kan., 1978, 82.

17. **Boring, J. G. and Farrar, M.,** Current review of the adulticidal activity of thiacetarsamide sodium, in *Proceedings of the Heartworm Symposium '77,* Otto, G. F., Ed., Veterinary Medicine Publ., Bonner Springs, Kan., 1978, 86.

18. **McCall, J. W., Lewis, R. E., Rawlings, C. A., and Lindermann, B. A.,** Re-evaluation of thiacetarsamide as an adulticide agent against *Dirofilaria immitis* in dogs, in *Proceedings of the Heartworm Symposium '80,* Otto, G. F., Ed., Veterinary Medicine Publ., Edwardsville, Kan., 1981, 141.

19. **Todd, K. S., DiPietro, J. A., Guterbock, W. M., Blagburn, B. L., and Noyes, J. D.,** Evaluation of thiacetarsamide therapy for *Dirofilaria immitis* infections in naturally infected dogs, in *Proceedings of the Heartworm Symposium '80,* Otto, G. F., Ed., Veterinary Medicine Publ., Edwardsville, Kan., 1981, 146.

20. **Palumbo, N. F., Perri, S. F., and Sylvester, M. S.,** Re-evaluation of thiacetarsamide therapy in canine heartworm disease, in *Proceedings of the Heartworm Symposium '80,* Otto, G. F., Ed., Veterinary Medicine Publ., Edwardsville, Kan., 1981, 149.

21. **Atwell, R. B.,** Canine dirofilariasis — assessment of a currently used, and of a proposed adulticide program, in *Proc. 22nd World Veterinary Congr.,* Perth, Western Australia, August 21, 1983, 117.

22. **Courtney, C. H., Sundlof, S. F., and Jackson, R. F.,** New dose schedule for the treatment of canine dirofilariasis with thiacetarsamide, in *Proceedings of the Heartworm Symposium '86,* Otto, G. F., Ed., American Heartworm Society, Washington, D.C., 1986, 49.

23. **Drudge, J. H.,** Arsenamide in the treatment of canine filariasis, *Am. J. Vet. Res.,* 13, 220, 1952.

24. **Holmes, R. A., McCall, J. W., Prasse, K. W., and Wilson, R. C.,** Thiacetarsamide sodium: pharmacokinetics, and the effects of decreased liver function on efficacy, in *Proceedings of the Heartworm Symposium '86,* Otto, G. F., Ed., American Heartworm Society, Washington, D.C., 1986, 57.

25. **Sundlof, S. F., Courtney, C. H., Bell, J. U., and Jackson, R. F.,** Pharmacokinetics of thiacetarsamide in relationship to therapeutic efficacy, in *Proceedings of the Heartworm Symposium '86,* Otto, G. F., Ed., American Heartworm Society, Washington, D.C., 1986, 65.

26. **Otto, G. F. and Jackson, R. F.,** Heartworm disease, in *Textbook of Veterinary Internal Medicine,* Vol. 2, Ettinger, S. J., Ed., W. B. Saunders, Philadelphia, 1975, 1014.

27. **Rawlings, C. A., Keith, J. C., Losonsky, J. M., Lewis, R. E., and McCall, J. W.,** Aspirin and Prednisolone modification of post-adulticide pulmonary arterial disease in heartworm infection: arteriographic study, *Am. J. Vet. Res.,* 182, 131, 1983.

28. **Jackson, R. F. and Otto, G. F.,** Thiacetarsamide re-evaluation: committee summary and interpretation, in *Proceedings of the Heartworm Symposium '80,* Otto, G. F., Ed., Veterinary Medicine Publ., Edwardsville, Kan., 1981, 153.

29. **Jackson, R. F.,** Complications during and following chemotherapy of heartworm disease, *J. Am. Vet. Med. Assoc.,* 154, 393, 1969.

30. **Morgan, H. C. and Rainey, C. T.,** Clinical aspects of canine heartworm disease, in *Canine Heartworm Disease: A Discussion of the Current Knowledge,* Bradley, R. E., Ed., University of Florida, Gainesville, Fla., 1972, 76.

31. **Osborne, C. A., Hammer, R. F., O'Leary, T. P., Pomeroy, K. A., Jeraj, K., Barlough, J. E., and Vernier, R. L.,** Renal manifestations of canine dirofilariasis, in *Proceedings of the Heartworm Symposium '80,* Otto, G. F., Ed., Veterinary Medicine Publ., Edwardsville, Kan., 1981, 67.

161

32. **Rawlings, C. A., Keith, J. C., Schaub, R. G., Losonsky, J. M., Lewis, R. E., and McCall, J. W.,** Post adulticide treatment: pulmonary disease and its modification with prednisolone and aspirin, in *Proceedings of the Heartworm Symposium '83,* Otto, G. F., Ed., Veterinary Medicine Publ., Bonner Springs, Kan., 1983, 122.
33. **Calvert, C. A. and Losonsky, J. M.,** Pneumonitis associated with occult heartworm disease in dogs, *J. Am. Vet. Med. Assoc.,* 186, 1097, 1985.
34. **Kelly, J. D.,** Canine heartworm disease, in *Current Veterinary Therapy VII: Small Animal Practice,* Kirk, R. W., Ed., W. B. Saunders, Philadelphia, 1980, 326.
35. **Jackson, R. F.,** The activity of levamisole against the various stages of *Dirofilaria immitis* in the dog, in *Proceedings of the Heartworm Symposium '77,* Otto, G. F., Ed., Veterinary Medicine Publ., Edwardsville, Kan., 1978, 111.
36. **Atwell, R. B.,** Canine dirofilariasis — clinical update, in *The Veterinary Annual,* 26th ed., Grunsell, C. S. G., Hill, F. W. G., and Raw, M. E., Eds., Scientechnica, Bristol, 1986, 259.
37. **Atwell, R. B., Carlisle, C., and Robinson, S.,** The effectiveness of levamisole hydrochloride in the treatment of adult *Dirofilaria immitis, Aust. Vet. J.,* 55, 531, 1979.
38. **Simpson, C. F., Gebhart, B. M., Bradley, R. E., and Jackson, R. F.,** Glomerulosclerosis in canine heartworm infection, *Vet. Pathol.,* 11, 506, 1974.
39. **Casey, H. W. and Splitter, G. A.,** Membranous glomerulonephritis in dogs infected with *Dirofilaria immitis, Vet. Pathol.,* 12, 111, 1975.
40. **Underwood, P. and Harwood, P. D.,** Survival and location of the microfilariae of *Dirofilaria immitis* in the dog, *J. Parasitol.,* 25, 23, 1939.
41. **Courtney, C. H. and Jackson, R. F.,** Recurrence of microfilariae of *Dirofilaria immitis* after microfilaricidal therapy without an adulticide, in *Proceedings of the Heartworm Symposium '74,* Morgan, H. C., Ed., Veterinary Medicine Publ., Bonner Springs, Kan., 1975, 87.
42. **Tulloch, G. S.,** A new approach to the treatment of heartworm disease in dogs, in *Proceedings of the Heartworm Symposium '74,* Morgan, H. C., Ed., Veterinary Medicine Publ., Bonner Springs, Kan., 1975, 85.
43. **Garlick, N. L.,** The management of canine dirofilariasis, *Canine Pract.,* 2, 22, 1975.
44. **Garlick, N. L., Beck, A. M., and Bryan, R. K.,** Canine dirofilariasis: 547 clinical cases treated first with dithiazanine iodide then with thiacetarsamide sodium, *Canine Pract.,* 3, 44, 1976.
45. **Courtney, C. H. and Boring, J. G.,** A comparison of the reaction to thiacetarsamide before or after the use of a microfilaricide, in *Proceedings of the Heartworm Symposium '77,* Otto, G. F., Ed., Veterinary Medicine Publ., Bonner Springs, Kan., 1978, 89.
46. **Simpson, C. F. and Jackson, R. F.,** Fate of microfilariae of *Dirofilaria immitis* following use of levamisole as a microfilaricide, *Z. Parasitenkd.,* 68, 93, 1982.
47. **Mills, J. N. and Amis, T. C.,** Levamisole as a microfilaricidal agent in the control of canine dirofilariasis, *Aust. Vet. J.,* 51, 310, 1975.
48. **Blair, L. S., Malatesta, P. F., and Ewanciw, D. V.,** Dose-response study of ivermectin against *Dirofilaria immitis* microfilariae in dogs with naturally acquired infections, *Am. J. Vet. Res.,* 44, 475, 1983.
49. **Plue, R. E., Seward, R. L., Acre, K. E., Cave, J. S., Schlotthauer, J. C., and Stromberg, B. E.,** Clearance of *Dirofilaria immitis* microfilariae in dogs using 200 mcg/kg ivermectin subcutaneously, in *Proceedings of the Heartworm Symposium '83,* Otto, G. F., Ed., Veterinary Medicine Publ., Edwardsville, Kan., 1983, 153.
50. **Jackson, R. F. and Seymour, W. G.,** Efficacy of avermectins against microfilariae of *Dirofilaria immitis,* in *Proceedings of the Heartworm Symposium '80,* Otto, G. F., Ed., Veterinary Medicine Publ., Edwardsville, Kan., 1981, 131.
51. **Campbell, W. C. and Benz, G. W.,** Ivermectin: a review of efficacy and safety, *J. Vet. Pharmacol. Ther.,* 7, 1, 1984.
52. **Seward, R. L.,** Reactions in dogs given ivermectin, *J. Am. Vet. Med. Assoc.,* 183, 493, 1983.
53. **Lindemann, B. A. and McCall, J. W.,** Microfilaricidal activity of invermectin against *Dipetalonema reconditum, J. Vet. Pharmacol. Ther.,* 6, 75, 1983.
54. **Sakamoto, T., Togoe, T., Yamamoto, Y., and Kitano, Y.,** Microfilaricidal effect on fenthion applied on the skin of the dog infected with *Dirofilaria immitis, Mem. Fac. Agric. Kagoshima Univ.,* 16, 75, 1980.
55. **Christie, R. J. and Harmon, R. K.,** Evaluation of fenthion dermal application in heartworm-infected dogs, *Canine Pract.,* 8, 33, 1981.
56. **Balbo, T. and Panichi, M.,** Comparative therapeutic tests in filariasis of the dog: preliminary results, *Atti Soc. Ital. Sci. Vet.,* 25, 153, 1971.
57. **Garlick, N. L.,** Intravenous and oral medication for the elimination of heartworms *(Dirofilaria immitis), J. Am. Vet. Med. Assoc.,* 159, 1435, 1971.
58. **Pacheco, G.,** Synopsis of Dr. Seiji Kume's reports at the first international symposium on canine heartworm diseases, in *Canine Heartworm Disease: The Current Knowledge,* Bradley, R. E., Ed., University of Florida, Gainesville, Fla., 1972, 137.

59. **Jackson, R. F.,** Off the cuff, *Am. Heartworm Soc. Bull.,* 8, 1, 1982.
60. **Lombard, C. W.,** Case report: reaction to DEC in a microfilaria-positive dog, *Am. Heartworm Soc. Bull.,* 8, 5, 1982.
61. **Courtney, C. H. and Nachreiner, R. F.,** The effect of diethylcarbamazine on fertility in the dog, in *Proceedings of the Heartworm Symposium '77,* Otto, G. F., Ed., Veterinary Medicine Publ., Bonner Springs, Kan., 1978, 95.
62. **Kume, S.,** Prophylactic therapy against *Dirofilaria immitis* with *dl*-tetramisole together with some observations on its microfilaricidal and adulticidal efficacy, in *Proceedings of the Heartworm Symposium '74,* Morgan, H. C., Ed., Veterinary Medicine Publ., Bonner Springs, Kan., 1975, 68.
63. **McCall, J. W., Cowgill, L. M., Plue, R. E., and Evans, T.,** Prevention of natural acquisition of heartworm infection in dogs by monthly treatment with ivermectin, in *Proceedings of the Heartworm Symposium '83,* Otto, G. F., Ed., Veterinary Medicine Publ., Edwardsville, Kan., 1984, 150.
64. **McCall, J. W., Lindemann, B. A., and Porter, C. A.,** Prophylactic activity of avermectins against experimentally induced *Dirofilaria immitis* infection in dogs, in *Proceedings of the Heartworm Symposium '80,* Otto, G. F., Ed., Veterinary Medicine Publ., Edwardsville, Kan., 1981, 126.
65. **Otto, G. F.,** Occurrence of the heartworm in unusual locations and in unusual hosts, in *Proceedings of the Heartworm Symposium '74,* Morgan, H. C., Ed., Veterinary Medicine Publ., Bonner Springs, Kan., 1975, 6.
66. **Parrott, T. Y., Greiner, E. C., and Parrott, J. D.,** *Dirofilaria immitis* infection in three ferrets, *J. Am. Vet. Med. Assoc.,* 184, 582, 1984.
67. **Wallach, J. D. and Boever, W. J.,** Pinnipeds, in *Diseases of Exotic Animals: Medical and Surgical Management,* W. B. Saunders, Philadelphia, 1983, 72.

Chapter 9

ADVERSE REACTIONS TO DIETHYLCARBAMAZINE IN THE TREATMENT OF DIROFILARIASIS

Peter F. L. Boreham and Richard B. Atwell

TABLE OF CONTENTS

I. INTRODUCTION

It became evident soon after the antifilarial properties of diethylcarbamazine (DEC) were discovered in 1947[1] that the drug could induce adverse reactions in microfilaremic dogs and humans. In the dog, the reaction was first described by Kume[2] thus: "some dogs develop allergic shock with the symptoms of vomiting, defecation, feeble pulse, pale mucous membranes, dyspnea, and collapse resulting in death." Several studies have now been undertaken to detail the clinical reactions, and, although ways of preventing it have been elucidated, the exact mechanism has not been determined. The dogma that the reaction is allergic in nature still remains and is found in many text books without critical appraisal of the available evidence. In this chapter, it is proposed to review the clinical and pathological nature of the reaction and to examine the available information on the mechanism together with a review of its prevention and treatment.

All authors agree that the reaction will only occur in microfilaremic dogs and not in dogs with occult infections. However, some authors report that the severity of the reaction is directly correlated with the number of blood microfilariae[3,4] while others dispute this conclusion.[5] In our own studies, those dogs with higher microfilaremias showed a greater proportion of reactions to DEC. However, it was not possible to predict accurately, based on microfilaremia, which dogs would react or to infer the severity of a reaction. It was thus suggested that another factor, such as the age of the microfilariae, which relates to the period of exposure to host serum factors may be important in its induction.[6]

II. DESCRIPTION OF THE REACTION

A. Prevalence

Several studies have reported the prevalence of reactions in dogs infected with *Dirofilaria immitis* and these are summarized in Table 1. The results appear to fall into two groups. Studies conducted in North America[4,5,7] reveal a higher prevalence and also a greater severity of the clinical reaction compared to results from Japan[2] or Australia.[8] Since outbred dogs have been used in most of these studies it is unlikely that genetic variability of the host accounts for the different morbidity results raising the intriguing possibility that different parasite demes exist in different geographic locations. In an attempt to try to answer this question the analysis of isoenzymes of adult *D. immitis* from different geographic locations is being undertaken. To date, no differences have been found in isoenzymes present in *D. immitis* collected in Gifu, Japan and from various centers within Australia[9] and, in fact, very little polymorphism was found at all. So far, material from North America has not been examined.

B. Clinical Reaction

Several descriptions of the reaction to DEC in dogs[2-5,8] and man[11-15] have been published. In the dog, three distinct phases occur following an oral dose of DEC (6 to 20 mg/kg) (Table 2):

Phase 1. The dogs become depressed after 30 to 60 min and are disinterested in their surroundings and less responsive to stimuli. The dogs often defecate and may have bloody diarrhea and prolonged tenesmus. Vomition of bile-colored material and/or digested food may occur.

Phase 2. A period of cardiovascular depression occurs with the cardiac apex beat becoming less palpable and cardiac sounds less audible on auscultation. Bradycardia may also be present during this phase with a 40 to 60% decrease in heart rate. The amplitude of the femoral arterial pulse may also be markedly reduced. In severe cases, the apex beat and the arterial pulse become nonpalpable and the heart sounds inaudible. This second phase is usually short and will not be detected if clinical appraisal is not constant.

Table 1
MORBIDITY AND MORTALITY IN DOGS
INFECTED WITH *D. IMMITIS* AND
TREATED WITH DEC

No. tested	Oral dose (mg/kg)	Morbidity (%)	Mortality	Ref.
65	5.5	11 (16.9)	3 (4.6)	1
22	10	15 (68.2)	5 (22.7)	7
10	6.6	10 (100)	1 (10.0)	5
28	10	25 (89.3)	NR[a]	4
8	6	2 (25.0)	0 (0)	8
6	20	2 (33.3)	1 (16.7)[b]	8
170	6	42 (24.7)	2 (1.2)	8

[a] NR = not recorded.
[b] In severe shock when killed.

Table 2
CLINICAL SIGNS SEEN IN DOGS REACTING
TO DEC THERAPY

Phase 1	Phase 2	Phase 3
Depression	Bradycardia	Pale mucosa
Lethargy	Soft heart sounds	Poor capillary refill
Defecation	Weak apex beat	Tachycardia
Urination	Low amplitude pulse	Tachypnea
		Hepatomegaly
		Abdominal cramping
		Recumbent
		Dyspnoea
		Cyanosis
		Terminal shock

Phase 3. The dogs show signs of shock and become recumbent with pale, cool, dry mucosae, a slow capillary refill time, cool periphery, and a very low amplitude femoral arterial pulse. The liver becomes enlarged and increasingly palpable beyond the right costal arch. Tachycardia and polypnea leading to dyspnea often occur associated with mild cyanosis. Forceful, tonic, abdominal wall contractions are present and seem to be associated with abdominal pain. Three hours after oral DEC the clinical signs have usually regressed, although depression may remain for about 6 hr. At this stage the animals can walk normally and in most cases will eat and drink when offered food and water.

C. Pathology

The major histopathological changes occurring during the DEC reaction are to be found in the liver[16-18] and are detailed in Chapter 6. At necropsy the liver is dark and enlarged with fibrin clots attached to the serosal surfaces, particularly on the surface of the diaphragm. The cut surface of the liver bleeds profusely which, together with fibrin deposition, suggests passive venous congestion of the liver with effusion of high-protein lymph onto the surface.

The most obvious histological feature is hepatic congestion and hemorrhage, mainly around the central veins, associated with occasional thrombus formation. The hepatic veins, especially the secondary veins, are constricted while the lymphatics associated with the larger hepatic veins are dilated and contain highly proteinaceous lymph. An increased number of irregular, scattered, inflammatory foci are present consisting predominantly of eosinophils, but with a few microfilariae lymphocytes and neutrophils. The number of microfilariae present in the liver does not correlate with the number present in the peripheral venous blood or with the severity of the inflammatory response. However, there does appear to be an association between the clinical severity of the reaction and vascular changes in the liver.[18] Mast cells with metachromatic granules associated mainly with the larger vessels are present in small to moderate numbers in the livers in both reactive and nonreactive dogs following DEC therapy. However, no evidence of mast cell degranulation is seen in liver sections using selective stains.[18]

Blood-tinged ascitic fluid often collects in the peritoneal cavity of dogs undergoing a DEC reaction. This fluid has a similar concentration of electrolytes and proteins to plasma suggesting that effusion of plasma occurs through the hepatic capsule causing the hydroperitoneum,[17] and is associated with the deposition of fibrin clots on the liver capsule.

In general, the portal areas are not affected, although occasional congestion of the veins and dilatation of the lymphatics is seen. The hepatic lymph nodes may become edematous[16] with hemorrhage into the space around the gall bladder, between the liver and diaphragm, and into the lumen of the intestine.

D. Pathophysiology

Conflicting data exist on the changes that occur in the blood of dogs undergoing DEC reactions. In a series of 18 dogs studied during the reaction in Hawaii[4] the most notable changes were a decrease in platelet numbers and fibrinogen levels, a rise in plasma transaminases (ALT), and the appearance of fibrin monomers. The thrombocytopenia and rise in transaminases have been confirmed.[5,6] However, in the dogs in Australia, major changes in the fibrinolytic system were not evident.[6] Changes in the cellular components of the blood are consistent with hypovolemic shock. A small but consistent drop in hematocrit, hemoglobin concentration, and erythrocyte counts occurs within the first 30 min after DEC administration followed by an increase above control values. These results suggest that erythrocytes are initially being sequestered, probably in the liver, and that subsequently hemoconcentration occurs. The consistent finding of a rise in ALT 1 to 2 hr post-treatment (Figure 1a) with levels raised for up to 7 days suggests that the liver is the target organ for the DEC reaction in dogs. The plasma concentrations of lactic dehydrogenase, creatinine phosphokinase, especially isoenzyme CK1 and CK3 (Figure 1b), and alkaline phosphatase enzyme concentrations are also raised.

During the clinical reaction in anesthetized dogs, changes in blood pressure have been measured.[19] A mild increase in portal venous pressure, together with a concurrent reduction in heart rate and in systolic and diastolic arterial pressures, has been seen, presumably associated with the hypovolemic shock. Wedged hepatic venous pressures rise during the reaction indicating an increase in postsinusoidal venous pressures. Portal venous angiography has shown partial cessation of blood flow and pooling of blood with dilation of vessels suggesting reduced portal venous return to the liver, probably due to venous congestion associated with the hepatomegaly.

All the available biochemical, histological, clinical, and pathological data indicate that the primary site of the reaction in the dog is the liver, in particular, the constriction of the hepatic veins. Such constriction would lead to hepatic venous congestion, hepatomegaly and secondary hypovolemic shock associated with reduced venous return to the heart, and possibly subsequent systemic hypotension.

III. MECHANISMS OF THE REACTION

In attempting to understand the mechanisms of the reaction to DEC in the dog, the key

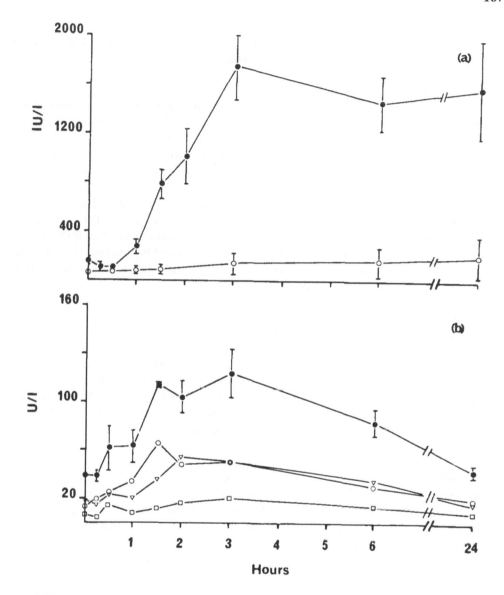

FIGURE 1. (a) Concentration of alanine aminotransferase (international units/liter, IU/ℓ) in the blood of dogs infected with *D. immitis* and treated with DEC (10 mg/kg orally). ●—● = reactive dogs (± SE), ○—○ = nonreactive dogs. (b) Creatinine phosphokinase concentration (units/liter, U/ℓ) in reactive dogs following DEC treatment. Total CK = ●—●, CK1 = ▽—▽, CK2 = □—□, CK3 = ○—○. (From Boreham, P. F. L. and Atwell, R. B., *Int. J. Parasitol.*, 13, 547, 1983. With permission.)

question to be answered is what is the mediator of the hepatic vein constriction. The possible mechanisms and available evidence will be reviewed.

A. Anaphylaxis

Although anaphylaxis (Type 1 hypersensitivity) is generally considered to be the cause of the DEC reaction, recent evidence would not support this conclusion. Histamine is the major autacoid released during anaphylaxis in the dog, and, although there are two reports of increases of histamine during the reaction (one based on a single dog[21] and the second a preliminary report[16]), a detailed study failed to detect any circulating histamine in peripheral venous blood during the reaction.[6] Two other pieces of evidence suggest that histamine and anaphylaxis are not the cause of the hepatic vein constriction. Neither H_1 or H_2 antagonists, given to 12 reactive dogs prior to DEC challenge, inhibited the reaction[20] and although mast cells are present in the liver they are not degranulated.[18]

B. Mast Cell Degranulation

It has been shown that when mast cells in the dogs are degranulated by an intravenous injection of compound 48/80 a similar series of events to the DEC reaction is seen.[3] However, clinical similarity with signs of shock does not necessarily imply similar causation. A mast cell-degranulating substance has been shown to be present in adult worms and microfilariae,[22] but the available evidence suggests that this is not the mechanism of the reaction.[23] Histamine and 5-hydroxytryptamine have not been detected in the circulation,[6] and inhibitors of both of these autacoids do not prevent the reaction.[20] One report does suggest that lodoxamide ethyl, a mast cell membrane-stabilizing drug will prevent the reaction.[22] This work could not be repeated and, additionally, two other mast cell membrane-stabilizing drugs, lodox-amide tromethamine and sodium chromoglycate, did not prevent the reaction.[20] In fact, the strongest evidence that mast cells are not involved in the reaction is provided from histological studies which show no evidence of degranulation.[18]

C. Immune Complex Activation of Biological Systems

Immune complexes, if formed in a host, can cause pathology in a number of ways, the most important of which is via the activation of complement. The activated products, C3a and C5a, are potential mediators of the reaction. No increase in circulating immune complexes was found during the DEC reaction, and concentrations of all the components of the classical pathway (C5 was not measured as this is technically difficult in the dog) and the total amount of the alternative pathway complement, as well as factors B and D, were all within the normal range for the dogs at all times during the reaction.[6] These findings make it very unlikely that immune complexes are involved in the reaction.

D. Disseminated Intravascular Coagulation (DIC)

Palumbo and co-workers[4] suggested that coagulopathy leading to DIC was one of the principle mechanisms of the reaction. Evidence to support this conclusion was provided by the decreases in platelet numbers, probably as the result of aggregation, and fibrinogen levels together with the appearance of fibrin monomers. These results could not be duplicated in Australia.[6] However, DIC could well be a secondary nonspecific event following the severe hypovolemic shock seen with DEC administration.

E. Endotoxic Shock

Some features similar to the DEC reaction in dogs do occur in endotoxin shock. However, dogs with severe endotoxin shock characteristically have mucosal hyperemia and hypoglycemia.[23] Neither of these has been seen in the DEC reaction.

F. Stimulation of the Hepatic Sympathetic Nerves

Stimulation of the pre- or postganglionic sympathetic nerves supplying the hepatic veins could result in catecholamine release and vasoconstriction. Although α adrenergic receptors are present in canine hepatic veins, neither α nor β adrenergic-blocking drugs will prevent the reaction when given to known DEC- reactive dogs prior to DEC administration.[20]

G. Endorphins

Recent studies have shown that the release of endorphins (natural opiates) is involved in hemorrhagic, endotoxic, and septic shock.[24] The DEC reaction is not prevented by the use of two endorphin inhibitors, naloxone or meptazinol, and, therefore, any such association is probably invalid.[20]

H. Release of Surface Antigens

Recent data have shown that host proteins may be bound to, and shed from, the surface of *D. immitis* microfilariae.[26] One possible explanation of the DEC reaction would be that DEC releases surface proteins from microfilariae which subsequently act directly or indirectly on the local hepatic veins. DEC is known to produce rapid changes in microfilarial surfaces, possibly involving the release of surface-acid mucopolysaccharide. A study of the surface proteins of both blood and uterine (obtained from adult, gravid, female worms) microfilariae using iodination and SDS-PAGE gel techniques showed that DEC has no effect on the surface proteins.[28]

I. Other Possible Mechanisms

Sections III.A-H above summarize current data and suggest that most of the mechanisms previously proposed to explain the DEC reaction are unlikely to be the major ones involved. The only drug which consistently prevents the reaction is dexamethasone.[20] This confirms observations in humans where steroids have been shown to prevent the Mazzotti reaction of onchocerciasis.[29,30] The observation that diazepam prevents the reaction[26] has not been confirmed.[20] Dexamethasone has many actions, but since it inhibits phospholipase A_2, which leads to the formation of leukotrienes and prostaglandins from arachidonic acid, these autacoids may be involved. Arachidonic acid metabolism via the cyclooxygenase pathway leads to the formation of prostaglandins. This pathway can be blocked by nonsteroidal, anti-inflammatory drugs. One such drug, Indemethacin, does not block the DEC reaction in dogs indicating that prostaglandins are unlikely to be involved. At present, there is little evidence that leukotrienes are important mediators of the DEC reaction or hypersensitivity reactions in the dog, unlike the guinea pig and man, but further studies on the possible involvement of these compounds are required.

Granulocytes have been demonstrated to be the key effector cells in DEC-induced clearance and killing of microfilariae. In the dogs infected with *D. immitis* the major cell is the neutrophil[31] while the eosinophil is important in human filarial infections.[32] These cells may provide the key to the DEC reaction. In humans, immediately following DEC treatment of onchocerciasis, eosinophils localize and degranulate around microfilariae in the skin and release granule major basic protein (MBP) onto or in close proximity to parasite surfaces.[33] Substances such as MBP or the biologically active compounds in neutrophils may, therefore, be important contributors to the reaction in the dog.

Another possible mechanism would be the release by DEC of an internally localized mediator from microfilariae. If this hypothesis were true these substances could be present in the blood at some stage during the reaction. Kume[2] demonstrated that the ability to react to DEC could be transferred by plasma transfusions from reactive to naive dogs. Our own unpublished studies have shown that infusion of reactive plasma (10 mℓ/kg) into DEC reactive dogs, but not nonreactive dogs, will sometimes induce a reaction. If this can be confirmed it should be possible to identify the ''mediator'' in the circulation using standard biochemical techniques.

IV. DIAGNOSIS AND CONTROL OF THE DEC REACTION

The diagnosis of the DEC reaction is based initially on a history of DEC administration to either a microfilaremic dog or one of unknown heartworm status. The signs of shock are not specific (except for the association with DEC administration), but the association of hepatomegaly, which can be easily palpated and, in some cases, is enlarged posteriorly to be in line with the prepuce, and clinical signs of shock are highly suggestive of the DEC reaction.

Clinical pathology can assist with diagnosis as microfilariae must be present for the reaction to occur and nonspecific hematological and biochemical signs associated with shock will be present. The most diagnostic factor is the increased concentrations of plasma ALT in clinical cases, these levels often persisting for several days after recovery.

It is necessary to treat dogs with high microfilaremias with caution as this is the only known predictive factor likely to be associated with a potential reaction. The value of giving an initial, small test dose of DEC is at present uncertain as dogs react at various dose rates (Table 1). Caution is particularly important with regard to postadulticide therapy and the removal of microfilariae. However, in occult cases DEC therapy can be introduced immediately so that more effective prophylaxis is ensured during the treatment course for the infection.

The DEC reaction can be reversed rapidly by infusing fluids intravenously to counteract the hemodynamic effects of shock. In experimental dogs, this response is extremely rapid and an apparently moribund dog may recover clinically in as little as 10 min following rapid and adequate fluid therapy. Fluids are administered rapidly by two catheters (if possible and if necessary) until a response is seen (reduced heart and respiration rates together with improved peripheral mucosal perfusion). Routine shock support therapy and nursing needs to be employed, e.g., ensuring diuresis, etc. Fluid therapy is reduced to maintenance levels once a response has been achieved.

Other therapy should also be considered. Dexamethasone, which has been shown to prevent the reaction experimentally when given prior to DEC administration, may be used at a dose of 1 mg/kg. However, fluid therapy is the most important aspect of treatment, and reliance on other therapy alone is not advised. Isoprenaline can also be used in the cardiac depressant stage (phase 2) of the DEC reaction, but early cases of the reaction are rarely seen in clinical practice.

In two experimental dogs, signs of caval syndrome developed after successful treatment and recovery from the DEC reaction.[34] This was presumably associated with the movement of adult worms to the right atrium during the DEC reaction, but no definite explanation for this association is available at this stage. Whether or not these occurrences were related to DEC treatment is similarly unknown. Delayed-onset shock following DEC has been infrequently observed both experimentally and in several clinical cases. One of our dogs reacted approximately 12 hr after DEC and, subsequently, was found dead the next day. Another dog reacted 6 hr after oral DEC dosage, but was successfully treated with fluid therapy.

It is probable that the true incidence of DEC reactions in the dog population is underdiagnosed as mild reactions are difficult to detect (unless constant clinical assessment is undertaken). Cases presented to a veterinary clinic will usually be the more severe cases in which the clients have observed enough of the signs of the reaction to be alarmed.

V. DEC-LIKE REACTIONS CAUSED BY OTHER DRUGS

A very important question which needs to be answered is whether therapy with other antifilarial compounds will induce similar reaction to DEC. The only drug so far studied systematically is ivermectin,[35] where dogs known to react to DEC did not react during a 3 hr period of observation to a single dose of ivermectin (50 μg/kg orally). Experience at the University of Queensland Veterinary Clinic suggested that DEC-like reactions do not occur with levamisole, although such reactions have been reported in man.[36] It has been postulated that cholinesterase inhibitors such as organophosphates and carbamates cannot be safely given to microfilaremic dogs because of the reactions they induce,[10] and intramuscular chloroquine produces a neurological disorder in microfilaremic dogs rather than the hypovolemic shock reaction.[37] At present, no mechanism of the chloroquine reaction has been determined. The metabolite of DEC, DEC-N-oxide, produces a reaction similar to those seen with DEC itself in experimental dogs.[38]

...

VI. SUMMARY

It is clear from the available evidence that adverse reactions to DEC are an important consideration in the clinical use of this drug in veterinary practice. The prevalence and severity of the reaction vary geographically, and the exact mechanism still remains a mystery. However, it is now possible to state categorically that it is not a Type 1 anaphylactic mechanism. The clinical reaction and exact site of the reaction are now well described in the dog, and adequate although nonspecific therapy is available. The design of specific therapy for the DEC reaction must await detailed elucidation of the mechanisms involved.

ACKNOWLEDGMENTS

The original work described in this chapter was undertaken with grants from the National Health and Medical Research Council of Australia and the University of Queensland.

REFERENCES

1. **Hewitt, R. I., Kushner, S., Stewart, H. W., White, E., Wallace, W. S., and SubbaRow, Y.,** Experimental chemotherapy of filariasis. III. Effect of 1-diethylcarbamyl-4-methylpiperazine hydrochloride against naturally acquired filarial infections in cotton rats and dogs, *J. Lab. Clin. Med.,* 32, 1314, 1947.
2. **Kume, S.,** Pathogenesis of allergic shock from the use of diethylcarbamazine to the canine dirofilariasis, in *Canine Heartworm Disease,* Bradley, R. E., Ed., University of Florida Press, Gainesville, Fla., 1970, 3.
3. **Garlick, N. L.,** Adverse drug effects precipitated by microfilariae of *Dirofilaria immitis, Clin. Toxicol.,* 9, 981, 1976.
4. **Palumbo, N. E., Desowitz, R. S., and Perri, S. F.,** Observations on the adverse reaction to diethylcarbamazine in *Dirofilaria immitis*-infected dogs, *Tropenmed. Parasitol.,* 32, 115, 1981.
5. **Powers, K. G., Parbuoni, E. L., and Furrow, R. D.,** *Dirofilaria immitis .* I. Adverse reactions associated with diethylcarbamazine therapy in microfilaremic dogs, in *Proceedings of the Heartworm Symposium '80,* Otto, G. F., Ed., Veterinary Medicine Publ., Edwardsville, Kan., 1981, 108.
6. **Boreham, P. F. L. and Atwell, R. B.,** Adverse drug reactions in the treatment of filarial parasites: haematological, biochemical, immunological and pharmacological changes in *Dirofilaria immitis* infected dogs treated with diethylcarbamazine, *Int. J. Parasitol.,* 13, 547, 1983.
7. **Palumbo, N. E., Perri, S. F., Desowitz, R. S., Una, S. R., and Read, G. W.,** Preliminary observations on adverse reactions to diethylcarbamazine (DEC) in dogs infected with *Dirofilaria immitis,* in *Proceedings of the Heartworm Symposium '77,* Otto, G. F., Ed., Veterinary Medicine Publ., Bonner Springs, Kan., 1978, 97.
8. **Atwell, R. B. and Boreham, P. F. L.,** Adverse drug reactions in the treatment of filarial parasites: clinical reactions to diethylcarbamazine therapy in dogs infected with *Dirofilaria immitis* in Australia, *J. Small Anim. Pract.,* 24, 695, 1983.
9. **Flockhart, H. A., Boreham, P. F. L., and Atwell, R. B.,** unpublished, 1984.
10. **Garlick, N. L. and Christy, K. E.,** Microfilaria-induced biochemical lesions in heartworm disease of canines, *Vet. Hum. Toxicol.,* 19, 14, 1977.
11. **Wilson, T.,** Hetrazan in the treatment of filariasis due to *Wuchereria malayi, Trans. R. Soc. Trop. Med. Hyg.,* 44, 49, 1950.
12. **Seo, B. S.,** Malayan filariasis in Korea, *Korean J. Parasitol.,* Suppl. 16, 1, 1978.
13. **Henson, P. M., Mackenzie, C. D., and Spector, W. G.,** Inflammatory reactions in onchocerciasis: a report on current knowledge and recommendations for further study, *Bull. W.H.O.,* 57, 667, 1979.
14. **Otto, G. F., Jachowski, L. A., Jr., and Wharton, J. D.,** Filariasis in American Samoa. III. Studies on chemotherapy against the nonperiodic form of *Wuchereria bancrofti, Am. J. Trop. Med. Hyg.,* 2, 495, 1953.
15. **Sutanto, I., Boreham, P. F. L., Munawar, M., Purnomo, and Partono, F.,** Adverse reactions to a single dose of diethylcarbamazine in patients with *Brugia malayi,* infection in Riau Province, West Indonesia, *Southeast Asian J. Trop. Med. Public Health,* 16, 395, 1985.

16. **Furrow, R. D., Powers, K. G., and Parbuoni, E. L.,** *Dirofilaria immitis.* II. Gross and microscopic hepatic changes associated with diethylcarbamazine citrate (DEC) therapy in dogs, in *Proceedings of the Heartworm Symposium '80*, Otto, G. F., Ed., Veterinary Medicine Publ., Edwardsville, Kan., 1981, 117.

17. **Atwell, R. B. and Boreham, P. F. L.,** Studies on the adverse reactions following diethylcarbamazine to microfilaria-positive *(D. immitis)* dogs, in *Proceedings of the Heartworm Symposium '83*, Otto, G. F., Ed., Veterinary Medicine Publ., Edwardsville, Kan., 1983, 105.

18. **Sutton, R. H., Atwell, R. B., and Boreham, P. F. L.,** Liver changes, following diethylcarbamazine administration, in microfilaremic dogs infected with *Dirofilaria immitis, Vet. Pathol.*, 22, 177, 1985.

19. **Atwell, R. B., Boreham, P. F. L., Euclid, J. M., and Carlisle, C.,** unpublished, 1984.

20. **Boreham, P. F. L., Atwell, R. B., and Euclid, J. M.,** Studies on the mechanism of the DEC reaction in dogs infected with *Dirofilaria immitis, Int. J. Parasitol.*, 15, 543, 1985.

21. **Desowitz, R. S., Palumbo, N. E., Perri, S. F., and Sylvester, M. S.,** Inhibition of the adverse reaction to diethylcarbamazine in *Dirofilaria immitis*-infected dogs by lodoxamide ethyl, *Am. J. Trop. Med. Hyg.*, 31, 309, 1982.

22. **Boreham, P. F. L.,** Activation *in vitro* of some biological systems by extracts of adult worms and microfilariae of *Dirofilaria immitis, J. Helminthol.*, 58, 207, 1984.

23. **Bradley, S. G.,** Cellular and molecular mechanisms of action of bacterial endotoxins, *Annu. Rev. Microbiol.*, 33, 67, 1979.

24. **Rees, M., Payne, J. G., and Bowen, J. C.,** Naloxone reverses tissue effects of live *Escherichia coli* sepsis, *Surgery*, 91, 81, 1982.

25. **Desowitz, R. S., Palumbo, N. E., and Tamashiro, W. K.,** Inhibition of the adverse reaction to diethylcarbamazine in *Dirofilaria immitis*-infected dogs by diazepam, *Tropenmed. Parasitol.*, 35, 50, 1984.

26. **Hammerberg, B., Rikihisa, Y., and King, M. W.,** Immunoglobulin interactions with surfaces of sheathed and unsheathed microfilariae, *Parasite Immunol.*, 6, 421, 1984.

27. **Gibson, D. W., Connor, D. H., Brown, H. L., Fuglsang, H., Anderson, J., Duke, B. O. L., and Buck, A. A.,** Onchocercal dermatitis: ultrastructural studies on microfilariae and host tissues, before and after treatment with diethylcarbamazine (Hetrazan®), *Am. J. Trop. Med. Hyg.*, 25, 74, 1976.

28. **Boreham, P. F. L. and Atwell, R. B.,** unpublished, 1984.

29. **Awadzi, K., Orme, M. L'E., Breckenridge, A. M., and Gilles, H. M.,** The chemotherapy of onchocerciasis. VII. The effect of prednisone on the Mazzotti reaction, *Ann. Trop. Med. Parasitol.*, 76, 331, 1982.

30. **Awadzi, K., Orme, M. L'E., Breckenridge, A. M., and Gilles, H. M.,** The chemotherapy of onchocerciasis. VI. The effect of indomethacin and cyproheptadine on the Mazzotti reaction, *Ann. Trop. Med. Parasitol.*, 76, 323, 1982.

31. **El-Sadr, W. M., Aikawa, M., and Greene, B. M.,** *In vitro* immune mechanisms associated with clearance of microfilariae of *Dirofilaria immitis, J. Immunol.*, 130, 428, 1983.

32. **Kephart, G. M., Gleich, G. J., Connor, D. H., Gibson, D. W., and Ackerman, S. J.,** Deposition of eosinophil granule major basic protein onto microfilariae of *Onchocerca volvulus* in the skin of patients treated with diethylcarbamazine, *Lab. Invest.*, 50, 51, 1984.

33. **Hammerberg, B.,** Sheathed and unsheathed microfilarial surfaces, interactions with host serum proteins in diethylcarbamazine, *Trop. Med. Parasitol.*, Suppl. 36, 5, 1985.

34. **Atwell, R. B. and Boreham, P. F. L.,** unpublished, 1983.

35. **Boreham, P. F. L. and Atwell, R. B.,** Absence of shock-like reactions to ivermectin in dogs infected with *Dirofilaria immitis, J. Helminthol.*, 57, 279, 1983.

36. **Mak, J. W. and Zaman, V.,** Drug trials with levamisole hydrochloride and diethylcarbamazine in Bancroftian and Malayan filariasis, *Trans. R. Soc. Trop. Med. Hyg.*, 74, 285, 1980.

37. **Desowitz, R. S., Palumbo, N. E., and Perri, S. F.,** The occurrence of an adverse reaction to chloroquine in *Dirofilaria immitis*-infected dogs, *Tropenmed. Parasitol.*, 34, 27, 1983.

38. **Hill, D. E.,** Studies on Various Aspects of Canine Heartworm Disease with Particular Reference to the DEC Reaction, B. Vet. Biol. thesis, University of Queensland, Brisbane, Australia, 1984.

Chapter 10

IMMUNOLOGY OF CANINE DIROFILARIASIS

Robert B. Grieve

TABLE OF CONTENTS

I. INTRODUCTION

The canine immune response to *Dirofilaria immitis* is one of the most intriguing aspects of canine dirofilariasis. Dogs which are otherwise immunologically competent will support live adult worms and microfilariae for several years. Furthermore, dogs can be continually infected and there is no evidence of protective immunity in adulticide-cured or older dogs. Although there is little evidence of an active protective immune response under natural conditions, infected animals are, nevertheless, immunologically responsive to infection.

An understanding of the various facets of the immune response to *D. immitis* enables exploitation of the immune response for the purpose of immunodiagnosis. Further, since immunologic mechanisms underlie much of the pathophysiology of canine heartworm disease, knowledge of these mechanisms may be important in development of new treatment strategies and the management of heartworm-infected dogs. Finally, there have been limited, but encouraging, studies which suggest that immunoprophylaxis against canine heartworm infection may be feasible.

II. IMMUNE RESPONSES IN EXPERIMENTALLY INFECTED DOGS

A. Antibody Responses

Humoral immune responses to experimental *D. immitis* infections in dogs have been studied by various investigators.[1,2] In those studies which measured antibody to adult worm antigens, specific antibody could be initially detected as early as 2 weeks after infection[1] and as late as 11 weeks after infection[3] depending upon the nature of the antigen and the serologic assay employed. Generally, antibody levels have been reported to increase until patency when, in many instances, antibody levels decrease transiently and either remain at constant levels or decrease over the course of study.[4-7] Secondary infections administered after the patency of the first infections do not produce anamnestic response, but may delay the decrease in antibody titers observed in dogs infected only once.[4]

There have been few studies concerned primarily with the kinetics of isotype-specific antibody responses following experimental infection.[6,8,9] Weiner and Bradley[8] demonstrated that IgM was active over the course of *D. immitis* infection. It is interesting that IgM, which is typically observed early and transiently in an immune response, persists at relatively high levels in chronic dirofilariasis. Reaginic antibody has been shown to appear at 65 days after infection, a time which corresponds to the fourth molt.[6,9] In one study, specific reaginic antibody levels persisted over the course of infection.[9] In a subsequent report from the same laboratory, reaginic antibody levels transiently disappeared between 97 days after infection and patency.[6]

Antibody responses to microfilarial surface antigens following experimental infection may vary considerably among different dogs.[7,10,11] Those dogs which are capable of generating an antibody-dependent immune-mediated killing of microfilariae typically produce antibody to microfilarial surface antigens which is evident in the circulation at a time which corresponds to the initial release of microfilariae.[7,10] In general, it has been reported that antibody to microfilarial surface antigens is not present in dogs during the prepatent period or in microfilaremic dogs.[10] However, in some dogs, this antibody may appear transiently concomitant with patency,[7] and the antibody may be present in as many as 18% of microfilaremic dogs.[12]

B. Cellular Responses

Hayasaki and co-workers assessed the ability of lymphocytes from infected dogs to elaborate soluble products which would inhibit macrophage migration in vitro.[9] They were unable to demonstrate any effect with lymphocyte culture supernatants from infected dogs as compared to uninfected dogs.

Peripheral blood lymphocytes (PBL) from experimentally infected dogs fail to respond to *D. immitis* antigens in in vitro lymphocyte transformation assays at a level greater than PBL from noninfected control dogs.[5,13] In fact, Weil et al. reported higher levels of antigen-induced blastogenesis in noninfected dogs as compared to infected dogs.[13] In addition to the nonresponsive state of PBL to *D. immitis* antigens, Grieve et al. observed significantly decreased mitogen-induced lymphocyte transformation in infected dogs.[5] Using similar methods, Weil et al. reported lowered, but not significantly decreased, responses of PBL from infected dogs to phytohemagglutinin.[13] The most recent study concerned with lymphocyte function in dogs with experimental infections demonstrated that B cells from PBL of microfilaremic dogs were impaired in their ability to respond to a T-dependent antigen.[14] Microfilaremic dogs immunized with burro erythrocytes (BE) had equivalent levels of BE-specific total antibody as compared to noninfected control dogs. However, BE-specific antibody in infected dogs was principally associated with persistently high levels of IgM, whereas BE-specific antibody responses in uninfected dogs followed classical IgM to IgG kinetics. Furthermore, in in vitro assays, primary IgM and IgG responses of B cells from microfilaremic dogs to BE were significantly suppressed. Numbers of B cells producing BE-specific IgG following secondary immunization were significantly depressed in microfilaremic dogs; however, the number of B cells producing BE-specific IgM was not different between infected and uninfected dogs. Results of this study interpreted in conjunction with the data of Weiner and Bradley[8] seem to suggest that dogs with chronic infection may have relatively high levels of parasite-specific IgM due to an impaired ability of infected dogs to shift from an IgM- to IgG-specific antibody following infection.

Various data summarized here suggest that dogs with microfilaremic *D. immitis* infection are immunologically impaired. These findings may explain, in part, the persistence of this infection in dogs. Furthermore, immune suppression may predispose dogs to other diseases and, therefore, may warrant consideration in the overall pathogenesis associated with *D. immitis* infection.

III. IMMUNODIAGNOSIS

A. Rationale

The principal reason for immunodiagnosis of *D. immitis* infection is the diagnosis of occult dirofilariasis. For the purposes of this chapter, occult dirofilariasis is defined as adult, immature or mature, *D. immitis* infection without discernible microfilaremia when standard, acceptable microfilaria detection techniques are used for diagnosis.

There are various reasons for occult infections. Rawlings and co-workers[15] described four types of occult infection. These different types included prepatent infections, unisexual infections, infections in which male worms were killed with anthelmintic, and infections rendered occult because of immune-mediated destruction of microfilariae. The prepatent infections described by these authors consisted of a group of seven dogs which all had severe alterations on pulmonary arteriograms prior to patency.

Unisexual infections would generally be expected to produce disease in proportion to the number of adult worms in an individual dog. Since there are usually few worms associated with naturally occurring unisexual infections there may not be a significant risk of extensive disease as a result of chronic infection in these dogs. However, up to 30% of dogs with occult infections in a hyperendemic area may have naturally occurring unisexual infections.[11] That fact, in conjunction with the possibility of spontaneous or drug-induced thromboembolic episodes, dictates that natually occurring unisexual infections be afforded significant consideration.

Drug-induced unisexual infections may be relatively common, and the extent of disease associated with these infections may be expected to be quite variable. Rawlings and co-workers[15] and Grieve and Knight[7] have recently reported instances where experimentally infected dogs were treated with adulticide and microfilaricide on multiple occasions and, at necropsy, dogs which had no evidence of microfilaremia were harboring adult female *D. immitis*. Research by Blair et al.[16] has demonstrated that thiacetarsamide is most effective against precardiac larvae, adults more than two years of age, and male worms. As suggested by Rawlings et al,[15] "clearance of microfilariae . . . following adulticide and microfilaricide treatment is usually interpreted as successful elimination of *D. immitis*." However, although clinical signs may resolve, many animals may still harbor infections which will continue to contribute to chronic disease as well as to the possibility of unexpected thromboembolic episodes.

Immune-mediated destruction of microfilariae acocunts for yet another type of occult infection. This type of occult infection is frequently associated with dogs which have considerable heartworm-related disease. The presence of circulating antibody to microfilarial surface antigens is considered to be indicative of this type of infection.[10,17] Immunologic mechanisms which account for stage-specific killing of microfilariae will be described below.

Finally, ectopic infections may be of significance in immunodiagnosis of dirofilariasis. These infections result from aberrant migration of developing larvae and, accordingly, adult worms may be observed in a variety of locations outside the cardiopulmonary system.[18] Disease associated with ectopic infections will be related to the site of infection; many ectopic infections, however, do not result in noticeable disease. The importance of these infections to immunodiagnosis of dirofilariasis is due to the belief that a single worm in an ectopic location could elicit a measurable antibody response or produce a demonstrable level of circulating antigen. Therefore, ectopic infections potentially confound interpretation of a positive immunodiagnostic result.

B. Assays for Antibody to *Dirofilaria immitis*

There have been numerous published reports on the development of immunodiagnostic assays for detection of antibody to *D. immitis*. Literature on immunodiagnosis of *D. immitis* infection is not comprehensively reviewed here. Alternatively, certain general characteristics of different tests will be described by reviewing the attributes of typical assays.

Antigens which have been used for serodiagnosis of canine dirofilariasis can, to a large extent, be categorized as adult or microfilarial antigens. To generalize further, the most widely used antibody-detection assays usually employ some derivative of an aqueous-soluble extract of adult worms or surface antigens on intact microfilariae.

Crude aqueous-soluble adult-worm extracts have been utilized in a variety of immunodiagnostic assays including assays of cell-mediated immune responses in dogs[5,9,13] and guinea pigs,[19-22] as well as antibody responses.[1,2,4,6,8,9,13,23-25] It has been recognized, however, that crude antigens from adult filarial worms are notoriously cross reactive with various other infections.[26] This fact has limited the specificity and, consequently, the usefulness of assays employing crude adult antigens for serodiagnosis of occult dirofilariasis. In assessing the complexity of crude aqueous-soluble extracts of *D. immitis* adults, Wheeling and Hutchison[27] demonstrated a minimum of 27 protein bands by disk electrophoresis. Research reported by Sawada and co-workers[28,29] and Takahashi and Sato,[30] for example, have demonstrated even more extensive biochemical and antigenic complexity in crude extracts of adult worms.

Various researchers have attempted to use purified fractions of crude aqueous-soluble adult-worm extracts to improve the specificity of serodiagnostic assays. Mantovani and Kagan[31] reported a high degree of specificity using an antigen purified with modifications of the scheme used by Sawada et al.[28] A trichloracetic acid-soluble portion of an aqueous-soluble somatic extract of adult worms was fractionated by gel filtration and ion exchange chromatography. A single fraction was obtained which, when used in indirect hemagglutination or skin-testing, permitted discrimination of *D. immitis*-infected dogs vs. *D. repens** or *Dipetalonema* sp.-infected dogs.[31] A similar antigen-fractionation method was used to obtain antigen for incorporation into inherently sensitive, quantitative, indirect immunofluorescence[32] and indirect enzyme-linked immunosorbent assays (ELISA)[3]; promising initial results were obtained in both assay systems. Further evaluation of this antigen in an ELISA has demonstrated that the ELISA employing this antigen is extremely sensitive and relatively specific.[11,33] Evaluation of the assay using sera from dogs from hyperendemic areas demonstrated the assay was capable of detecting all dogs with occult dirofilariasis[11] and that there was no pattern of cross reactivity with intestinal parasitism.[33] However, cross reactivity with sera from a portion of dogs with *Di. reconditum* was noted.[33]

Recently Scott and co-workers[34] have employed a different scheme for fractionating crude aqueous-soluble extracts of adult worms. They employed lectin chromatography to purify an antigen which showed both sensitivity and specificity when it was incorporated into an ELISA.

The detection of antibody to microfilarial surface antigens has been widely used for diagnosis of occult infections which result from immune-mediated destruction of microfilariae.[10,17] Typically, intact microfilariae are recovered from the circulation of microfilaremic dogs, subjected to chemical fixation, and used as antigen in an indirect fluorescent antibody assay (IFA). There is an excellent correlation between the presence of circulating antibody to microfilarial surface antigens and the presence of immune-mediated, occult heartworm disease in the same dog.[10,17] Although this assay tends to be specific, this IFA is relatively insensitive.[11,35] Presumably, this lack of sensitivity is due to the fact that most dogs with occult heartworm infection, as defined above, do not have occult infections as the result of immune-mediated destruction of microfilariae. Furthermore, although a positive test may generally be considered as conclusive evidence of occult dirofilariasis, it should be noted that sera from some dogs may be transiently positive concomitant with the onset of patency.[7]

C. Assays for Circulating *Dirofilaria immitis* Antigen

Evidence for *D. immitis* antigen in body fluids was first reported 20 years ago.[36] Soluble antigens in the urine of dogs reacted with rabbit antibody generated against a saline extract of *D. immitis*.[36] The fraction which was antigenic was partially isolated and characterized as a glycosylated protein which appeared to share properties with an antigen recovered from in vitro culture of *D. immitis*. Unfortunately, it is not possible to determine if the in vitro culture system consisted of adult worms, microfilariae, or both.[36] In subsequent work, Yamanouchi fractionated urine from microfilaremic dogs by gel filtration; he reported that all of the antigenic activity was present in the first of two fractions eluted from a Sephadex® G-50 column.[37]

Desowitz and Una were able to demonstrate soluble antigen in the sera of two out of five microfilaremic dogs examined.[38] Circulating antigen was evident when sera from microfilaremic dogs and rabbit antisera to soluble *D. immitis* extracts were reacted together by counterimmunoelectrophoresis.

Hamilton and Scott developed an immunoradiometric assay for the detection of soluble circulating *D. immitis* antigen.[39] Although the assay was capable of detecting antigen in 42% of microfilaremic dogs, the ability of the assay to detect antigen was limited by the presence of antibody in serum, presumably due to the formation of immune complexes.

Weil and co-workers[40] used methods similar to those reported by Desowitz and Una[38] to demonstrate circulating antigen in both microfilaremic and occult dogs. They did not observe this circulating antigen in dogs with *Di. reconditum* microfilaremia. Further, there was a

---

* When abbreviated, *D.* will refer to *Dirofilaria* and *Di.* to *Dipetalonema*.

correlation observed between the serum parasite antigen content and the number of adult worms present in individual dogs. In subsequent work, Weil et al.[41] developed monoclonal antibodies to the circulating antigens detected in their counterimmunoelectrophoresis system. They demonstrated that they had generated monoclonal antibodies to two epitopes shared by two major circulating antigens. Interestingly, the epitopes recognized were shown to be carbohydrate which is consistent with the previous observations of Saito[36] on *D. immitis* antigens in the urine of infected dogs. Antigenemia, as detected by these two monoclonal antibodies, was initially detected 6 months after infection. This observation and localization of the corresponding epitopes within the adult worms indicated that the circulating antigens were specific to the adult stage. The epitopes were also observed within microfilariae, however. The two monoclonal antibodies were incorporated into a direct ELISA for detection and measurement of circulating antigen. This ELISA was evaluated for sensitivity with sera from 9 dogs with occult *D. immitis* infection and with sera from 37 dogs with microfilaremic *D. immitis* infection. All but one of the sera, which was from a microfilaremic dog, was positive by this assay. There was no cross reactivity observed with sera from 26 uninfected dogs or with sera from 20 dogs with *Di. reconditum* microfilaremia. The quantity of circulating antigen and the worm burden showed direct proportionality similar to the correlation described previously by Weil et al.[40]

D. Evaluation of Immunodiagnostic Assays

While there have been numerous advances within the past several years in the development of new immunodiagnostic capabilities for canine heartworm infection, evaluation of assays has, in general, been inadequate. In many instances, claims for test accuracy are based upon very modest numbers of sera from animals with *D. immitis* and from uninfected animals. In some instances, evaluation includes sera from animals with gastrointestinal parasitism or microfilaremic *Di. reconditum* infections. Evaluations of this extent may provide interesting preliminary information and indicate whether the test is sufficiently accurate for further investigation.

It is difficult to recreate every scenario in the laboratory which may produce an animal with cross-reacting antibody. Numerous infections and neoplastic disease, alike, must be evaluated for cross-reactive potential. Furthermore, certain disease agents should be very thoroughly evaluated. For example, sera from *Di. reconditum* microfilaremic dogs is often used in evaluating the specificity of an assay. It is recognized, however, that cross reactivity to *D. immitis* antigens in sera from dogs with experimental *Di. reconditum* infection assayed by one particular serologic test may be restricted to a limited time during the course of infection.[42] Therefore, evaluations which employ sera from dogs with chronic microfilaremic *Di. reconditum* infections may underestimate the potential for cross reactivity with that infection. Similarly, evaluations employing sera from dogs with gastrointestinal parasitic infections may underestimate cross reactivity associated with a particular infection because sera are collected when the infection is patent rather than when larvae are migrating in tissues during prepatent infection.

Any test which is advocated for use in immunodiagnosis of dirofilariasis, therefore, should be rigorously evaluated using adequate numbers of sera from animals which live outside the controlled laboratory environment. Appropriate field evaluations, however, are difficult. Frequently, adequate numbers of sera are available from live dogs living in endemic areas, but without information on definitive infection status obtained by necropsy it is very difficult to evaluate the performance of an assay.[24,25,33] Only recently, the validity of certain widely used assays has been investigated using numerous sera from dogs with necropsy examination.[11,35,43,44] Furthermore, routine necropsy examination is not always able to detect recently terminated infections, prepatent infections, or ectopic infections.

The evaluation of an immunodiagnostic test, therefore, should incorporate several important measures once an assay has been initially developed. Performance of a test should be assessed in the laboratory using parameters such as precision, reproducibility, and parallelism.[45] Modest numbers of sera from dogs with *D. immitis* infection and from dogs with potentially cross-reactive infections should be assayed with the test and, if results are favorable, large-scale seroepidemiologic evaluation using well-characterized sera is essential. Finally, results of the evaluation of an assay should be reported in terms of sensitivity and specificity, measures which are familiar to most readers. In addition, whenever possible, results should be reported to include the quality of an assay within the population tested based upon positive and negative predictive values.[11] Positive and negative predictive values are the probabilities that a positive or a negative result, respectively, is correct within the population tested. The sensitivity and specificity of an assay and the prevalence of infection are all taken into account in determining predictive values.[46,47] For example, the positive predictive value is calculated as follows:

$$\text{Positive predictive value} = \frac{(\text{Sensitivity})(\text{Prevalence})}{(\text{Sensitivity})(\text{Prevalence}) + (1 - \text{Specificity})(1 - \text{Prevalence})}$$

Simply stated, a given assay with certain inherent sensitivity and specificity characteristics could provide a result which could be interpreted very differently (i.e., be more or less valuable) depending upon the prevalence of infection in the local population. While this concept is important to all diagnostic testing, it is especially important in immunodiagnosis of canine dirofilariasis because characteristics of sensitivity and specificity are widely variable among the most commonly used assays and because degrees of endemism are often very different among different locales within a relatively discrete geographic region.[48]

E. Potential New Approaches for Development of Immunodiagnostic Assays

In addition to the diagnosis of occult dirofilariasis, in general, there are other specific diagnostic questions which could potentially be answered with immunologic assays. There is good evidence to indicate that standard anthelmintic regimens for treatment of dirofilariasis are often less than totally effective.[7,15,16] It has been demonstrated that, depending upon the assay which is used, treatment success can be determined by monitoring levels of specific antibody following adulticide treatment.[7] Since levels of certain circulating antigens are directly proportional to the adult worm burden,[40,41] it may be possible to determine the relative efficacy of adulticide treatment by use of antigen detection assays. Further, since circulating antigen, in the absence of living worms, may disappear from the circulation faster than antibody, it may be possible to determine treatment success faster by antigen detection assays than by antibody measurement.

More accurate antibody detection assays are entirely feasible. Relatively little research has been concerned with purification of antigens for use in immunodiagnosis. Scott et al.[34] have shown encouraging preliminary data on the sensitivity and specificity of lectin-purified antigens. Research on assays employing a trichloracetic acid-soluble antigen purified by gel filtration and ion-exchange chromatography has demonstrated that the semipurified antigen may have value in antibody detection assays.[7,11,33] That antigen, however, has only been partially purified; additional fractionation may provide a more specific antigen preparation. Figure 1 illustrates this first-generation semipurified antigen (DISA-I) analyzed by gradient pore polyacrylamide gel electrophoresis (5 to 17.5%) in the presence of sodium dodecyl sulfate under reducing and nonreducing conditions. By comparison to a crude, soluble, somatic extract of male *D. immitis* it is clear that the antigen is relatively noncomplex; however, additional fractionation is still possible.

Various researchers have documented the occurrence of a parasite-specific IgE response

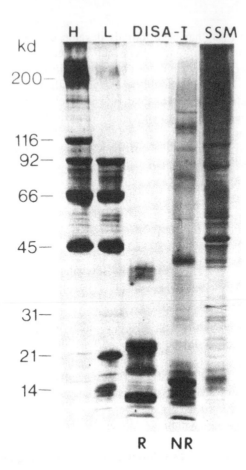

FIGURE 1.   Results of sodium dodecyl sulfate gradient pore polyacryl-amide gel electrophoresis of *Dirofilaria immitis* antigens. The gradient employed was 5 to 17.5% polyacrylamide. H = heavy molecular weight standards, L = light molecular weight standards, DISA-I = adult-derived semipurified antigen (R = reducing conditions, NR = nonreducing conditions), SSM = crude aqueous soluble somatic extract of adult male *D. immitis*. Constituent proteins were silver stained.

in *D. immitis*-infected dogs.[6,13,49] In addition, there have been a series of reports which document the properties of a characterized, purified allergen obtained from *D. immitis*.[50-56] However, there have been no published accounts of the development of an allergen-IgE-based assay for immunodiagnosis of filariasis. This approach would seem particularly worthy in view of the recent report by Weiss and co-workers[57] which established that IgE antibodies are more specific than IgG antibodies in human onchocerciasis and lymphatic filariasis.

Antigens with potential for antibody-detection assays could be isolated from other "compartments" not usually investigated, such as surface-associated carbohydrates,[58] lipids,[59] surfactant-soluble antigens[60] and excretory-secretory products of microfilariae,[61] adults,[11,62,63] and developing larvae[64-66] maintained in vitro. Antigens solubilized with Triton® X-100 from aqueous-insoluble material, for example, were reactive with sera from dogs with occult infections, but were unreactive with sera from microfilaremic dogs.[60]

In addition to separating antigens on the basis of allergenic reactivity and physicochemical characteristics, selection of relevant antigens on the basis of immune reactivity is a promising approach to isolation of antigens for antibody detection assays. The immunoblotting technique which permits identification of antibody-reactive antigens which have been separated according to molecular weight or isoelectric focusing has been extremely valuable in identification of relevant antigens.[11,67] Immunoblotting was used to demonstrate that different antigen-recognition patterns were evident in antibody obtained at different times during infection and in antibody collected from microfilaremic dogs vs. that collected from dogs with occult infection.[11,67] Boto and co-workers[68] used immunoprecipitation methods to demonstrate that several antigens derived from *D. immitis* microfilariae were recognized by antibody from infected dogs as well as from human patients with onchocerciasis. Both immunoblotting and immunoprecipitation can be valuable in identifying antigens recognized by certain sera and for identifying sera/antibody which has definable antigen-recognition characteristics. If an antibody source which recognizes relevant antigens can be identified, affinity chromatography can be used to recover workable quantities of the desired antigen. This method has been used to a limited extent to purify *D. immitis* antigens for use in diagnosis of zoonotic dirofilariasis.[69,70] Although the use of polyclonal antibody as a ligand in affinity chromatography may select for both cross reactive and specific antigens, the use of monoclonal antibodies in affinity purification schemes offers real promise.

Antigen-detection assays should prove to be extremely powerful tools in diagnosis of canine dirofilariasis. It should also be emphasized, however, that antigen-detection assays in general are not superior in specificity to antibody-detection assays *a priori*. The assay described by Weil et al.[41] appears to be very promising in terms of sensitivity and specificity. One distinct advantage of that assay is stability of the antigen following treatment to disassociate immune complexes. Other assays have been less successful with problems associated with immune complexes[39] or particular assay formats.[71] Development of additional antigen-detection assays could be specifically concerned with investigations of certain parasite metabolites. In vitro-derived proteins and glycoproteins would certainly be probable candidates for soluble circulating antigens. Development of assays for parasite-derived volatile fatty acids in body fluids has also been proposed.[72]

Finally, in addition to improved antigen- and antibody-detection assays, it seems likely that an additional immunologic phenomenon, namely the idiotype/anti-idiotype network, which has been exploited for immunodiagnosis of malaria[73] could be used for diagnosis of dirofilariasis. Jerne proposed that the immune system consists of a network of idiotypes and anti-idiotypes which are involved in immune regulation.[74] This hypothesis dictates that an antibody with an idiotype in the binding site may evoke an anti-idiotypic antibody which would behave structurally as the original antigen. The presence of anti-idiotypic antibodies has been documented in various systems. For example, spleen cells producing anti-idiotypic antibody related to parasite antigen have been observed in mice infected with *Schistosoma mansoni*, another metazoan parasite.[75] Similarly, the demonstration of *D. immitis* antigen-related auto-anti-idiotypic antibody could theoretically be diagnostic of infection. In addition, anti-idiotypic antibodies could be used instead of antigen to probe for *D. immitis*-specific antibody. These assays are potentially very sensitive, and the theoretical specificity afforded by this type of assay is extraordinary.

IV. IMMUNE-MEDIATED KILLING OF MICROFILARIAE

Although dogs can apparently be reinfected with *D. immitis* over the course of a life-span and no evidence presently exists of a naturally occurring complete protective immune response against developing larvae or adults, certain dogs are capable of effective immune-mediated destruction of microfilariae. The phenomenon is dog-dependent with only a portion of an infected population developing this immunity. Interestingly, if dogs with occult infection are treated and reinfected, the resultant infection will also be an occult infection.[10] Immunity is stage specific: apparently, microfilariae are killed, but developing larvae and adults remain unharmed. Considerable lung pathology is associated with this phenomenon,[76] and dogs with this form of immune-mediated occult heartworm infection frequently present with a great deal of disease.[77,78]

There is no obvious reason to explain why some animals will spontaneously develop the ability to immunologically kill microfilariae in a parasite stage-dependent manner. Wong and co-workers[17] have suggested that transplacental transfer of microfilariae into pups from bitches with high levels of microfilaremia could be a mechanism for sensitization of dogs to microfilarial antigens prior to infection. It would also seem possible, and consistent with the hypothesis, that soluble circulating microfilarial antigen could achieve the same result. Furthermore, since pups may acquire anti-*D. immitis* antibody with colostrum,[79] passive transfer of microfilaria antigen-related auto-anti-idiotypic antibody may also sensitize dogs to the corresponding microfilarial antigen. It is also tempting to speculate that the ability of certain dogs to respond to intact microfilariae, soluble microfilarial antigens, or anti-idiotypic antibody is somehow genetically associated.

In lieu of an understanding of what mechanism predisposes certain dogs to the ability to immunologically kill microfilariae there is some information about the effect of mechanisms for immune killing. The phenomenon of immune-mediated destruction of microfilariae in dogs with adult worms of both sexes can be experimentally induced. Wong[80] demonstrated that heartworm-naive dogs could be effectively immunized against microfilariae by repeatedly inoculating dogs with living microfilariae. Further, sera from those immune dogs would suppress microfilaria production in adult worms maintained in vitro, and, upon intravenous immunization, the immune serum would decrease the levels of microfilaremia in infected, microfilaremic dogs. While this study could not definitively demonstrate that microfilariae were being killed rather than being sequestered somewhere out of the peripheral circulation, subsequent work showed that heartworm-naive dogs immunized the same way which were challenged with infective larvae would develop occult dirofilariasis with histologic evidence of microfilarial death.[17] Rawlings and co-workers[15] removed adult female worms from dogs with immune-mediated occult dirofilariasis and transplanted the worms into uninfected dogs. Subsequent to transplantation, microfilariae were observed in the circulation of three of four dogs which received transplanted worms. This work demonstrated that female worms, which constitute this type of infection, are capable of producing microfilariae and further validates the concept that an active process in the dogs with immune-mediated occult infections is responsible for the fact that microfilariae are not evident in the circulation.

Weil and co-workers[81] inoculated radiolabeled microfilariae into uninfected, microfilaremic and occult dogs which had antibody to microfilarial surface antigens. Microfilariae rapidly disappeared from the circulation following inoculation into dogs with occult infections. By histopathologic examination, by determination of radioisotope concentrations in various tissues, and by external radioisotopic scanning, the authors concluded that microfilariae were killed in the lungs.

In vitro studies on immune mechanisms associated with *D. immitis* microfilarial killing demonstrate that neutrophils, antibody, and complement may all be important effectors in immune-mediated microfilarial killing.[82,83] El-Sadr et al.[82] have reported that microfilariae are optimally killed by neutrophils in the presence of complement and the IgM fraction of serum from a dog with occult infection. In a similar study, Rzepczyk and Bishop[83] corroborated the findings of El-Sadr et al.[82] that neutrophils were the principal effector cell involved in microfilarial killing. Further, they demonstrated that neutrophils which adhered to microfilariae mediated toxicity by both hydrogen-peroxide release and degranulation. Although they observed both IgM and IgG on the surface of microfilariae, IgM appeared to bind preferentially to the crypts, or annulae, on the cuticle surface. Correspondingly, they noted that neutrophils appeared to adhere preferentially in the same area on the microfilariae. By contrast, Weil and co-workers[81] reported that the IgG fraction of serum from dogs with immune-mediated occult infection was responsible for in vitro leukocyte binding to microfilariae. While each of these reports[81,83] consistently observed antibody binding to microfilarial surface antigens when microfilariae were incubated in sera from dogs with occult infection, they did not find these antibodies in sera from uninfected or microfilaremic dogs.

In an interesting series of experiments, Hammerberg and co-workers[84] were able to expand upon previous observations of antibody binding to microfilarial surfaces. They demonstrated that microfilariae recovered from dogs had innately bound immunoglobulin on the surface. In addition, microfilariae which were free of immunoglobulin were capable of binding IgM and IgG from noninfected dogs, microfilaremic dogs, and from dogs immunized with living microfilariae. Perhaps the most exciting finding in this study was that microfilariae were able to shed antibody from their surfaces. Furthermore, immunoglobulins were shed in a different manner depending upon the source of antibody. IgM and IgG from uninfected dogs were shed within 2 to 5 min, whereas IgM was lost in less than 5 min, but IgG was lost at a more gradual rate when sera from microfilaremic dogs was added to live microfilariae. Immunoglobulins of both classes derived from dogs immunized with live microfilariae remained attached to microfilarial surfaces for at least 1 hr of in vitro maintenance. These authors note that it was not determined whether immunoglobulin was binding antigen or was attached by an Fc receptor on the microfilarial surface. Either situation is possible and both may occur depending upon the source of the serum, i.e., from microfilaremic or immune-mediated occult infections. This work, through the use of live, metabolically inhibited microfilariae, has clearly demonstrated the necessity of considering the dynamics of the microfilarial surface as well as its ability, at any given time, to bind antibody and, subsequently, leukocytes. The presence of antibody from microfilaremic dogs which is shed relatively fast provides possible explanation of why other researchers did not observe antibody on microfilarial surfaces using serum from microfilaremic dogs.

There is very little additional information on microfilarial surface properties. It has been demonstrated histochemically that *D. immitis* microfilariae have carbohydrate on the surface.[58,84] Microfilariae derived from the uterus of female worms as well as from peripheral circulation will bind antibody from dogs with immune-mediated occult dirofilariasis.[15] However, there may be differences in the extent of antigenicity observed on the surface of intrauterine microfilariae as opposed to circulating microfilariae.[85] Complement has been shown to bind directly to the surface of microfilariae in the form of C3,[82,86] properdin,[86] and C5[86] and to be depleted from serum by intact microfilariae and microfilarial products obtained in vitro.[86] Carbohydrates on the surface of microfilariae, and subsequently shed by microfilariae maintained in vitro, were reported to be associated with complement reactivity.[86]

In addition to the lung pathology associated with immune-mediated destruction of microfilariae in occult dirofilariasis, glomerular damage in dogs has been shown to be directly related, at least in part, to the immune response to *D. immitis*.[87-89] Deposits of IgG have been observed within the glomerular basement membrane,[87] and it has been reported that the amount of immune-complex material present in the glomerulus is proportional to the intensity and duration of microfilaremia.[88] Microfilariae have been shown to be connected with capillary endothelial cells by narrow cytoplasmic bands which suggests that immune complexes in the glomerular basement membrane may form *in situ* subsequent to the deposition of antigen by microfilariae.[88] Antibodies reactive with adult-worm antigen could be eluted from kidneys of dogs experimentally infected with *D. immitis*.[89] Although it is unlikely that the antigens used to demonstrate these antibodies were adult-stage specific, it is entirely possible that adult-associated antigens are deposited in the glomerular basement membrane with subsequent *in situ* complex formation or that soluble circulating immune complexes with adult- and microfilaria-associated antigens[41] accumulate during the course of infection.

Future attempts to elucidate the nature of the antigens involved in these complexes could begin with more thorough investigation of the antigens present in the urine of infected dogs.[36,37] The source of antigens which contribute to immune complexes may be determined by using immunocytochemical methods[41] or by studying immune complexes in rodents with infections constituted only of microfilariae[90-92] or adults.[93,94]

V. PROTECTIVE IMMUNE RESPONSES

Dogs with previous or existing adult *D. immitis* infections can be reinfected; however, there is limited evidence of concomitant immunity. Total average adult-worm numbers in dogs are remarkably similar to dogs living in endemic areas, and infection levels in relation to exposure to infective larvae do not appear to be additive.[48] Although it is necessary to exercise caution when interpreting data from dogs obtained from pounds, it appears that some mechanism, perhaps immunologic, serves to limit the population of adult worms in individual dogs.

In addition to limited epidemiologic data which suggest that a level of protective immunity may occur under natural conditions, there is a body of experimental evidence which demonstrates that developing larvae can be killed by immunologic mechanisms. Wong and co-workers used gamma radiation-attenuated infective larvae to immunize dogs.[95] Dogs were challenged with intact, nonirradiated infective larvae, and immunized dogs showed variable but marked levels of protection when compared to nonimmunized controls.

Blair and Campbell[96] immunized ferrets by infecting animals with third-stage larvae and terminating infections 2 months later with ivermectin treatment. This regimen was repeated twice. Following challenge with infective larvae, four immunized ferrets yielded an average of 0.5 worms each, whereas 14 nonimmunized ferrets permitted development of an average of 6.6 worms each. In a subsequent study conducted by Blair and co-workers,[97] dogs were immunized against developing larvae using a regimen of two chemically abbreviated infections similar to that used in ferrets.[96] Dogs immunized in that manner averaged 82% fewer worms than nonimmunized controls. Although neither chemically abbreviated infections nor immunization with radiation-attenuated larvae provide a thorough or practical means for immunoprophylaxis, the data demonstrate that protective immunization against dirofilariasis is possible.

Research with other parasitic nematodes would suggest that surface antigens on developing larval stages may be relevant to protective immune responses. In a very recent study, surface antigens of *D. immitis* third-stage larvae were investigated.[105] One major protein, with a molecular weight of 35 kdaltons, was evident following immunoprecipitation of radiolabeled surface proteins from live larvae. Both rabbit antisera generated to third-stage larvae and sera from dogs immunized with irradiated larvae were effective in precipitating this protein. This single, major immunoreactive protein was evidently not glycosylated and could be resolved into two constituents with two-dimensional gel analysis. This work suggests that analysis of surface antigens on third- and fourth-stage larvae may be relatively straightforward. Further, since the antigen appeared to be exclusively protein, its production by biotechnological methodologies may be easier than production of glycoproteins.

In initial studies utilizing a mouse model for early larval development, immunization with live or radiation-attenuated larvae resulted in 25 to 40% reduction in challenge larvae.[98] In contrast, immunization with killed larvae elicited no protection. Attempts to passively immunize mice with a stage-specific monoclonal antibody reactive with an epitope on the surface of infective larvae typically resulted in ≤25% reduction in challenge larvae. Although only modest levels of protection were achieved in these experiments, certain observations were possible. First, active immunization against developing larvae could be achieved at significant levels using the murine host. Second, live larvae, either intact or radiation-attenuated, were necessary to elicit protection; dead larvae would not produce protective immune responses. Third, monoclonal antibody directed against a surface antigen on infective larvae could effect some larval destruction. This was encouraging in view of the idea that effector mechanisms other than antibody-mediated mechanisms may contribute to protection. An important conclusion reached in this study was that since the third-stage larvae is a relatively short-lived stage in the dog,[99] i.e., 96 hr or less, an appropriate target for potentially effective immune responses is the fourth-stage larvae. Accordingly, the fourth-stage larva warrants additional research.

Since there is little evidence of protective immune responses to *D. immitis* occurring in naturally infected dogs in endemic regions, a cursory appraisal of the potential for immunoprophylaxis may be negative. However, dogs can be immunized against infection using at least two different, albeit impractical, methods. With sound research approaches involving an intricate appreciation for the biology and immunology of *D. immitis* infection and, perhaps, contemporary biotechnological approaches which could potentially circumvent problems of procurement and production of relevant antigens, the development of vaccines against dirofilariasis is a real possibility.

VI. IMMUNE RESPONSE OF THE VECTOR

In addition to numerous mechanical and physiological attributes which may dictate vector competence,[48] the encapsulation-melanizaton response of mosquitos to microfilariae of *D. immitis* is important in neutralizing the developing parasite in the vector.[100-102] This response, although not an immune response in the context of mammalian immunology, is an effective defense response. The response to microfilariae is remarkably different in magnitude between different mosquito species and between strains of the same species.[100] Hemocyte lysis and melanization occur immediately after microfilariae are exposed to hemolymph.[101] Although classical hemocyte encapsulation is not observed with microfilariae exposed to hemolymph via intrathoracic inoculation, hemocytes appear to be necessary for activation of the immune response.[101] Interestingly, when melanization appeared complete, a double membrane-like structure appeared to enclose the melanized microfilariae and hemocyte debris; it was hypothesized that this membrane may function to isolate the melanized parasite preventing additional hemocyte activation.[101] Figure 2 is a transmission electron micrograph which illustrates a melanized *D. immitis* microfilaria at 3 to 4 days after inoculation into the hemocoel of *Aedes trivittatus*. Observations on the melanization of *D. immitis* larvae located intracellularly within the Malpighian tubules suggest that melanization can occur intracellularly within the tubules without direct contact with hemocytes.[102]

While the melanization response is responsible for neutralizing microfilariae and larvae developing in the mosquito, recent studies have suggested that the defense reaction of *Aedes trivittatus* against *D. immitis,* for example, may reduce the parasite burden and increase the chances for survival for both vector and the parasite.[103,104] Therefore, the implications of effective immune responses in the vector are complex. Further investigation in this area will be important to a more thorough understanding of the host-vector-parasite relationship and may provide insight toward novel approaches for interrupting parasite transmission.

ACKNOWLEDGMENTS

The author gratefully acknowledges Drs. David Abraham and Mario Philipp for permitting the discussion of their unpublished research results. Dr. Bruce Christensen and Mr. Keith Forton kindly allowed the use of their transmission electron micrograph of a melanized *D. immitis* microfilaria.

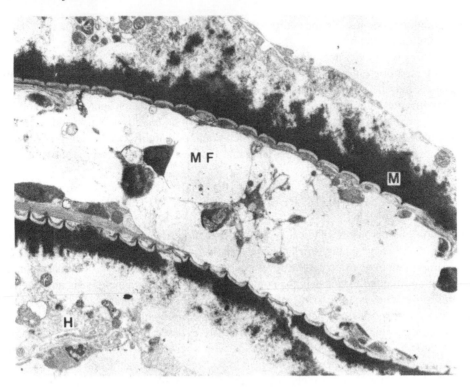

FIGURE 2.   Melanized *Dirofilaria immitis* microfilaria at 3 to 4 days postinoculation into the hemocoel of *Aedes trivittatus*. M = melanin deposits; MF = microfilaria; H = hemocyte remnants. (Photograph provided by Dr. B. Christensen and Mr. K. Forton.)

REFERENCES

1. **Pacheco, G.,** Progressive changes in certain serological responses to *Dirofilaria immitis* infection in the dog, *J. Parasitol.,* 52, 311, 1966.
2. **Tulloch, G. S., Pacheco, G., Casey, H. W., Bills, W. E., Davis, I., and Anderson, R. A.,** Prepatent clinical, pathologic, and serologic changes in dogs infected with *Dirofilaria immitis* and treated with diethylcarbamazine, *Am. J. Vet. Res.,* 31, 437, 1970.
3. **Grieve, R. B., Mika-Johnson, M., Jacobson, R. H., and Cypess, R. H.,** Enzyme-linked immunosorbent assay for measurement of antibody responses to *Dirofilaria immitis* in experimentally infected dogs, *Am. J. Vet. Res.,* 42, 66, 1981.
4. **Weiner, D. J. and Bradley, R. E.,** Serologic changes in primary and secondary infections of beagle dogs with *Dirofilaria immitis,* in *Canine Heartworm Disease: The Current Knowledge,* Bradley, R. E., Ed., University of Florida, Gainesville, Fla., 1972, 77.
5. **Grieve, R. B., Gebhardt, B. M., and Bradley, R. E.,** *Dirofilaria immitis:* cell-mediated and humoral immune responses in experimentally-infected dogs, *Int. J. Parasitol.,* 9, 275, 1979.
6. **Hayasaki, M.,** Reaginic and hemagglutinating antibody production in dogs infected with *Dirofilaria immitis,* *Jpn. J. Vet. Sci.,* 44, 63, 1982.
7. **Grieve, R. B. and Knight, D. H.,** Anti-*Dirofilaria immitis* antibody levels before and after anthelmintic treatment of experimentally infected dogs, *J. Parasitol.,* 71, 56, 1985.
8. **Weiner, D. J. and Bradley, R. E.,** The 2-mercaptoethanol labile immunoglobulin response of beagles experimentally-infected with *Dirofilaria immitis, J. Parasitol.,* 59, 696, 1973.
9. **Hayasaki, M., Nakagaki, K., Kobayashi, S., and Ohishi, I.,** Immunological response of dogs to *Dirofilaria immitis* infection, *Jpn. J. Vet. Sci.,* 43, 909, 1981.
10. **Wong, M. M. and Suter, P. F.,** Indirect fluorescent antibody test in occult dirofilariasis, *Am. J. Vet. Res.,* 40, 414, 1979.

11. **Tamashiro, W. K., Powers, K. G., Levy, D. A., and Scott, A. L.,** Quantitative and qualitative changes in the humoral response of dogs through the course of infection with *Dirofilaria immitis, Am. J. Trop. Med. Hyg.,* 34, 292, 1985.

12. **Grieve, R. B., Glickman, L. T., Bater, A. K., Mika-Grieve, M., Thomas, C. B., and Patronek, G. J.,** Canine *Dirofilaria immitis* infection in a hyperzootic area: examination by serologic and parasitologic findings at necropsy and by two serodiagnostic methods, *Am. J. Vet. Res.,* 47, 329, 1986.

13. **Weil, G. J., Ottesen, E. A., and Powers, K. G.,** *Dirofilaria immitis:* parasite-specific humoral and cellular immune responses in experimentally infected dogs, *Exp. Parasitol.,* 51, 80, 1981.

14. **Grieve, R. B., Brooks, B. O., Babish, J. G., Jacobson, R. H., and Cypess, R. H.,** Lymphocyte function in experimental dirofilariasis: B-cell responses to heterologous antigen, *J. Parasitol.,* 68, 341, 1982.

15. **Rawlings, C. A., Dawe, D. L., McCall, J. W., Keith, J. C., and Prestwood, A. K.,** Four types of occult *Dirofilaria immitis* infection in dogs, *J. Am. Vet. Med. Assoc.,* 180, 1323, 1982.

16. **Blair, L. S., Malatesta, P. F., Gerckens, L. S., and Ewanciw, D. V.,** Efficacy of thiacetarsamide in experimentally infected dogs at 2, 4, 6, 12, or 24 months post-infection with *Dirofilaria immitis,* in *Proceedings of the Heartworm Symposium '83,* Otto, G. F., Ed., Veterinary Medicine Publ., Edwardsville, Kan., 1983, 130.

17. **Wong, M. M., Suter, P. F., Rhode, E. A., and Guest, M. F.,** Dirofilariasis without circulating microfilariae: a problem of diagnosis, *J. Am. Vet. Med. Assoc.,* 163, 133, 1973.

18. **Otto, G. F.,** Occurrence of the heartworm in unusual locations and in unusual hosts, in *Proceedings of the Heartworm Symposium '74,* Morgan, H. C., Ed., Veterinary Medicine Publ., Bonner Springs, Kan., 1975, 6.

19. **Kobayakawa, T., Ishiyama, H., and Senda, F.,** Delayed hypersensitivity to *Dirofilaria immitis.* I. Migration inhibition test, *Jpn. J. Parasitol.,* 22, 369, 1973.

20. **Kobayakawa, T. and Ishiyama, H.,** Delayed hypersensitivity to *Dirofilaria immitis.* II. Blast transformation test, *Jpn. J. Parasitol.,* 23, 226, 1974.

21. **Kobayakawa, T., Kobayashi, T., and Ishiyama, H.,** Delayed hypersensitivity to *Dirofilaria immitis.* III. The *in vitro* cytotoxic activity of sensitized lymphocytes and their effect upon the mortality of microfilariae in diffusion chambers implanted intraperitoneally into guinea pigs, *Jpn. J. Parasitol.,* 23, 300, 1974.

22. **Kobayakawa, T.,** Cell-mediated immunity to *Dirofilaria immitis, Jpn. J. Med. Sci. Biol.,* 28, 11, 1975.

23. **Hayasaki, M.,** Indirect hemagglutination test for diagnosis of canine filariasis, *Jpn. J. Vet. Sci.,* 43, 21, 1981.

24. **Scholtens, R. G. and Patton, S.,** Evaluation of an enzyme-linked immunosorbent assay for occult dirofilariasis in a population of naturally exposed dogs, *Am. J. Vet. Res.,* 44, 861, 1983.

25. **Gillis, J. M., Smith, R. D., and Todd, K. S., Jr.,** Diagnostic criteria for an enzyme-linked immunosorbent assay for occult heartworm disease: standardization of the test system in naturally exposed dogs, *Am. J. Vet. Res.,* 45, 2289, 1984.

26. **Voller, A. and DeSavigny, D.,** Diagnostic serology of tropical parasitic disease, *J. Immunol. Methods,* 46, 1, 1981.

27. **Wheeling, C. H. and Hutchison, W. F.,** Disc electrophoresis and immunoelectrophoresis of soluble extracts of *Dirofilaria immitis, Jpn. J. Exp. Med.,* 41, 171, 1971.

28. **Sawada, T., Takei, K., Katamine, D., and Yoshimura, T.,** Immunological studies on filariasis. III. Isolation and purification of antigen for intradermal skin test, *Jpn. J. Exp. Med.,* 35, 125, 1965.

29. **Sawada, T., Sato, K., and Sato, S.,** The further studies on the separation of responsible protein in the skin test antigen FST by column chromatography, disc electrophoresis and isoelectric focusing technique, in *Recent Advances in Researches on Filariasis and Schistosomiasis in Japan,* Sasa, M., Ed., University of Tokyo Press, Tokyo, 1970, 169.

30. **Takahashi, J. and Sato, K.,** Studies on the hemagglutination test on filariasis. I. The fractionation and purification of antigens by column chromatography and disc electrophoresis, *Jpn. J. Exp. Med.,* 46, 7, 1976.

31. **Mantovani, A. and Kagan, I. G.,** Fractionated *Dirofilaria immitis* antigens for the differential diagnosis of canine dirofilariasis, *Am. J. Vet. Res.,* 28, 213, 1967.

32. **Gittleman, H. J., Grieve, R. B., Hitchings, M. M., Jacobson, R. H., and Cypess, R. H.,** Quantitative fluorescent immunoassay for measurement of antibody to *Dirofilaria immitis* in dogs, *J. Clin. Microbiol.,* 13, 309, 1981.

33. **Glickman, L. T., Grieve, R. B., Breitschwerdt, E. B., Mika-Grieve, M. M., Patronek, G. J., Domanski, L. M., Root, C. R., and Malone, J. B.,** Serologic pattern of canine heartworm *(Dirofilaria immitis)* infection, *Am. J. Vet. Res.,* 45, 1178, 1984.

34. **Scott, A. L., Parquette, S. C., and Levy, D. A.,** Lectin affinity purification of antigens from *Dirofilaria immitis,* in *Proceedings of the Heartworm Symposium '83,* Otto, G. F., Ed., Veterinary Medicine Publ., Edwardsville, Kan., 1983, 63.

35. **Sisson, D., Dilling, G., Wong, M. M., and Thomas, W. P.,** Sensitivity and specificity of the indirect-fluorescent antibody test and two enzyme-linked immunosorbent assays in canine dirofilariasis, *Am. J. Vet. Res.,* 46, 1529, 1985.

36. **Saito, T.,** Studies on the antigenicity of metabolites of the canine filariae in urine, *Kurume Med. J.,* 13, 1, 1966.

37. **Yamanouchi, S.,** Immunological studies on filariasis. III. Antigenic substance in the urine of dogs infected with *Dirofilaria immitis, Kurume Med. J.,* 19, 117, 1972.

38. **Desowitz, R. S. and Una, S. R.,** The detection of antibodies in human and animal filariasis by counter-immunoelectrophoresis with *Dirofilaria immitis* antigens, *J. Helminthol.,* 50, 53, 1976.

39. **Hamilton, R. G. and Scott, A. L.,** Immunoradiometric assay for quantitation of *Dirofilaria immitis* antigen in dogs with heartworm infections, *Am J. Vet. Res.,* 45, 2055, 1984.

40. **Weil, G. J., Malane, M. S., and Powers, K. G.,** Detection of circulating parasite antigens in canine dirofilariasis by counterimmunoelectrophoresis, *Am. J. Trop. Med. Hyg.,* 33, 425, 1984.

41. **Weil, G. J., Malane, M. S., Powers, K. G., and Blair, L. S.,** Monoclonal antibodies to parasite antigens found in the serum of *Dirofilaria immitis*-infected dogs, *J. Immunol.,* 134, 1185, 1985.

42. **Lindemann, B. A. and McCall, J. W.,** Current status of canine dipetalonemiasis in the United States, in *Proceedings of the Heartworm Symposium '83,* Otto, G. F., Ed., Veterinary Medicine Publ., Edwardsville, Kan., 1983, 24.

43. **Sheen, D., Sweeny, K., and Jones, C. K.,** Heartworm testing: comparing results of ELISA with filter technique and necropsy findings, *Vet. Med. Small Anim. Clin.,* 1, 49, 1985.

44. **Martin, T. E., Collins, G. H., Griffin, D. L., and Pope, S. E.,** An evaluation of 4 commercially available ELISA kits for the diagnosis of *Dirofilaria immitis* infection in dogs, *Aust. Vet. J.,* 62, 166, 1985.

45. **Hamilton, R. G., Scott, A. L., D'Antonio, R., Levy, D. A., and Adkinson, N. F., Jr.,** *Dirofilaria immitis:* performance and standardization of specific antibody immunoassays for filariasis, *Exp. Parasitol.,* 56, 298, 1983.

46. **Galen, P. S. and Gambino, S. R.,** *Beyond Normality: The Predictive Value and Efficiency of Medical Diagnosis,* John Wiley & Sons, New York, 1975.

47. **Fletcher, R. H., Fletcher, S. W., and Wagner, E. H.,** *Clinical Epidemiology: The Essentials,* Williams & Wilkins, Baltimore, 1982, chap. 3.

48. **Grieve, R. B., Lok, J. B., and Glickman, L. T.,** Epidemiology of canine heartworm infection, *Epidemiol. Rev.,* 5, 220, 1983.

49. **Hsu, C., Melby, E. C., Jr., and Farwell, A. E., Jr.,** Demonstration and interspecies cross-sensitization of reaginic antibodies in dogs infected with *Dirofilaria immitis, Am. J. Trop. Med. Hyg.,* 23, 619, 1974.

50. **Fujita, K.,** Separation of *Dirofilaria immitis* allergen from the IgG-inducing antigens, *Jpn. J. Med. Sci. Biol.,* 28, 139, 1975.

51. **Fujita, K., Ikeda, T., and Tsukidate, S.,** Immunological and physicochemical properties of a highly purified allergen from *Dirofilaria immitis, Int. Arch. Allergy Appl. Immunol.,* 60, 121, 1979.

52. **Fujita, K. and Tsukidate, S.,** Preparation of a highly purified allergen from *Dirofilaria immitis:* reaginic antibody formation in mice, *Immunology,* 42, 363, 1981.

53. **Fujita, K., Tsukidate, S., and Ikeda, T.,** *Dirofilaria immitis:* physicochemical properties of IgG-inducing antigen with special reference to the comparison with highly purified allergen, *Trop. Med.,* 23, 193, 1981.

54. **Fujita, K. and Tsukidate, S.,** Allergen concentration in *Dirofilaria immitis* (Nematoda), *J. Parasitol.,* 70, 313, 1984.

55. **Fujita, K., Ikeda, T., and Tsukidate, S.,** Amino acid composition of the highly purified allergen from *Dirofilaria immitis, Int. Arch. Allergy Appl. Immunol.,* 73, 184, 1984.

56. **Fujita, K. and Tsukidate, S.,** A highly purified allergen from excretory and secretory products of *Dirofilaria immitis, Int. J. Parasitol.,* 14, 547, 1984.

57. **Weiss, N., Hussain, R., and Ottesen, E. A.,** IgE antibodies are more species-specific than IgG antibodies in human onchocerciasis and lymphatic filariasis, *Immunology,* 45, 129, 1982.

58. **Cherian, P. V., Stromberg, B. E., Weiner, D. J., and Soulsby, E. J. L.,** Fine structure and cytochemical evidence for the presence of polysaccharide surface coat of *Dirofilaria immitis* microfilariae, *Int. J. Parasitol.,* 10, 227, 1980.

59. **Hutchison, W. F., Turner, A. C., Grayson, D. P., and White, H. B., Jr.,** Lipid analysis of the adult dog heartworm, *Dirofilaria immitis, Comp. Biochem. Physiol.,* 53B, 495, 1976.

60. **Grieve, R. B., DeGregory, K. T., and Lindmark, D. G.,** Extraction and partial characterization of surfactant-soluble antigens from adult female *Dirofilaria immitis, Acta Trop.,* 42, 63, 1985.

61. **Devaney, E. and Howells, R. E.,** Culture systems for the maintenance and development of microfilariae, *Ann. Trop. Med. Parasitol.,* 73, 139, 1979.

62. **April, M., D'Antonio, R., Scott, A. L., Malick, A., Roberts, E. P., and Levy, D. A.,** Development of a chemotherapeutic model for microfilaricidal drugs to *Dirofilaria immitis, Acta Trop.,* 41, 383, 1984.

63. **Earl, P. R.,** Filariae from the dog *in vitro, Ann. N.Y. Acad. Sci.,* 77, 163, 1959.

64. **Wong, M. M., Knighton, R., Fidel, J., and Wada, M.,** *In vitro* cultures of infective-stage larvae of *Dirofilaria immitis* and *Brugia pahangi, Ann. Trop. Med. Parasitol.,* 76, 239, 1982.

65. **Lok, J. B., Mika-Grieve, M., Grieve, R. B., and Chin, T. K.,** In vitro development of third- and fourth-stage larvae of *Dirofilaria immitis:* comparison of basal culture media, serum levels and possible serum substitutes, *Acta Trop.,* 41, 145, 1984.

66. **Devaney, E.,** *Dirofilaria immitis:* the moulting of the infective larva *in vitro, J. Helminthol.,* 59, 47, 1985.

67. **Boto, W. M. O., Powers, K. G., and Levy, D. A.,** Antigens of *Dirofilaria immitis* which are immunogenic in the canine host: detection by immuno-staining of protein blots with the antibodies of occult dogs, *J. Immunol.,* 133, 975, 1984.

68. **Boto, W. M. O., D'Antonio, R., and Levy, D. A.,** Homologous and distinctive antigens of *Onchocerca volvulus* and *Dirofilaria immitis:* detection by an enzyme-linked immuno-inhibition assay, *J. Immunol.,* 133, 981, 1984.

69. **Welch, J. S. and Dobson, C.,** Immunodiagnosis of parasitic zoonoses: comparative efficacy of three immunofluorescence tests using antigens purified by affinity chromatography, *Trans. R. Soc. Trop. Med. Hyg.,* 72, 282, 1978.

70. **Welch, J. S. and Dobson, C.,** Immunodiagnosis of parasitic zoonoses: sensitivity and specificity of *in vitro* lymphocyte proliferative responsiveness using nematode antigens purified by affinity chromatography, *Trans. R. Soc. Trop. Med. Hyg.,* 75, 5, 1981.

71. **Hamilton, R. G., Hussain, R., Alexander, E., and Adkinson, N. F.,** Limitations of the radioimmunoprecipitation polyethylene glycol assay (RIPEGA) for detection of filarial antigens in serum, *J. Immunol. Methods,* 68, 349, 1984.

72. **Soprunov, F. F., Lurje, A. A., Laynis, Yu. Ya., Soprunova, N. Ya., and Alieva, H. H.,** Metabolic end products of helminths: their degradation and excretion by the host, *Acta Trop.,* 38, 449, 1981.

73. **Potocnjak, P., Zavala, F., Nussenzweig, R., and Nussenzweig, V.,** Inhibition of idiotype-anti-idiotype interaction for detection of a parasite antigen: a new immunoassay, *Science,* 215, 1637, 1982.

74. **Jerne, N. K.,** Towards a network theory of the immune system, *Ann. Immunol., (Paris),* 125, 373, 1974.

75. **Powell, M. R. and Colley, D. G.,** Demonstration of splenic auto-anti-idiotypic plaque-forming cells in mice infected with *Schistosoma mansoni. J. Immunol.,* 134, 4140, 1985.

76. **Castleman, W. L. and Wong, M. M.,** Light and electron microscopic pulmonary lesions associated with retained microfilariae in canine occult dirofilariasis, *Vet. Pathol.,* 19, 355, 1982.

77. **Knight, D. H.,** Heartworm heart disease, *Adv. Vet. Sci. Comp. Med.,* 21, 107, 1977.

78. **Knight, D. H.,** Heartworm disease, in *Textbook of Veterinary Internal Medicine, Diseases of the Dog and Cat,* Vol. 1, 2nd ed., Ettinger, S. J., Ed., W. B. Saunders, Philadelphia, 1983, chap. 54.

79. **Hayasaki, M.,** Passive transfer of anti- *Dirofilaria immitis* hemagglutinating antibody from the mother dog to its offspring, *Jpn. J. Vet. Sci.,* 44, 781, 1982.

80. **Wong, M. M.,** Studies on microfilaremia in dogs. II. Levels of microfilaremia in relation to immunologic responses of the host, *Am. J. Trop. Med. Hyg.,* 13, 66, 1964.

81. **Weil, G. J., Powers, K. G., Parbuoni, E. L., Line, B. R., Furrow, R. D., and Ottesen, E. A.,** *Dirofilaria immitis.* VI. Antimicrofilarial immunity in experimental filariasis, *Am. J. Trop. Med. Hyg.,* 31, 477, 1982.

82. **El-Sadr, W. M., Aikawa, M., and Greene, B. M.,** *In vitro* immune mechanisms associated with clearance of microfilariae of *Dirofilaria immitis, J. Immunol.,* 130, 428, 1983.

83. **Rzepczyk, C. M. and Bishop, C. J.,** Immunological and ultrastructural aspects of the cell-mediated killing of *Dirofilaria immitis* microfilariae, *Parasite Immunol.,* 6, 443, 1984.

84. **Hammerberg, B., Rikihisa, Y., and King, M. W.,** Immunoglobulin interactions with surfaces of sheathed and unsheathed microfilariae, *Parasite Immunol.,* 6, 421, 1984.

85. **Hayasaki, M.,** Antigenicity of microfilarial and adult *Dirofilaria immitis* in indirect fluorescent antibody test, *Jpn. J. Vet. Sci.,* 45, 113, 1983.

86. **Staniunas, R. J. and Hammerberg, B.,** Diethylcarbamazine-enhanced activation of complement by intact microfilariae of *Dirofilaria immitis* and their in vitro products, *J. Parasitol.,* 68, 809, 1982.

87. **Casey, H. W. and Splitter, G. A.,** Membranous glomerulonephritis in dogs infected with *Dirofilaria immitis, Vet. Pathol.,* 12, 111, 1975.

88. **Aikawa, M., Abramowsky, C., Powers, K. G., and Furrow, R.,** *Dirofilariasis.* IV. Glomerulonephropathy induced by *Dirofilaria immitis* infection, *Am. J. Trop. Med. Hyg.,* 30, 84, 1981.

89. **Abramowsky, C. R., Powers, K. G., Aikawa, M., and Swinehart, G.,** *Dirofilaria immitis.* V. Immunopathology of filarial nephropathy in dogs, *Am. J. Pathol.,* 104, 1, 1981.

90. **Ohgo, T.,** Basic studies on the *Dirofilaria immitis* and the experimental infection with larval filariae in mice, *J. Osaka City Med. Cent.,* 29, 745, 1980.

91. **Zielke, E.,** On the longevity and behaviour of microfilariae of *Wuchereria bancrofti, Brugia pahangi* and *Dirofilaria immitis* transfused to laboratory rodents, *Trans. R. Soc. Trop. Med. Hyg.,* 74, 456, 1980.

92. **Grieve, R. B. and Lauria, S.,** Periodicity of *Dirofilaria immitis* microfilariae in canine and murine hosts, *Acta Trop.,* 40, 121, 1983.

93. **Zielke, E.**, Preliminary studies on the transplantation of adult *Dirofilaria immitis* into laboratory rodents, *Ann. Trop. Med. Parasitol.*, 71, 243, 1977.

94. **Grieve, R. B., Griffing, S. A., Goldschmidt, M. H., and Abraham, D.**, Transplantation of adult *Dirofilaria immitis* into Lewis rats: parasitologic and serologic findings, *J. Parasitol.*, 71, 391, 1985.

95. **Wong, M. M., Guest, M. F., and Lavoipierre, M. M. J.**, *Dirofilaria immitis:* fate and immunogenicity of irradiated infective stage larvae in beagles, *Exp. Parasitol.*, 35, 465, 1974.

96. **Blair, L. S. and Campbell, W. C.**, Immunization of ferrets against *Dirofilaria immitis* by means of chemically abbreviated infections, *Parasite Immunol.*, 3, 143, 1981.

97. **Blair, L. S., Jacob, L., and Ewanciw, D. V.**, Immunization of dogs against *Dirofilaria immitis* by means of chemically abbreviated infections, *Mol. Biochem. Parasitol.*, Suppl., 678, 1982.

98. **Abraham, D., Grieve, R. B., Mika-Grieve, M. M., and Seibert, B. P.**, Active and passive immunization of mice against infection with *Dirofilaria immitis, J. Parasitol*, submitted.

99. **Kotani, T. and Powers, K. G.**, Developmental stages of *Dirofilaria immitis* in the dog, *Am. J. Vet. Res.*, 43, 2199, 1982.

100. **Christensen, B. M., Sutherland, D. R., and Gleason, L. N.**, Defense reactions of mosquitoes to filarial worms: comparative studies on the response of three different mosquitoes to inoculated *Brugia pahangi* and *Dirofilaria immitis* microfilariae, *J. Invertebr. Pathol.*, 44, 267, 1984.

101. **Forton, K. F., Christensen, B. M., and Sutherland, D. R.**, Ultrastructure of the melanization response of *Aedes trivittatus* against inoculated *Dirofilaria immitis* microfilariae, *J. Parasitol.*, 71, 331, 1985.

102. **Bradley, T. J. and Nayar, J. K.**, Intracellular melanization of the larvae of *Dirofilaria immitis* in the malpighian tubules of the mosquito, *Aedes sollicitans, J. Invertebr. Pathol.*, 45, 339, 1985.

103. **Christensen, B. M.**, *Dirofilaria immitis:* effect on the longevity of *Aedes trivittatus, Exp. Parasitol.*, 44, 116, 1978.

104. **Christensen, B. M.**, Observations on the immune response of *Aedes trivittatus* against *Dirofilaria immitis, Trans. R. Soc. Trop. Med. Hyg.*, 75, 439, 1981.

105. **Philipp, M., and Davis, T. B.**, Biochemical and immunologic characterization of a major surface antigen of *Dirofilaria immitis* infective larvae, *J. Immunol.*, 136, 2621, 1986.

Chapter 11

CAVAL SYNDROME

Richard B. Atwell and Ibrahim B. J. Buoro

TABLE OF CONTENTS

I. INTRODUCTION

Caval syndrome has many synonyms (liver failure syndrome,[1] venae cavae embolism,[2] dirofilarial hemoglobinuria,[3]) and occurs in some dogs infected with *Dirofilaria immitis,* being characterized by the acute development of shock and signs of intravascular hemolysis. In Australia, it has been reported to occur mostly in dogs averaging 5 years of age,[4] while in North America[1,5] it reportedly occurs in dogs 3 to 5 years of age. However, caval syndrome is not age dependent and has been reported in one dog as old as 18 years of age.[6] Similarly, there appears to be no breed or sex predisposition,[2] and the development of the syndrome is closely related to the presence of filariae in the proximity of the right atrial chamber. As with all diseases, there does exist variation in the clinical signs seen with cases of this syndrome. It is proposed that this is due chiefly to the variable extent of underlying pathology present when the set of circumstances that induces the syndrome in different cases is developed.

II. PATHOPHYSIOLOGY

Jackson et al., while speculating on the pathophysiology of caval syndrome, questioned why large numbers of *D. immitis* were found in dogs presented with caval syndrome.[5,7] They stressed the relative young age of the dogs involved and concluded that such dogs probably had a massive first exposure to *D. immitis* receiving a large number of infected mosquito bites at one time, or over a short period of time, so that a large number of *D. immitis* matured together, probably before any protective immunity could develop to limit the survival of infective larvae. Protective immunity seems to occur in heartworm infections where the subsequent infection rate in the individual dog is limited, so that the worm counts of the middle-aged to older dogs are usually quite low.

The contention that large numbers of filariae are necessary would seem to find support in the work of Sawyer and Weinstein[8] who, in 1963, while performing transmission studies with *D. immitis,* inadvertently induced caval syndrome in two experimental dogs following inoculation of 200 *D. immitis* larvae (L3) subcutaneously in the inguinal region. At necropsy the distribution of *D. immitis* was similar to that for natural cases of caval syndrome. Sawyer later repeated this experiment and obtained similar results.[9]

In 1962, von Lichtenberg, Jackson, and Otto[10] suggested that local vein obstruction by adult *D. immitis* played a major role in the pathogenesis of caval syndrome; they ascribed the pathological changes in the liver to the presence of *D. immitis* in the posterior vena cava and suprahepatic veins. Similarly, Jackson found enlarged and congested livers in caval syndrome cases with the hepatic lymphatics visible in some cases.[1]

Ishihara and his colleagues made similar observations pointing out that embolism of the venae cavae by adult *D. immitis* was the trigger for caval syndrome and cited as supporting evidence the fact that prompt recovery ensues once the *D. immitis* are surgically extracted.[3] They associated the hepatic injury with passive venous congestion of the liver caused by the obstructive presence of adult *D. immitis* in the posterior vena cava.

In contrast to this, Atwell and Farmer[4] drew attention to the fact that the right atrium was the cardiac chamber most involved in reported cases of caval syndrome where the average number of filariae in the right atrium and venae cavae per dog at necropsy was 64.[2,3,11] They also hypothesized that movement of filariae occurred from the distal pulmonary arteries and right ventricle into the right atrial area to induce the syndrome.[12] This movement of *D. immitis* from the right ventricle into the right atrium would be facilitated by any concurrent regurgitation through the tricuspid valve as a result of *D. immitis* physically interfering with the valve and/or expansion of the valve annulus as a result of right ventricular dilation in response to chronic pulmonary hypertension caused by dirofilariasis.[13] Movement of *D. immitis* into the venae cavae could thus be facilitated by severe regurgitation and, possibly,

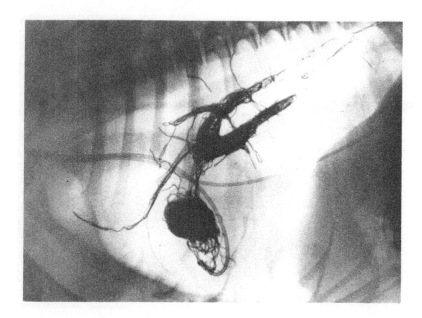

FIGURE 1. Lateral radiograph of an experimentally induced case of caval syndrome with radiopaque threads in the right atrium and in the pulmonary arteries. (From Buoro, I. B. J. and Atwell, R. B., *Aust. Vet. J.*, 61, 267, 1984. With permission.)

directly into the posterior vena cava by the reverse flow of blood seen in the posterior vena cava in normal dogs during atrial contraction.[14]

Such movement between the right atrium and right ventricle has been confirmed with the use of echocardiography in dogs in Japan where movement in both directions across the valve is seen in severely affected dogs.[15,16] Echocardiographic evidence also shows the right atrium to be obviously enlarged and the most involved cardiac chamber in caval cases. These cases typically have more worms present in the right atrium and right ventricle than do infected dogs without signs of caval syndrome, supporting earlier observations.[1,4,17]

Insertion of up to 30 worms obtained from clinical cases of caval syndrome into dogs infected with *D. immitis* but without signs of caval syndrome did not result in the dogs developing clinical signs of caval syndrome.[18] When considering the number of worms seen in caval cases this would not be expected to induce the syndrome as an average of 64 worms were present in the atrial area of reported caval cases.

Atwell and Boreham[19] when comparing the levels of circulating immune complexes of *D. immitis*-infected dogs and cases of caval syndrome, found no significant difference between the two groups. Similarly, the levels of total hemolytic complement in *D. immitis*-infected dogs, caval cases, and normal dogs free from infestation were not significantly different. Plasma from caval-syndrome dogs did not lyse dog or sheep erythrocytes, and neither the supernatant nor the sediment of a preparation of worms, taken from caval syndrome cases and from infected dogs without caval syndrome, had any detectable hemolytic activity. These findings suggested that the hemolysis was more likely mechanical in nature.

While using an experimental model designed to assess the role played by different physical factors, it was demonstrated that occupation of the right atrial chamber by a critical number (n = 60) of filamentous structures similar to *D. immitis* was enough to precipitate the clinical signs of caval syndrome[18] (Figure 1). The resulting syndrome was shown to be clinically and pathologically similar to natural caval syndrome. Initially, it was found that surgical extraction of the filamentous structures from the right atrium resulted in cessation of hemolysis and resolution of the clinical signs. However, it was subsequently demonstrated that

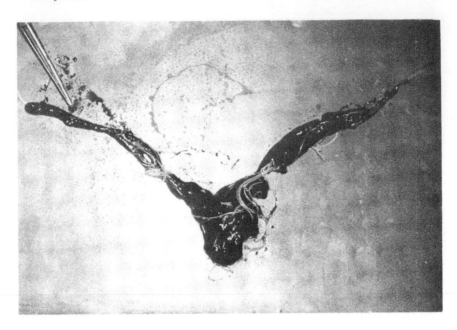

FIGURE 2.   Filariae found *in situ* in the venae cavae and right atrium at necropsy from a natural case of caval syndrome.

their complete extraction from the cardiovascular system was not necessary. A simple translocation of the filamentous structures from the right atrium to the anterior vena cava was sufficient to stop the hemolysis and for the clinical signs to abate. This latter finding is of interest as it would help to explain the cessation of clinical signs seen in some clinical cases of caval syndrome after apparently unsuccessful worm extraction. What probably happens in those cases is that a rearrangement or relocation of the "hemolytic" mass or "knot of worms" in the right atrial region occurs, reducing obstruction (the cause of the shock component) and turbulence (the cause of hemolysis) and so abolishing the clinical signs.[13]

Based on results obtained using the above model, and taking into consideration the clinical response to surgical treatment and necropsy findings of natural cases of caval syndrome, it is suggested that occupation of the right atrium by filamentous structures plays a central role in the pathophysiology of caval syndrome.

In the natural disease, as the number of filariae increases, it is suggested that movement of filariae occurs from the pulmonary arteries back towards the right ventricle and right atrium. The cause of this movement is presumed to be part of the natural life cycle.[12,20] If there is a large number of worms and/or tricuspid valve interference and regurgitation, accentuation of this movement would occur into the right atrium and beyond. If many filariae mature and arrive at a similar time then a critical number of worms (of the order of 60) could be present in the right atrium and venae cavae at one period thus inducing the acute signs of caval syndrome (Figure 2). Support for this is based on the age variation of filariae with older worms being seen in the right atrial area.[21] In fact, worms extracted from this area during surgery often appear old, flaccid, and yellowed[22] which differ from other reports[17] but are possibly related to the age of the actual infection in each case with geriatric worms more likely in the atrial area in older infections.

The presence of a mass of *D. immitis* in the right atrium, a high velocity, low pressure chamber, apart from obstructing venous flow and return, could be expected to be associated with an increase in nonlaminar flow and a consequential increase in shearing forces on the erythrocyte membrane. Since erythrocytes from caval-syndrome cases have increased susceptibility to mechanical hemolysis and are known to have elevated levels of cholesterol,[23] which impairs their elasticity and predisposes them to hemolysis, their subjection to augmented shearing stresses could be expected to result in lysis.

Selective, plasma-hemoglobin sampling above and below the site of artificial threads in the model of caval syndrome showed that hemolysis was associated with the flow of blood through the mass of threads.[18]

Additionally, it has been established that shear stress affects all formed blood elements in a similar way. In the case of platelets, induced aggregation and release reactions[24] are associated with platelets being transformed into bizzare shapes, becoming sticky, and, presumably, generating the formation of fibrin on the surfaces they adhere to. This deposition of fibrin strands (on the surface of filariae and within the knot of filariae) would be expected to contribute to, and enhance the level of, intravascular hemolysis similar to that process seen in microangiopathic hemolysis.

In the experimental model, dogs with caval syndrome did develop thrombocytopenia and did develop bizzare platelet forms suggestive of platelet activation and aggregation.[18] This is presumably associated with the turbulence generated by the nonlaminar flow elevating shear forces and so leading to platelet activation. Such platelet activity could also lead to further downstream arterial pathology as outlined in Chapter 7.

While being reported to occur rarely in cases of dirofilariasis,[25] disseminated intravascular coagulation was not shown to be associated with the hemolysis in 21 cases of caval syndrome.[26] However, decreased platelet counts were observed which would further support the experimental findings mentioned above and the belief that turbulence-induced platelet activation and secondary fibrin deposition are factors in the hemolysis (microangiopathic) in addition to the direct effect of increased shear force causing mechanical lysis of the erythrocytes.

Following successful extraction of filariae in 19 cases, average central venous pressures fell, the systolic murmur and jugular pulse were reduced, and hemolysis abated.[27] These findings support the fact that filarial mass in the right atrial area is associated with reduced venous return and obstructed flow (high venous pressures), the development of a systolic murmur (turbulence and/or tricuspid regurgitation), and hemolysis. However, tricuspid regurgitation alone will not produce hemolysis unless there is an altered shear force due to filarial-induced turbulence. Dogs in which turbulence is not reduced by extraction will not improve and usually die.

It has also been shown that partial obstruction of the posterior vena cava does not cause caval syndrome, whereas complete blockage leads to death within 20 to 60 min with no hemolysis being induced.[18] Placing a turbulence-generating device in the posterior vena cava similarly did not induce caval syndrome. Only the placement of a critical mass of artificial filariae in the right atrium has been shown to induce caval syndrome.[18] Caval syndrome has also been seen following DEC shock reactions. The cause is not known, but presumably filariae have moved to the right atrial area in association with the hypovolemic shock and/or reduced cardiac output seen in the DEC reaction.[22]

Thus, it would appear that dogs with caval syndrome have large numbers of mature worms which can easily induce clinical signs of shock due to obstruction and reduced venous return. The presence of worms in the atria is due either to crowding pressure and movement of worms towards the ventricle and thus the atria, or being moved in that direction by the development of "undefined cardiac abnormalities". These could occur secondarily to the effects of cor pulmonale from severe pulmonary arterial pathology, increased right ventricular afterload and subsequent dilation of the tricuspid annulus, or directly due to valvular interference by the worm mass. The subsequent turbulence induced by the presence of worms is sufficient to alter shear forces and induce platelet activation and mechanical hemolysis

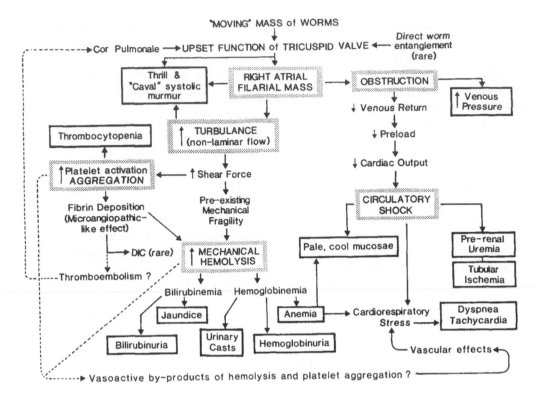

FIGURE 3.   Proposed pathophysiology of caval syndrome.

(associated with angiopathic hemolysis) which is due to the direct rupture of erythrocytes already shown to be susceptible to mechanical trauma.

Figure 3 outlines the proposed pathophysiology of caval syndrome.

III. CLINICAL SIGNS

The affected dog is usually presented with the history of a sudden onset of anorexia and depression.[7] Clinical examination usually reveals a distressed, dyspneic dog with tachycardia, tachypnea, abdominal respiration, pale mucosae (probably due to the effects of shock more than from anemia), and a reduced capillary refill time.[4] The body-temperature response is variable,[2,3] and jaundice and hemoglobinuria may or may not be grossly evident.[4] Petechial hemorrhages[4] may also be evident in the mucosa and suggest thrombocytopenia believed to be due to turbulence-induced platelet activation and aggregation.

A loud, holosystolic, low-frequency heart murmur is usually present over the tricuspid area. An engorged jugular and a jugular pulse are present and a precordial thrill is sometimes palpable over the tricuspid area. The murmur of caval syndrome usually disappears after filarial extraction,[6,27,28] this being associated with either a reduction in the degree of tricuspid regurgitation and/or reduced turbulence associated with the removal of a knot of worms that had previously caused audible turbulence.

The first recognizable clinical description of what is now known as caval syndrome was in 1956 by Adams[29] when he described a clinical condition in a 9-year-old, male pointer presented in a moribund state, with hyperthermia, subcutaneous edema, and intensely icteric mucous membranes. Six years later, Jackson reported cases of canine dirofilariasis which presented with a sudden onset of weakness, hemoglobinuria and bilirubinuria, and which died shortly thereafter.[17] In the same year, von Lichtenberg, Jackson, and Otto described in some detail the hepatic pathology in 12 dogs with caval syndrome presented with weakness, anorexia, hemoglobinuria, and bilirubinuria.[10] The dogs died within a few days of presentation. A year later, Sawyer and Weinstein experimentally induced caval syndrome in two dogs and observed similar clinical signs[8] to those previously reported.

In 1969, Jackson noted that canine dirofilariasis is usually a slowly progressing disease, so that clinical signs are more often seen in older than younger dogs.[1] He contrasted this with caval syndrome in which there was usually a lack of history of prior clinical signs with the associated sudden onset of weakness, anorexia, hemoglobinuria, and bilirubinuria. Since then, caval syndrome has been well characterized clinically.[2-5]

IV. CLINICAL PATHOLOGY

The biochemical and hematological changes in caval syndrome have been extensively studied.[3,4,23] While not diagnostic, they are indicative of the severity of the case at hand. The values for erythrocyte count, hematocrit, and mean corpuscular hemoglobin are found to be significantly lower than control values. However, while plasma hemoglobin levels varied from 100 to 300 mg/ℓ there appears to be no correlation between plasma hemoglobin levels and the presence or absence of hemoglobinuria.[3] However, plasma hemoglobin levels do fall with successful extraction of filariae.[13,27] Hemosiderinuria is sometimes seen, but this is more suggestive of the chronic, low-grade hemolysis seen in most dogs with heartworm disease (see Chapter 6). In fact, there is most likely a metabolic anemia (normochromic, normocytic) in moderate to severe cases, as well as the low-grade, mechanical hemolysis mentioned above. Both of these are separate to the acute hemolysis seen in caval syndrome which shows signs of regeneration based on the values of the reticulocytic production index.[18]

When dogs with caval syndrome and those with severe dirofilariasis were compared to noninfected dogs, the infected group had lower levels of lecithin cholesterol acyltransferase than control values, and the erythrocyte, mechanical fragility value, and serum-free cholesterol concentration were both higher than control values.[3] In addition, it was noted that there was an association of the biochemical changes with the variable presence of "spur cells", target cells, spherocytes, and damaged erythrocytes, which were noted in up to 9% of caval syndrome cases examined.[3,18] Leukocytosis with a neutrophilia and eosinopenia was also evident in caval syndrome cases.[3] Similar but less severe findings were reported in cases of severe dirofilariasis in which no clinical signs of caval syndrome were apparent.

Examination of the osmotic fragility curve in caval syndrome cases indicated that, while the median corpuscular fragility was normal, the minimum resistance (1% hemolysis) was lower and the maximum resistance (99% hemolysis) higher than those of the control group, suggesting the presence of "younger" and more resistent erythrocytes in the caval group.[3,30] Creatinine values were found to be normal in a group of caval cases,[5] while values for plasma aspartate aminotransferase, plasma alanine aminotransferase, and isoenzyme one of lactic dehydrogenase were raised in cases of caval syndrome and in severe cases of dirofilariasis.[3] Plasma urea levels (prerenal factors) and bilirubin levels will be elevated in relation to the degree of shock and hemolysis seen clinically.

In a study on plasma lipids from cases of caval syndrome, plasma-free cholesterol levels were elevated, while lecithin cholesterol acyltransferase activity was depressed.[23] Lipids account for about 50% of the weight of the mammalian erythrocyte; cholesterol and phospholipids being the major lipids.[23] As the erythrocyte membrane and plasma exchange their cholesterol content utilizing an exchange equilibrium, elevated plasma-cholesterol levels, such as are found in *D. immitis*-infected dogs, could cause a corresponding elevation of cholesterol content in the erythrocytic membrane. Such cells have a reduced elasticity and are known to be easily injured.[23]

V. NECROPSY

Caval cases do not always have filariae in the posterior vena cava and, similarly, naturally infected dogs will have live filariae in the venae cavae without signs of caval syndrome. It is emphasized that, while the venae cavae can be involved, this does not have to be, and, as such, the term caval syndrome is misleading.

Adams found the majority of *D. immitis* in the posterior vena cava and hepatic veins in a case of caval syndrome.[29] In 1962, von Lichtenberg, Jackson, and Otto described hepatic lesions in 12 dogs presented with caval syndrome.[10] The majority of the dogs had from 30 to 150 adult *D. immitis* and, while the distribution of the *D. immitis* varied, all dogs had worms in the posterior vena cava and either in the pulmonary artery or the right ventricle. They attributed the liver lesions to passive venous congestion as a result of cor pulmonale and interference with function of the tricuspid valve by *D. immitis*. (See Chapter 6 for further detail.)

When the necropsy results of caval syndrome cases and cases of severe dirofilariasis were compared, approximately twice as many *D. immitis* (101) were in the dogs of the caval-syndrome group as there were in cases of the severe, dirofilariasis group (58).[11] In addition, in dogs with caval syndrome, over half of the worms present were in the venae cavae and right atrium in about equal numbers.[11] This observation has been subsequently confirmed.[21] It was also reported that half of the dogs with severe dirofilariasis and about a quarter of asymptomatic dogs had worms in these locations as well.[11] However, these two latter groups of dogs had fewer adult *D. immitis* in these locations when compared to cases of caval syndrome.[11]

Fujii, in a comparison of cases of caval syndrome with ''acute symptoms'' and those with ''subacute symptoms'', found that, while the overall *D. immitis* numbers were similar to both categories, they tended to be located more in the pulmonary artery and right ventricle in the subacute category.[2] Jackson found that the distribution of *D. immitis* varies: whereas in some cases they were found distributed throughout the length of the venae cavae and as far anteriorly as the jugular vein and posteriorly as the iliac veins, some of the cases had *D. immitis* confined to the right-atrial area alone.[1] He also noted that dogs with caval syndrome had enlargement of the right atrium with or without enlargement of the right ventricle.

In contrast to the typical caval case, 3 dogs with clinical signs similar to caval syndrome had only 9 to 15 adults *D. immitis* in the right ventricle and pulmonary artery at necropsy, the right atrium, anterior and posterior venae cavae of the 3 dogs being free from infestation.[31] However, calcified worm bodies were coiled around the chorda tendinea of the tricuspid valve, the apex of which was thickened and gelatinous, and two of the dogs had ruptured chordae tendineae. While these findings are not typical of caval syndrome, the cause of the hemolysis was supposedly due to the increased blood turbulence generated by the entanglement of the valve, similar to the turbulence generated around a cardiac-valve prosthesis when inserted into people. Cases similar to this have been seen in our practice when dogs are presented with acute development of RSCHF associated with a marked thrill and loud murmur over the tricuspid area (Figure 4). These dogs have signs similar to caval syndrome, including hemoglobinuria, and usually have marked valve entanglement by worms believed to be the cause of the turbulence and associated hemolysis.[27] Acute tricuspid regurgitation due to valve entanglement is responsible for the acute onset of RSCHF, and the associated turbulence is enough to produce a marked thrill and murmur and, presumably, lead to a secondary elevation of shear stress high enough to induce hemolysis.[22]

Thus, at necropsy of caval syndrome cases large numbers of worms are seen, usually in the right-atrial area, the dogs have evidence of intravascular hemolysis (renal discoloring) and have a variable degree of routine pathology due to *D. immitis*.

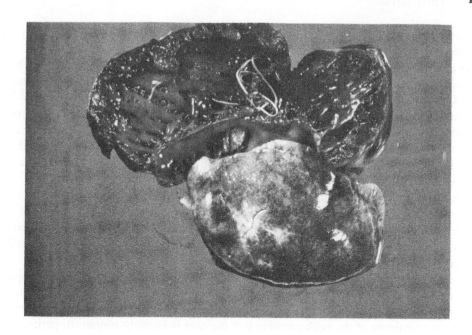

FIGURE 4.   Entanglement of the tricuspid valve from a case of acute RSCHF with signs similar to caval syndrome. (Courtesy of Dr. W. R. Kelly.)

VI. DIAGNOSIS, PROGNOSIS, AND TREATMENT

Obvious cases of caval syndrome are considered an emergency and must be surgically treated as soon as is possible.

Diagnosis is based on the sudden development of clinical signs of severe cardiorespiratory stress (tachypnea, tachycardia), signs of shock (associated with the reduced preload due to obstructed venous return), and the development of hemoglobinuria (due to hemolysis associated with worm-induced turbulence). Age, sex, and breed are not diagnostic factors, and the association of the above clinical signs in a dog not being treated prophylactically from a known, infected area is very supportive of the diagnosis.

Some cases have severe signs of heartworm disease,[2] but most are younger dogs with mild signs of heartworm disease.[5] In considering the differential diagnoses, other causes of acute intravascular hemolysis and shock need to be considered.

Prognosis is always guarded, but, where possible, at least renal assessment (plasma urea and creatinine levels, urinalysis) should be performed, particularly if the signs have been present for several days, suggesting that secondary renal involvement due to reduced perfusion and tubular ischemia may have occurred. A routine hemogram will enable assessment of the degree of anemia, and plasma bilirubin, hemoglobin, haptoglobin, and urinary hemoglobin and hemosiderin levels can be used to assess the degree of hemolysis. Liver enzyme levels will usually be elevated, but very high values have been seen in moderately affected dogs and, thus, their significance is difficult to interpret.[22]

The appearance of urinary hemoglobin casts is a poor prognostic sign and usually indicates hemolysis has been in progress for 24 hr or more, thus heightening the chance of renal impairment associated with iron-induced nephropathy and ischemia due to reduced peripheral perfusion.

Although procedures will vary, at the University of Queensland the following procedure is adopted: clinical examination, assess the clinical severity, take blood for basic profile (in particular renal assessment), collect urine (catheterize if necessary) for urinalysis and to confirm hemoglobinuria (separate from hematuria), and proceed to radiography for a jugular venotomy.

It is emphasized that the hemodynamics of the right-atrial area have to be altered to both improve venous return and reduce turbulence-induced hemolysis. Thus, a considerable number of filariae have to be removed to effectively treat the animal and to improve its cardiovascular status. Experimentally, 60 artificial threads were needed to induce the syndrome and, once those threads in the right-atrial area were removed, cases that would normally have died would recover. However, all 60 "filariae" did not have to be recovered for survival to occur. In natural cases it is obvious that filariae are left in inaccessible areas of the heart and pulmonary arteries and even in the venae cavae and atrium, but dogs still recover as the critical, turbulent mass of filariae have been displaced or rearranged, thus quickly altering venous return to the heart.

Once clinical signs have become evident, the only treatment which has been found to be effective is surgical and is based on a technique of jugular venotomy followed by extraction of adult worms from the right-atrial region using special alligator forceps.[2,5]

The dogs are usually placed in left, lateral recumbency and the right, external, jugular vein exposed as close as is possible to the right, thoracic inlet. The dogs can be anesthetized (preferably using a fentanyl-droperidrol combination) or the procedure may be performed under local anesthesia of the neck ensuring that the case is securely restrained, particularly once the forceps are intracardiac to avoid cardiac trauma due to sudden movement. Usually, in severe cases the dogs are so depressed that restraint is not a problem.

Practice on a cadaver, with the right chest wall resected and the right lung reflected and packed dorsally, allows the clinicians to "feel and look" their way from the venotomy site to the right atrium and the posterior vena cava.

The exposed vein is anchored with stay sutures and then alligator forceps are inserted carefully toward the heart. It is essential to be gentle in inserting the forceps as the vena cava can be easily torn, although, surprisingly, this seems to rarely occur. Radiographic guidance makes the procedure easier but also tends to encourage less patience and gentleness with advancement of the forceps. It is important to "feel your way" slowly down the cava and if in doubt "spot check" if you have radiographic support. The forceps should be long enough to reach the posterior vena cava and, after practice on different-sized cadavers, the operator should be able to gauge the length required for different-sized dogs to reach the atrial area. During insertion, the neck may need to be manipulated to facilitate entrance to the thorax and, subsequently, the heart. Often the forceps need to be medially directed at the level of the heart to enter the atrium. Care is needed once clasping begins in the heart to avoid entrapping the atrial wall in the forceps and causing atrial tears. It is not difficult to penetrate the atrial wall, particularly if the dog moves suddenly. Dogs will survive mild atrial tears, but complete rupture induces a terminal hemopericardium.

Once filariae are grasped they are extracted and the procedure repeated until a considerable knot of worms (enough to cause obstruction to venous return and turbulence) is removed. This situation can be assessed by the associated abatement of the systolic murmur as the turbulent mass is removed. Usually five to six negative extractions are performed before the procedure is stopped. The venotomy can be repaired or the vein ligated both above and below the incision.

In Japan, a flexible two-directional forcep (see Chapter 4, Reference 61) has been developed to enable filariae to be extracted prophylactically both from routine cases and from caval cases, from both the venae cavae and the right atrium, and from the pulmonary arteries and ventricle. These are used with ultrasonic and fluoroscopic guidance and could revolutionize the way we treat routine heartworm cases. That is, it could be possible to mechanically, rather than chemically, remove a large percentage of filariae and so avoid the problems of embolism of dead worms and the potential severity of thromboembolism and resultant arterial compromise, particularly in severely affected dogs. Selection of such routine cases could be based on antigen levels to predict large worm numbers.

Fluid therapy is important once the surgical procedure has been finished to help ensure renal function and avoid shutdown. However, intense fluid therapy prior to overcoming the shock status of these dogs will not improve them, the basic defect being one of obstruction to venous return and venously administered fluid will therefore not help overcome the basic defect. Additionally, in severe cases immediate surgery should be performed as that will relieve the primary cause of the shock. Valuable time could be lost attempting unnecessary fluid therapy instead of concentrating on preparing for surgery.

Large numbers of filariae have been recovered from individual dogs. Sometimes dogs will be represented months or years later with similar signs due to the presence of more filariae in the atrial area. Sometimes repeat procedures (two or three) have been performed on the same dogs 12 to 24 hr after the initial operation when the clinical signs had not improved.[32] Successful outcome eventuated with removal of further filariae, presumably to a level below that necessary to form the critical "obstructive" mass. If few filariae are extracted at surgery and/or if plasma creatinine and bilirubin values are high, a grave prognosis is given.

Successful surgery is usually associated with clinical improvement of the patient, the reduction of venous distension of the hind legs, the reduction of the heart murmur, and the clearing of the urine. Presumably, also the reduction of tachycardia and tachypnea is expected once the effects of the surgical procedure have been overcome. The associated *D. immitis* infection still needs to be treated, and rest and a good quality diet is recommended to aid with the recovery from anemia and from the effects of caval syndrome and surgery. It is advisable to prescribe an antibiotic cover as occasionally in acute cases there is the possibility of wound contamination during the procedure. Cases are usually routinely treated for dirofilariasis 1 to 2 months after extraction.

The papers by Jackson and colleagues[5,11] and by Fujii,[2] reporting 512 cases are considered excellent, basic papers on this syndrome and its treatment.

VII. SUMMARY

The literature on caval syndrome has highlighted a central role that the presence of large numbers of adult *D. immitis* in the thoracic, posterior vena cava was thought to play in the pathogenesis of caval syndrome. It has been shown instead that the right atrium is the more likely site of hemolysis. The conclusion reached is that, for "mechanical", intravascular hemolysis and shock to occur, a vascular obstruction to blood flow (to reduce venous return) has to occur at a point where blood velocity is high (to induce turbulence), a condition fulfilled within the right atrium in cases of caval syndrome.

Additionally, it has been established that shear stress affects in a similar way all formed, blood elements and, in the case of platelets, induces aggregation and release reactions with the platelets being transformed into bizzare shapes, becoming sticky, and presumably generating the formation of fibrin on the surfaces to which they adhere. This deposition of fibrin strands would be expected to then contribute to, and enhance the level of, intravascular hemolysis occurring due to direct, shear-induced, mechanical factors.

Thus, it would seem the factors involved in caval syndrome relate to (1) some degree of venous obstruction, (2) right-atrial turbulence associated with a presence of large numbers of filariae, and (3) the possibility of shear-induced, mechanical hemolysis associated with some preexisting potential for lysis and, perhaps, microangiopathic hemolysis associated with platelet activation and secondary fibrin deposition.

Thus, the syndrome consists mainly of shock due to reduced venous return caused by filarial obstruction and the development of hemolysis associated with platelet activation and turbulence in a younger dog with a large number of filariae, presumably infected during a short but massive exposure period. The extent of associated pulmonary pathology, dependent on the number of filariae and on the duration of infection and associated, secondary, arterial disease, would seem to be unrelated to the induction of the syndrome.

REFERENCES

1. **Jackson, R. F.**, The venae cavae or liver failure syndrome in heartworm disease, *J. Am. Vet. Med. Assoc.*, 154, 384, 1969.
2. **Fujii, I.**, A clinical study of venae cavae embolism by heartworms of dogs, *Bull. Azabu Vet. Coll.*, 30, 105, 1975.
3. **Ishihara, K., Kitagawa, H., Ojima, M., Yagata, Y., and Suganuma, Y.**, Clinicopathological studies on canine dirofilarial hemoglobinuria, *J. Vet. Sci.*, 40, 525, 1978.
4. **Atwell, R. B. and Farmer, T. S.**, Clinical pathology of the 'caval syndrome' in canine dirofilariasis in northern Australia, *J. Small Anim. Pract.*, 23, 675, 1982.
5. **Jackson, R. F., Seymor, W. G., Growney, P. J., and Otto, G. F.**, Surgical treatment of the caval syndrome of canine heartworm disease, *J. Am. Vet. Med. Assoc.*, 171, 1065, 1977.
6. **Kitani, S., Tagawa, M., Nishida, N., Uematsu, K., Sako, T., Koyoma, H., Uchino, T., Motoyoshi, S., and Kurokawa, K.**, A case of caval syndrome (tricuspid valvular insufficiency) of heartworm disease in a senile dog, *Bull. Nippon Vet. Zootech. Coll. Nihon Jui Chikusan Daigaku Kiyo*, 30, 136, 1981.
7. **Jackson, R. F.**, The venae cavae syndrome, in *Proceedings of the Heartworm Symposium '74*, Morgan, H. C., Ed., Veterinary Medicine Publ., Bonner Springs, Kan., 1975, 48.
8. **Sawyer, T. K. and Weinstein, P. P.**, Experimentally induced canine dirofilariasis, *J. Am. Vet. Med. Assoc.*, 143, 975, 1963.
9. **Sawyer, T. K.**, The venae cavae syndrome in dogs experimentally infected with *Dirofilaria immitis*, in *Proceedings of the Heartworm Symposium '74*, Morgan, H. C., Ed., Veterinary Medicine Publ., Bonner Springs, Kan., 1975, 45.
10. **Von Lichtenberg, F., Jackson, R. F., and Otto, G. F.**, Hepatic lesions in dogs with dirofilariasis, *J. Am. Vet. Med. Assoc.*, 141, 121, 1962.
11. **Jackson, R. F., Otto, G. F., Bauman, P.M., Peacock, F., Hinrichs, W. L., and Maltby, J. H.**, Distribution of heartworms in the right side of the heart and adjacent vessels, *J. Am. Vet. Med. Assoc.*, 149, 515, 1966.
12. **Atwell, R. B.**, Early stages of disease of the peripheral pulmonary arteries in canine dirofilariasis, *Aust. Vet. J.*, 56, 157, 1980.
13. **Buoro, I. B. J. and Atwell, R. B.**, Development of a model of caval syndrome in dogs infected with *Dirofilaria immitis*, *Aust. Vet. J.*, 61, 267, 1984.
14. **Franklin, K. L. and Janker, R.**, Effects of respiration upon the venae cavae of certain mammals, as studied by means of X-ray cinematography, *J. Physiol. London*, 81, 434, 1934.
15. **Yamada, E.**, Experimental and clinical studies of echocardiography for the diagnosis of canine heartworm disease. II. Clinical application of canine heartworm disease, *Bull. Azabu Vet. Coll.*, 4, 281, 1979.
16. **Hashimoto, H.**, personal communication, 1985.
17. **Jackson, R. F., Von Lichtenberg, F., and Otto, G. F.**, Occurrence of adult heartworm in the venae cavae of dogs, *J. Am. Vet. Med. Assoc.*, 141, 117, 1962.
18. **Buoro, I. B. J.**, Aspects of Natural and Experimental Caval Syndrome, Ph.D. thesis, University of Queensland, Brisbane, Australia, in press.
19. **Atwell, R. B. and Boreham, P. F. L.**, Possible mechanisms of the caval syndrome in dogs infected with *Dirofilaria immitis*, *Aust. Vet. J.*, 59, 161, 1982.
20. **Knight, D. H.**, Heartworm disease, in *Veterinary Internal Medicine*, 2nd ed., Ettinger, S. J., Ed., W. B. Saunders, Philadelphia, 1983, 1099.
21. **Nomura, Y., Ibaraki, J., and Saito, Y.**, Pathological observation in the venae cavae syndrome of canine dirofilariasis, *Bull. Azabu Vet. Coll.*, 3, 129, 1982.
22. **Atwell, R. B.**, unpublished data, 1982.
23. **Ishihara, K., Kitagawa, H., Yokoyama, S., and Ohashi, H.**, Studies on haemolysis in canine dirofilarial hemoglobinuria lipid alterations in blood serum and red cell membrane, *Jpn. J. Vet. Sci.*, 43, 1, 1981.
24. **Jen, C. J. and McIntire, L. V.**, Characteristics of shear-induced aggregation in whole blood, *J. Lab. Clin. Med.*, 103, 115, 1984.

25. **Kociba, G. J. and Hathaway, J. E.,** Disseminated intravascular coagulation associated with heartworm disease in the dog, *J. Am. Anim. Hosp. Assoc.,* 10, 373, 1974.
26. **Kitagawa, H., Sasaki, Y., and Ishihara, K.,** Clinical studies on canine dirofilarial hemoglobinuria: blood coagulation system findings, *Jpn. J. Vet. Sci.,* 47, 575, 1985.
27. **Kitagawa, H., Sasaki, Y., and Ishihara, K.,** Clinical studies on dirofilarial hemoglobinuria: central venous pressure before and after heartworm removal, *Jpn. J. Vet. Sci.,* 47, 691, 1985.
28. **Ohno, I.,** personal communication, 1982.
29. **Adams, E. W.,** A case of dirofilariasis, with obstruction of hepatic veins, *North Am. Vet. Assoc.,* 37, 229, 1956.
30. **Jain, N. C.,** Osmotic fragility of erythrocytes of dogs and cats in health and in certain hematological disorders, *Cornell Vet.,* 63, 411, 1973.
31. **Tagawa, M., Kim, B. S., Nakawishi, A., Nakamura, T., Kobori, H., Ejima, H., Kurokawa, K., and Mitani, S.,** Hemoglobinuria caused by tricuspid valvular lesion with special reference to three dogs with dirofilaria infection, *Bull. Nippon Vet. Zootech. Coll. Nihon Jui Chikusan Daigaku Kiyo,* 31, 114, 1982.
32. **Farmer, T. S.,** personal communication, 1985.

Chapter 12

FELINE HEARTWORM DISEASE

Ray Dillon

TABLE OF CONTENTS

I. INTRODUCTION

Dirofilaria immitis infection in cats is a recognized clinical problem with increasing incidence and public awareness. Heartworm disease in cats has been reported worldwide and is consistently diagnosed in heartworm endemic areas.[1-14] The increased awareness of the disease has made ante mortem diagnosis more common.[15-19] The prevalence of heartworm infection in the cat is generally accepted to be similar to that of the dog population of the area, but at a lower rate of infection.

Experimental production of heartworm disease is more difficult in cats than in dogs,[12,20-26] and the percentage of infective larvae (L_3) developing into adult worms is significantly less in cats (1 to 25%) than in dogs (40 to 90%).[3,22,26] Experimentally, when adequate infective larvae are used, the percentage of infective larvae developing into adult *D. immitis* is low, but the percentage of cats from which adult worms are recovered is high (66 to 90%). Thus, the cats at risk would be those present in a heavily endemic area where repeated bites by infected mosquitoes would be common. Further, it would appear that infective larvae in the cat are perhaps poorly oriented and, therefore, ectopic sites for the adult worms (brain, subcutaneous tissue) are more common than in the dog. After experimental inoculations of L_3 larvae in the cat, abnormal migration patterns and the resultant death of migrating larvae would appear to cause subcutaneous nodules and granuloma formation.[21,22]

Correspondingly, the worm burden is less in the cat (range usually one to nine worms) compared to the dog, but up to 19 adults have been experimentally produced in one cat. Although the adult worms reach significant size in the cat (female > 21 cm, male > 12 cm), their development seems to be slower than in the dog.[22,26] Once infected by adult heartworms either by infective larvae (L_3) or transplantation (from superinfected dogs), the natural resistance of the cat induces a shortened period of patency and a lower concentration of or absence of microfilariae.[21,22] The average time for a patent microfilaremia following the experimental transfer of infective larvae into a cat is about 8 months compared to the 5 to 6 1/2 months in the dog. Since the microfilaremia is transient and is of very low numbers, the use of concentration tests is recommended for diagnosis. A negative test certainly does not exclude heartworm disease. Microfilaremia is uncommon (< 20% of spontaneous clinical cases) and inconsistant or transient when present.[3] Infective larvae developed in about 1% of *Anopheles quadrimaculatus*, *Aedes aegypti*, and *Ae. togoi* that fed on cats with patent infections.[22] Thus, the cat is a potential but probably insignificant source as a reservoir for the parasite.

Evidence tends to support the premise that adult heartworms in the cat have a shorter lifespan (probably less than 2 years) compared to those in the dog (approximately 5 years).[22,26,27] Survival of adult worms after transplantation or after L_3 infections have indicated that the cat does not harbor the adult for as long as the dog and that spontaneous recovery is much more likely in the cat. A shortened longevity would contribute to an underestimation of the true incidence of heartworm disease in the cat based on routine necropsies. A gradual decrease in the number of adult worms found in the heart has been noted when cats are chronologically studied.[22,26] Thus compared to the dog, the cat has a lower susceptibility to *D. immitis* with a more transitory nature to the disease.

II. PATHOLOGY

The general pulmonary pathology in the cat is similar to that of the dog.[22,26] Arterial dilation and thickening, villous endarteritis, and cellular infiltrates of the adventitia are typically more severe in the caudal pulmonary arteries. A typical reaction of the cat is the development of severe muscular hypertrophy in the smaller arteries. The host's response to

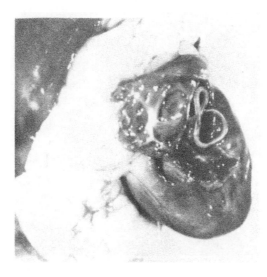

FIGURE 1. The heart of an asymptomatic cat that died peracutely with two adult heartworms. The lungs were severely congested. The right ventricular wall is dissected free.

the parasite is intense as demonstrated by enlarged pulmonary arteries within 1 week of transplantation.[3] Embolization of pulmonary arteries is a major contributing factor to the initiation of clinical signs. Although pulmonary hypertension does occasionally occur, right-axis electrocardiographic changes, radiographic evidence of right-sided cardiac enlargement, and right-sided heart failure are infrequent, suggesting that severe cor pulmonale is uncommon in heartworm disease of the cat.

Although in chronic cases perivascular reactions and evidence of thrombus formation with recanalization are noted, it would appear that the physical presence of the worm causes more pathology in the cat than in the dog (Figures 1 and 2). Thrombotic obstruction of blood flow, especially in the caudal pulmonary arteries, is associated with acute clinical signs. The lung lobe involved becomes edematous with areas of necrosis and hemorrhage believed to be associated with the release of serotonin and bradykinin, affecting the perfusion of the collateral circulation and intimal permeability. If the cat survives the initial embolic lesion, recanalization around the obstruction occurs rapidly and the lung is markedly improved within days.

III. CLINICAL DISEASE

A. Infection Rates

There is no age prediliction to *D. immitis* infection in cats and a wide age range of clinically infected cats has been reported (1 to 17 years).[13,15,16] Indoor and outdoor cats are represented, but the disease is more common in outdoor cats. The higher incidence in males compared to females may represent a sex susceptibility, but more probably represents increased exposure. Feline leukemia virus (FeLV) infection is not a predisposing factor, and heartworm infection is not a common incidental finding at necropsy of cats with FeLV.[3]

B. Clinical Signs

Infected cats may die acutely, exhibit chronic signs, or have no clinical signs at all (Table 1). Based on cardiopulmonary changes and experimental studies, most infected cats even with severe heartworm disease are asymptomatic. In the acute cases, death may be so rapid

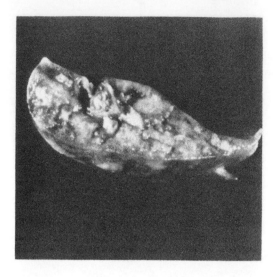

FIGURE 2. A cross-sectional cut of the caudal lung lobes of the cat in Figure 1 demonstrates that, although the cat had been asymptomatic, chronic changes of fibrosis and severe hypertrophy of the pulmonary vasculature were obvious.

Table 1
CLINICAL SIGNS
ASSOCIATED WITH
FELINE DIROFILARIASIS

Acute	Chronic
Collapse	Coughing
Dyspnea	Vomiting
Convulsions	Dyspnea
Vomiting/diarrhea	Lethargy
Blindness	Anorexia
Tachycardia	Weight loss
Syncope	

as to preclude diagnosis or treatment. Sudden death has been attributed to circulatory collapse and respiratory failure from acute pulmonary, arterial, and parenchymal infarction.[1,13,28] Acute collapse may occur with or without previous clinical signs. Cats which die from heartworm disease can be clinically normal 1 hr before death. All cats in endemic areas with periacute death should be examined for heartworm disease. In such cases as few as two worms have been found accompanied by severe pulmonary congestion, infarction, and edema. The worms in the acute syndrome are not always found embolizing the main pulmonary arteries.

The most common complaints in cats with clinical signs are coughing, dyspnea, vomiting, lethargy, anorexia, and weight loss. Vomiting or respiratory signs are the predominate clinical complaint in chronic cases. It is unusual for an infected cat to have both respiratory and gastrointestinal signs. Vomiting tends to be sporadic and can be related to eating. The vomitus generally contains food or foam and is rarely bile stained. Retching and severe paroxysmal vomiting is a rare finding. Heartworm disease in endemic areas should be included in the differential diagnosis of chronic emesis in the cat.[3]

The respiratory complaints most common are coughing and intermittent dyspnea. The dyspnea may represent acute embolism. On occasions, occlusion of a caudal pulmonary artery is accompanied by radiographic appearance of lung lobe consolidation and the development of life-threatening, acute dyspnea. Hemoptysis is occasionally noted. Coughing can occur in severe paroxysmal attacks. Periods which may last several days when the animal appears normal are often seen between coughing episodes. Coughing is usually temporarily responsive to corticosteroid therapy with exacerbations occurring during therapy. Such a clinical presentation and response to therapy often lead to a tentative diagnosis of bronchial asthma in the infected cat.

The nonspecific clinical signs are similar to those seen with many feline diseases. Anorexia and/or lethargy can be the only presenting signs in some cases of heartworm disease. Similarly, heartworm disease is often an incidental finding on thoracic radiographs during diagnostic screening.

Cats with worms found in abnormal locations may have signs attributable to local pathology. Neurological signs are uncommon, but can occur in infected cats with or without worms in the central nervous system (CNS).[2,5,7,29,30]

IV. DIAGNOSIS

A. Physical Examination

The physical examination is normal in most infected cats. A systolic murmur and, occasionally, a gallop rhythm can be present, but, as a general rule, are uncommon. Harsh lung sounds are the most frequent abnormal finding and can be present in cats without respiratory signs. Ascites, exercise intolerance, and signs of right-sided heart failure are rare. There does not seem to be a correlation between the clinical signs, physical findings, and radiographic findings in cases of feline heartworm disease.

B. Clinical Pathology

Routine complete blood counts may demonstrate a mild anemia (22 to 33% hematocrit), occasional nucleated erythrocytes, and an eosinophilia and basophilia. The anemia is present in about one third of infected cats and is nonregenerative based on reticulocyte counts. Peripheral eosinophilia, present in about one third of cats at diagnosis, is an inconsistent finding even on serial samples in the same cat and is dependent on the stage of infection. The eosinophilia occurs 4 to 7 months postinfection and intermittently thereafter.[21,22] Absence of eosinophilia does not exclude a diagnosis of feline dirofilariasis. As in the dog, the presence of basophilia is highly suggestive of heartworm disease.[3]

Blood chemistries and urinalysis in infected cats are usually normal. Although hyperglobulinemia does occur in some animals,[15] it is not consistent nor predictable and should not be used to rule out feline heartworm disease.[3,27] Normal serum globulins and normal electrophoresis are found in cats that are heartworm positive based on the results of the Knott test, the indirect fluorescent antibody test (IFA), and/or the enzyme-linked immunosorbent assay (ELISA) (Table 2).

Cats experimentally infected either with L_3 larvae or transplanted adults usually produce a transient microfilaremia if both sexes of worms are present. The microfilaremia, however, is of short duration and of low concentration. Thus, a positive blood test for microfilariae is unlikely, but diagnostic if present. The odds are increased by repeated testing (three to four tests) and using larger quantities of blood (5 mℓ) for each test. Concentration techniques such as the Knott test or millipore filter techniques are best. Even with repetitive testing, occult infections represent over 80% of feline heartworm disease.

C. Serology

There are three methods which have been used in the diagnosis of feline heartworm

Table 2
DIAGNOSTIC TESTS AVAILABLE FOR
SUSPECTED FELINE DIROFILARIASIS

Complete blood count	IFA-microfilarial antibody
Knott test	ELISA-adult antibody
Thoracic radiographs	ELISA-adult antigen
Fecal examination	Tracheal wash
EKG	Arteriogram

disease: immunofluorescence tests for microfilarial antibody, ELISA for antibody to adult worms, and ELISA for adult antigen. The variability of the methodology of individual laboratories and confusion over the interpretation of results have provided the practicing veterinarian with serious problems in making a definitive diagnosis of dirofilariasis based on serology alone. In view of the high incidence of occult disease in the cat, a good serological test to detect *D. immitis* antigen would be a valuable asset to diagnosis.

The IFA (detecting antibodies to microfilarial cuticular antigen) is positive in about 33% of cats with occult heartworm infections. However, in other occult cats (e.g., presence of sterile worms, worms of only one sex) or in the absence of host response to cuticular microfilarial antigen a diagnostic titer will not occur.[16] The use of the somatic IFA (detecting antibodies to microfilarial somatic antigen) is nonspecific.[26]

ELISA tests which are designed to detect antibodies to adult heartworm antigens show promise, but false-positive reactions related to cross reactivity remain a potential complication.[26,31] The use of such ELISA tests (as adapted from canine ELISA tests)[31,32] in the cat to confirm a clinical diagnosis has been very helpful. Initial studies of cats that have eliminated adult parasites naturally or after adulticidal treatment reveal that a negative ELISA antibody titer develops when host antibody gradually decreases to negative concentrations (4 to 6 months).

An ELISA test for the detection of adult antigen in the circulation (Filarochek®) has been developed which can be used on any species of animal.[33] This test is positive in cats within 2 to 4 weeks of transplantation of mature adult worms from dogs.[27] Since the antigen being detected seems to be derived primarily from the adult reproductive tract (especially of the female), a low worm burden, or a mainly male infection, or sexually immature worms may not produce enough antigen to make the test positive.[33] Cats infected with L_3 larvae became positive in the test after 8 months following larval infection. However, cats with clinical heartworm disease and high ELISA-antibody titers have been negative on ELISA-antigen testing. Thus, because of the nature of the low number and slow maturation of adult worms typically noted in cats, it would be prudent at this time to consider a positive ELISA-antigen test diagnostic, but a negative test does not rule out dirofilariasis. After elimination of adult worms, the ELISA-antigen test becomes negative and, as for the dog, may be predictive of the complete removal of filariae.

D. Electrocardiogram

Although subtle signs of right ventricular enlargement are occasionally noted (with unipolar chest leads),[15,34] a right axis vector ($>$120 degrees) on a standard 6-lead electrocardiogramm (EKG) is rare.[3] Ectopic ventricular beats and other arrhythmias have rarely been seen after adulticide therapy in asymptomatic cats.

E. Radiography

Radiography is one of the best screening tests for feline heartworm infection. The pulmonary parenchymal changes are nonspecific and can change rapidly in infected cats. These changes include diffuse or coalescing infiltrates, perivascular densities, and lung atelec-

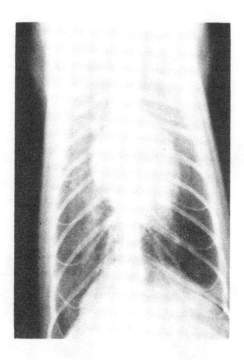

FIGURE 3. Although the pulmonary parenchymal pattern varies greatly, the most consistent lesion in cats with heartworm disease is the marked prominence of the caudal pulmonary artery. However, severe lung densities can sometimes "hide" this right caudal artery.

tasis.[4,17,35,36] The most distinctive radiographic sign is enlarged pulmonary arteries with ill-defined margins (Figure 3). This is most prominent in the caudal lung lobes on the ventro-dorsal view. Truncation and tortuosity of the pulmonary arteries are occasionally seen, but not as commonly as in the dog. An enlarged main pulmonary arterial segment extending beyond the cardiac border on the ventro-dorsal or dorso-ventral view is not a classic feature of feline heartworm infection.

Arteriograms as a diagnostic tool may more clearly demonstrate the enlarged pulmonary arteries and embolic obstruction of the vessel.[4,32,35] A nonspecific angiocardiogram is a simple and safe method for confirming a tentative diagnosis of dirofilariasis in the cat. A radiographic exposure 5 to 6 sec after injection of contrast medium into the jugular vein will provide good visualization of worms. There does not seem to be a correlation between the severity of the arterial disease as seen with arteriography and the severity of clinical signs or of postadulticide reactions (Figure 4).

F. Tracheal Cytology

The finding of eosinophils on a tracheal wash is common in heartworm disease, asthma, and other parasitic lung diseases. In feline heartworm disease, the number of eosinophils in the wash seems to correlate with the stage of infection, being common 4 to 7 months after L_3 infection and often absent at other times even when adult worms are present. Results typical of chronic inflammation may be present after the eosinophilic reaction resolves. Careful fecal examination should be performed before the tracheal wash to determine whether other parasitic lung infections are the cause of the eosinophilia. This may reveal the large operculated eggs of *Paragonimus kellicotti* or the larvae of *Aelurostrongylus abstrusus* or of *Strongyloides felis*.

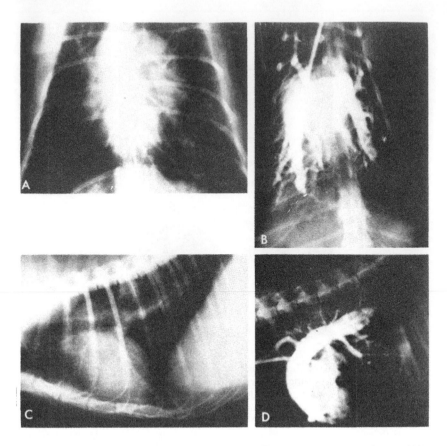

FIGURE 4.   A cat was presented for chronic vomiting. The arteriogram was used as an aid in diagnosis because of the nondiagnostic radiographic pattern. (A) Ventro-dorsal radiograph with some prominence of the caudal pulmonary arteries. Notice the absence of an enlarged right ventricle and main pulmonary arterial segment. (B) Ventro-dorsal radiographs 4 sec after the injection of positive contrast medium demonstrating the enlargement and truncation of the pulmonary arteries. With the use of contrast, a bulge in the main pulmonary arterial segment can be seen. (C) Lateral radiograph with minimal changes except for slight increase in pulmonary density in the caudal lung lobes. (D) Lateral radiographs 4 sec after the injection of positive contrast demonstrating truncation of the caudal pulmonary arteries.

G. Differential Diagnosis

In the cat with respiratory signs, heartworm disease must be differentiated from *A. abstrusus* or *P. kellicotti* infection, asthma, and cardiomyopathy. Although each disease in various stages can mimic the clinical and radiographic pulmonary parenchymal changes of heartworm, the pulmonary arterial changes of heartworm disease are unique and can be enhanced by contrast radiography procedures. Peripheral eosinophilia, eosinophilic tracheal cytology, and the chronic cough of feline dirofilariasis can be misdiagnosed as "bronchial asthma". However, a higher prevalence of feline asthma has not been reported in heartworm endemic areas. The enlarged pulmonary arteries and muscular hypertrophy seen in experimental infections of *A. abstrusus* and *Toxocara cati* infection are clinically uncommon in heartworm disease.

V. THERAPY

A. Infection

Treatment of feline heartworm cases with thiacetarsamide sodium (2.2 mg/kg intravenously twice a day for 2 days) is safely tolerated by cats without immediate complications of hepato- or nephrotoxicity.[3,15,27] Clinical signs tend to improve after therapy. However, chronically anorexic cats may have to be hyperalimentated. Although the presence of circulating microfilariae is uncommon, dithiazanine iodide and levamisole hydrochloride have both been used successfully and safely as microfilaricides.

Complications after therapy are related to embolization. Sudden death can occur especially within the first 10 days postadulticide. Although severe secondary thrombocytopenia has not been noted, thromboembolism can induce severe pulmonary infarction causing hemoptysis and dyspnea. Embolism most often affects the caudal lung lobes, and thoracic radiographs may demonstrate a consolidated lung lobe. Oxygen therapy is indicated if hypoxemia or cyanosis occur. High doses of corticosteroids (0.5 to 1 mg/kg prednisolone three times a day) with careful i.v. fluid (lactated ringer) therapy will often support the cat through the crisis. The routine use of corticosteroids is not recommended before or after thiacetarsamide in cats. Aspirin has not been evaluated in feline heartworm disease and thrombocytopenia does not seem to be a common sequela in severe heartworm disease. Although cats are sensitive to platelet aggregation in aortic embolism and aspirin has been recommended to improve the collateral blood flow, the pulmonary vasculature of the cat in the acute crisis relative to aspirin therapy is unknown. Aspirin inhibits prostaglandin formation and thus may increase leukotriene production in the lung. This could result in increased bronchospasm and pulmonary hypertension. Until more information is available, it is not currently recommended that aspirin be used in feline heartworm disease and it may even be contraindicated in the acute crisis.

The periacute nature of the postadulticide reaction dictates that the cat be under constant attention especially during the first 2 weeks.[27] Even though clinical and radiographic signs of acute embolism can resolve over 1 to 2 days, death can occur before therapy can be instituted. The client should be made aware that the risk and severity of complications in the cat are more severe than they are in the dog.

Currently, in the asymptomatic cat with the incidental diagnosis of dirofilariasis, the potential severity of the post adulticidal reaction poses a dilemma for the veterinarian where the risk of postadulticide complications is probably greater than is the potential risk from spontaneous disease. Although *D. immitis* adults may not live as long in cats as in dogs, the development of clinical signs and of death (low probability) are also of potential risk if cats are left untreated. Up until now, the efficacy of thiacetarsamide could not be evaluated in client cats because of the occult nature of the disease. However, with cats that have had a postadulticide microfilaremia, repeated attempts to eliminate microfilariae have failed and repeated adulticidal therapy was required. Problems associated with adulticide efficacy were alluded to in other microfilaremic cats.[18,19] However, current research seems to indicate that the adulticide is effective and that clinical signs usually abate during the initial weeks after thiacetarsamide.

B. Prophylaxis

Due to the relative resistance of the cat to infective larvae and perhaps the resistance of the cat to mosquito bites, it is not currently recommended that diethylcarbamazine be used as a prophylactic in the cat. If in a highly endemic area a client requests preventative measures, diethylcarbamazine can be safely administered to cats. The value of ivermectin in the cat is unknown at this stage.

REFERENCES

1. **Bernard, M. A.,** Feline dirofilariasis, *Can. Vet. J.,* 11, 190, 1970.
2. **Cusick, P. K., Todd, K. S., Jr., Blake, J. A., and Daly, W. R.,** *Dirofilaria immitis* in the brain and heart of a cat from Massachusetts, *J. Am. Anim. Hosp. Assoc.,* 12, 490, 1976.
3. **Dillon, A. R.,** Feline heartworm disease: clinical evaluation, in *Proceedings of the Heartworm Symposium '83,* Otto, G. F., Ed., Veterinary Medicine Publ., Edwardsville, Kan., 1983, 31.
4. **Donahoe, J. M., Kneller, S. K., and Lewis, R. E.,** *In vivo* pulmonary arteriography in cats infected with *Dirofilaria immitis, J. Am. Vet. Radiol. Soc.,* 17, 147, 1976.
5. **Donahoe, J. M. R. and Holzinger, E. A.,** *Dirofilaria immitis* in the brains of a dog and cat, *J. Am. Vet. Med. Assoc.,* 164, 518, 1974.
6. **Lillis, W. G.,** *Dirofilaria immitis* in dogs and cats from south-central New Jersey, *J. Parasitol.,* 50, 802, 1964.
7. **Mandelker, L. and Brutus, R. L.,** Feline and canine dirofilarial encephalitis, *J. Am. Vet. Med. Assoc.,* 159, 776, 1971.
8. **Noyes, J. D.,** Illinois state veterinary medical association six year survey, in *Proceedings of the Heartworm Symposium '77,* Otto, G. F., Ed., Veterinary Medicine Publ., Bonner Springs, Kan., 1978, 1.
9. **Seymore, D. N.,** Dirofilariasis in a cat, *Mod. Vet. Pract.,* 61, 251, 1980.
10. **Sherman, W. A. and Wechsler, S. J.,** Unusual case of heartworm disease in a cat, *Vet. Med. Small Anim. Clin.,* 70, 1320, 1973.
11. **Stackhouse, L. L. and Clough, E.,** Clinical report: five cases of feline dirofilariasis, *Vet. Med. Small Anim. Clin.,* 67, 1309, 1972.
12. **Soifer, F. K.,** Dirofilariasis in a cat, *Vet. Med. Small Anim. Clin.,* 71, 484, 1976.
13. **Teske, R. H.,** Dirofilariasis in a cat, *J. Am. Vet. Med. Assoc.,* 159, 891, 1971.
14. **Todd, K. S., Jr., Byerly, C. S., Small, E., and Krone, J. V.,** Heartworm infections in Illinois cats, *Feline Pract.,* 6, 41, 1976.
15. **Calvert, C. A. and Mandell, C. P.,** Diagnosis and management of feline heartworm disease, *J. Am. Vet. Med. Assoc.,* 180, 550, 1982.
16. **Dillon, R., Sakas, P. S., Buxton, B. A., and Schultz, R. D.,** Indirect immunofluorescence testing for diagnosis of occult *Dirofilaria immitis* infection in three cats, *J. Am. Vet. Med. Assoc.,* 181, 80, 1982.
17. **Hawe, R. S.,** The diagnosis and treatment of occult dirofilariasis in a cat, *J. Am. Anim. Hosp. Assoc.,* 14, 577, 1979.
18. **Harlton, B. W.,** Treatment of dirofilariasis in a domestic cat (a clinical report), *Vet. Med. Small Anim. Clin.,* 69, 1440, 1974.
19. **Schwartz, A.,** Two cases of feline heartworm disease, *Feline Pract.,* 5, 20, 1975.
20. **Byerly, C. S., Donahoe, J. M. R., and Todd, K. S., Jr.,** Histopathological changes in cats experimentally infected with *Dirofilaria immitis,* in *Proceedings of the Heartworm Symposium '77,* Otto, G. F., Ed., Veterinary Medicine Publ., Bonner Springs, Kan., 1978, 79.
21. **Donahoe, J. M. R., Kneller, S. K., and Lewis, R. E.,** Hematologic and radiographic changes in cats after inoculation with infective larvae of *Dirofilaria immitis, J. Am. Vet. Med. Assoc.,* 168, 413, 1976.
22. **Donahoe, J. M. R.,** Experimental infection of cats with *Dirofilaria immitis, J. Parasitol.,* 61, 599, 1975.
23. **Fowler, J. L., Matsuda, K., and Fernau, R. C.,** Experimental infection of the domestic cat with *Dirofilaria immitis, J. Am. Vet. Med. Assoc.,* 8, 79, 1972.
24. **Pacheco, G.,** Synopsis of Dr. Seije Kume's reports at the first international symposium on canine heartworm disease, in *Canine Heartworm Disease: The Current Knowledge,* Bradley, R. E., Ed., University of Florida, Gainesville, Fla., 1972, 137.
25. **Rawlings, C. A., Losonsky, J. M., Lewis, R. E., Hubble, J. J., and Priestwood, A. K.,** Response of the feline heart to *Aelurostrongylus abstrusus, J. Am. Anim. Hosp. Assoc.,* 16, 573, 1980.
26. **Wong, M. M., Pedersen, N. C., and Cullen, J.,** Dirofilariasis in cats, *J. Am. Anim. Hosp. Assoc.,* 19, 855, 1983.
27. **Dillon, A. R.,** Feline heartworm disease, *Grace Kemper Research Project,* in progress.
28. **Griffiths, J. H., Schlotthauer, J. C., and Gehrman, F. W.,** Feline dirofilariasis, *J. Am. Vet. Med. Assoc.,* 140, 577, 1979.
29. **Otto, G. F.,** Abnormal host — abnormal locations, *Am. Heartworm Soc. Bull.,* 4, 14, 1978.
30. **Otto, G. F.,** Occurrence of the heartworm in unusual locations and in unusual hosts, in *Proceedings of the Heartworm Symposium '74,* Morgan, H. C., Ed., Veterinary Medicine Publ., Bonner Springs, Kan., 1975, 6.
31. **Grieve, R. B., Mika-Johnson, M., Jacobson, R. H., and Cypess, R. H.,** Enzyme-linked immunosorbent assay for measurement of antibody responses to *Dirofilaria immitis* in experimentally infected dogs, *Am. J. Vet. Res.,* 42, 66, 1981.
32. **Weil, G. J., Lord, P. F., and Grieve, R. B.,** Occult feline dirofilariasis confirmed by angiography and serology, *J. Am. Anim. Hosp. Assoc.,* 19, 847, 1983.

33. **Weil, G. J., Malane, M. S., Powers, K. G., and Blair, L. S.,** Monoclonal antibodies to parasite antigens found in the serum of *Dirofilaria immitis*-infected dogs, *J. Immunol.,* 134, 1185, 1985.
34. **Calvert, C. A. and Coulter, D. B.,** Electrocardiographic values for anesthetized cats in lateral and sternal recumbencies, *Am. J. Vet. Res.,* 42, 1453, 1981.
35. **Owens, J. M. and Twedt, D. C.,** Nonselective angiocardiography in the cat, *Vet. Clin. North Am.,* 7, 309, 1977.
36. **Suter, P. F. and Lord, P. F.,** Radiographic differentiation of disseminated pulmonary parenchymal diseases in dogs and cats, *Vet. Clin. North Am.,* 4, 687, 1974.

Chapter 13

DIROFILARIASIS IN MAN

Peter F. L. Boreham

TABLE OF CONTENTS

I. INTRODUCTION

Human dirofilariasis is a rare, nonlife-threatening zoonotic disease found throughout the world where *Dirofilaria* sp. are common. The incidence of reports of infections is increasing, partly because of a heightened awareness of the disease by the medical profession and partly because of its spread in the definitive host in recent years. For example, in Australia canine dirofilariasis is spreading southwards,[1] while in the U.S. there has been a recent spread into the Mississippi River valley.[2] Several species belonging to two subgenera, *Dirofilaria* and *Nochtiella*,[3] of the genus *Dirofilaria* have been recorded infecting man. The dog heartworm, *D. immitis*, is usually seen in the lungs as a "coin lesion", but has been recorded in the cardiovascular system,[4] subcutaneous tissue,[5] eye,[6] abdominal cavity,[7] and bladder.[8] *D. repens*, a subcutaneous parasite of the dog, and *D. tenuis* and *D. ursi*, natural parasites of the raccoon and bear, respectively, have been found in a variety of subcutaneous tissues of man. The name *D. conjunctivae* has been used to describe the subcutaneous parasite of man, but it is now accepted that *D. conjunctivae* is not a separate species but has been used to describe human infections with *D. repens* or *D. tenuis*.[9,10]

II. HUMAN DIROFILARIASIS DUE TO THE SUBGENUS *DIROFILARIA*

The first report of a human infection with a member of the subgenus *Dirofilaria* was in 1887 when de Magalhaes reported finding one male and one female worm in the left ventricle of a male child from Rio de Janeiro, Brazil.[11] The parasite was identified as *D. magalhaesi* which is now considered to be a synonym of *D. immitis*. Subsequently, *D. immitis* was identified in the pulmonary artery in 1940[12] and in a pulmonary infarction in 1961.[13] Since that time, approximately 100 cases have been recorded in the literature. In no instance have microfilariae been found circulating in the blood. Man is a dead-end host who may acquire an infection when bitten by infected mosquitoes. It is believed that *D. immitis* does not normally survive in the subcutaneous tissues but occasionally is able to migrate to the right ventricle where it develops into a sexually immature worm. When death of the worm occurs, it is washed into the pulmonary artery and embolization occurs.

A. Clinical Features

The typical picture of a patient infected with *D. immitis* is one who has a spherical nodule 1 to 3 cm in diameter (a coin lesion) in the lungs discovered on routine radiography (Figure 1) or accidentally at autopsy.[7,14] Lesions are usually found in the peripheral portion of the lung as a single nodule with a random distribution between the lobes. Occasionally, two nodules have been reported in the same patient.[15]

Robinson and co-workers[11] reviewed the clinical symptoms present in 47 patients with pulmonary dirofilariasis. Over half the patients (57%) were asymptomatic, while the most common symptoms were cough (23%), chest pains (17%), eosinophilia (15%), hemoptysis (9%), and fever (6%). All infections were found in adults. It is believed that clinical symptoms may occur when the worm dies in the heart and passes into the pulmonary artery and an infarct occurs.

B. Pathology

Macroscopically, the pulmonary lesion appears as a well-circumscribed, grayish-yellow nodule, 1 to 3 cm in diameter, surrounded by normal lung parenchyma. Microscopically, there is a central zone of necrosis surrounded by a narrow, granulomatous zone composed of epithelial cells, plasma cells, lymphocytes, and an occasional, giant cell. The lesion is demarcated peripherally by fibrous tissue. The lung parenchyma around the lesion contains scattered collections of lymphocytes, macrophages, and eosinophils.[16] A single worm, usu-

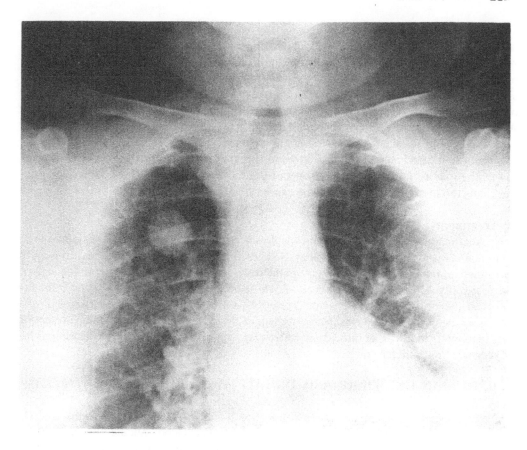

FIGURE 1. Coin lesion of *Dirofilaria immitis*. (Original radiograph kindly provided by Dr. D. E. Moorhouse, University of Queensland.)

ally necrotic and often fragmented and focally calcified, is present in the lumen of a small artery within the necrotic tissue. The worm itself does not completely fill or obstruct the lumen of the artery, but blockage is completed by intense fibroblastic proliferation. In almost all cases, the parasites are associated with thrombus formation within the pulmonary artery. It has been suggested that the worms release some kind of toxin which results in the disproportionately large globular area of necrosis which is not consistent with embolic infarction.[17] Possibly the clotting factors described by Crawford and colleagues[18] are involved and the process may be similar to that seen in pulmonary aspergillomata.[19]

C. Diagnosis

The most likely person to acquire an infection is a Caucasian male 40 to 60 years of age who lives in an area endemic for canine dirofilariasis.[15] Dirofilariasis should be considered as a possible diagnosis in patients who develop a solitary pulmonary nodule 3 cm or less in diameter in any lobe of the lung and who are asymptomatic or have an associated, mild respiratory illness. There are many causes of well-defined pulmonary lesions which must be considered in a differential diagnosis, including carcinoma, tuberculosis, fungal infections, and hamartomas.[20] Radiography and tomographic scans are nonspecific.

Serological tests for filariasis have not been very helpful in diagnosis. In proven positive cases of pulmonary dirofilariasis in North America, the indirect hemagglutination test was positive in six out of nine patients while the bentonite flocculation test was positive in three out of seven patients.[15] In one patient where a complement fixation test was used, a negative result was obtained.[13] The sera of four proven cases in Australia all had raised fluorescent antibody titers.[21] Similarly in Japan, a patient with pulmonary dirofilariasis had a raised antibody titer as demonstrated by an enzyme-linked immunosorbent assay (ELISA) test.[22] Angiocardiography of the right heart and pulmonary artery for adult worms in a single patient, transthoracic needle aspiration in four patients, examination of bronchial washings and biopsy, and sputum cytology have not proved to be of any value in diagnosis.[15] On two occasions, diagnosis of dirofilariasis has been made without surgical removal of the nodule. A fine needle-aspiration biopsy of a noncalcified mass in the right, upper lobe of the lung of a 62-year-old female revealed in sections the cross section of a worm identified as *D. immitis*.[23] In the second case, computerized tomography enabled a small lesion in a 52-year-old male to be aspirated with a percutaneous needle. Fragments of parasite were seen with the characteristic appearance of a lesion caused by *D. immitis*.[24]

D. Treatment

No treatment other than surgery is available since it is the dead worm that is found in the lungs that is responsible for the pathology. A thoracotomy, with wedge resection, is normally performed immediately after a solitary pulmonary nodule has been identified, because of the possibility of carcinoma. However, should it be possible to develop a specific immunological diagnostic test, surgery would probably not be indicated in most cases of dirofilariasis as the evidence suggests that no growth of the granuloma occurs and impairment of lung function is minimal. In those cases where diagnosis was confirmed by needle aspiration, surgery was not undertaken.[23,24]

III. HUMAN DIROFILARIASIS DUE TO THE SUBGENUS *NOCHTIELLA*

Three parasites are important causes of infection in man — *D. repens*, *D. tenuis*, and *D. ursi* (Table 1). Almost all human cases appear as subcutaneous nodules which may be migratory. *D. repens* is a natural parasite of dogs and cats and has been found in man in Europe, Asia and Africa, especially in Mediterranean countries, Russia, and Japan.[9,10] *D. tenuis* and *D. ursi* have only been recorded in North America. *D. tenuis* is a parasite of the subcutaneous tissues of the raccoon (*Procyon lotor*) and human infections from *D. tenuis* have been found mainly in the southeastern states of the U.S.[25,26] *D. ursi* is a parasite of the peritracheal and perirenal tissues of bears in North America and Asia. Human infections have been reported in the states and provinces along the U.S.-Canadian border.

A. Clinical Features

D. tenuis and *D. repens* have been found in subcutaneous nodules excised from many parts of the body, especially the breast, arms, legs, scrotum, eyelid, and conjunctiva.[16] Usually a nodule or swelling 1 to 2 cm in diameter is evident a few weeks before the affected area becomes painful. Occasionally, the nodule may be migratory. During the migratory phase of the worm through subcutaneous tissue, a slight inflammatory reaction may develop. After death of the worm, a severe granulomatous reaction may occur around the worm. The main features of ocular dirofilariasis are pain, swelling, edema, hyperemia of the conjunctiva, and itching and pruritus.

B. Pathology

The early nodule is an abscess with mainly polymorphonuclear leukocytes and eosinophils. In the chronic stage, the reaction is characterized by granulomatous inflammation consisting of epithelioid cells, foreign-body giant cells, histiocytes, and eosinophils.[16] Usually a single, coiled worm is present which is often dead and degenerating. Occasionally, adult female worms containing microfilariae have been seen.[10]

Table 1

CHARACTERISTICS OF MEMBERS OF THE GENUS *DIROFILARIA* FOUND IN MAN

	D. immitis	*D. repens*	*D. tenuis*	*D. ursi*
Subgenus	*Dirofilaria*	*Nochtiella*	*Nochtiella*	*Nochtiella*
Natural hosts	Dog, cat, wolf, fox, dingo, and other carnivores	Dog, cat	Raccoon	Bear
Distribution	Worldwide, especially U.S., Australia, Japan	Europe, Asia, Africa, especially Mediterranean countries and Russia	Southeast U.S.	U.S.-Canadian border
Adult male	12—20 cm	5—7 cm	4—4.8 cm	7 cm
Adult female	25—31 cm	10—17 cm	8—13 cm	16 cm
Diameter	150—330 μm	220—660 μm	280—330 μm	<200 μm
Cuticular ridges	Internal, except at caudal tip of male where external ridges are present	Conspicuous, sharp, external ridges separated by a distance greater than the width of the ridges, 6 to 12 μm by 1.5 to 3 μm high; approx. 100 ridges	External ridges low and rounded with wavy, broken, and branching pattern; the space between ridges is narrower than the ridges themselves	Tall, sharp, narrow, external ridges separated by a distance 3 to 4 times wider than the ridges themselves; approx. 65 ridges
Main pathology	Lungs (''coin lesion'')	Subcutaneous nodule may be migrating; eye, neck, breast, arms, abdomen, lower extremities	Similar to *D. repens*	Subcutaneous nodule

C. Diagnosis

Diagnosis is dependent upon identification of the worm either in tissue section or after the intact worm has been removed surgically. The major characteristics for species identification are given in Table 1. Serology has not proved to be a useful technique.

D. Treatment

The only treatment available is surgical removal of the nodule or the worm from the conjunctiva. A patient will normally only consult a medical practitioner with dirofilariasis when nonspecific symptoms occur resulting from the death of the worm.

IV. IDENTIFICATION OF *DIROFILARIA* SPECIES IN TISSUE SECTIONS

In cross section, members of the genus *Dirofilaria* have a thick, multilayered cuticle 5 to 15 μm thick which may be smooth or have longitudinal ridges. The lateral chords are usually prominent and the musculature is well developed with numerous cells. Internal organs consist of an intestine and two uteri in the female and a single reproductive tube in the male.[26] Specific diagnostic features of *Dirofilaria* sp. are given in Table 1 and illustrated in Figures 2 and 3. Although not yet proven to occur in man, other species of *Dirofilaria* may prove to be zoonotic, e.g., *D. subdermata* found in the subcutaneous tissues of the Canadian porcupine (*Erethizon dorsatum*).[27]

V. SERODIAGNOSIS OF DIROFILARIASIS

The only attempts at determining the incidence and distribution of dirofilariasis in the human population have been undertaken in Australia in both the Caucasian and aboriginal populations.[21,28,29] The incidence of serum antibodies in man to purified *Dirofilaria* antigen, using fluorescent antibody tests, was found to be proportional to the incidence of *D. immitis* in the canine population. The incidence of antibodies was higher in the aboriginal population (20.1%) than in the Caucasian population (2.6%). It was suggested that the low incidence of detection of human dirofilariasis in the areas where these studies were undertaken indicates that most larvae fail to establish themselves in the human host and die. Caucasians with eosinophilia were found to have positive sera in 22.2% of cases.[28] It was also noted that six out of nine patients with eosinophilia of unknown etiology who had *D. immitis* antibodies in their serum showed symptoms of cranial involvement suggesting the possibility that *D. immitis* is an agent of eosinophilic meningitis.[30] There is controversy as to whether helminth parasites, including *D. immitis*, can cause allergic asthma. In Oahu, Hawaii where approximately 50% of the dogs are infected with *D. immitis*, a significantly higher incidence of specific IgE antibodies was found in asthmatic patients than in control patients, suggesting the possible involvement of this parasite in allergic asthma.[31] There is also a suggestion that exposure to animal filariae may reduce the severity of the disease caused by *Wuchereria bancrofti* and *Brugia* sp. and could even confer some protection.[32]

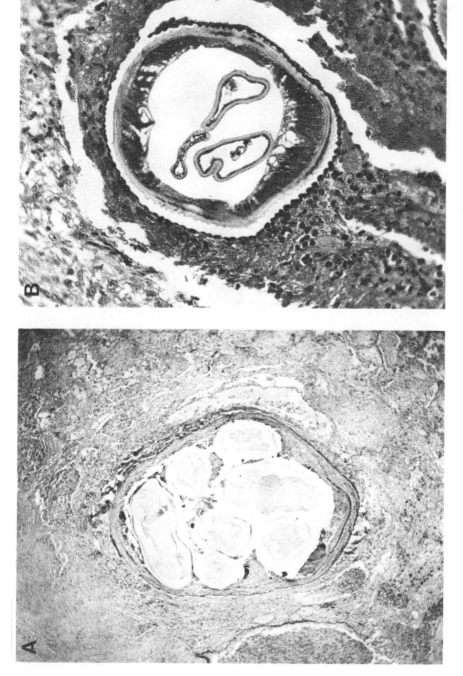

FIGURE 2. (A) Immature male *D. immitis* in a pulmonary vessel. (Movat; magnification × 50.) (B) Female *D. tenuis* in subcutaneous tissue. (Movat; magnification × 80.) (Photomicrographs kindly provided by Dr. R. C. Neafie, Armed Forces Institute of Pathology, Washington, D.C.)

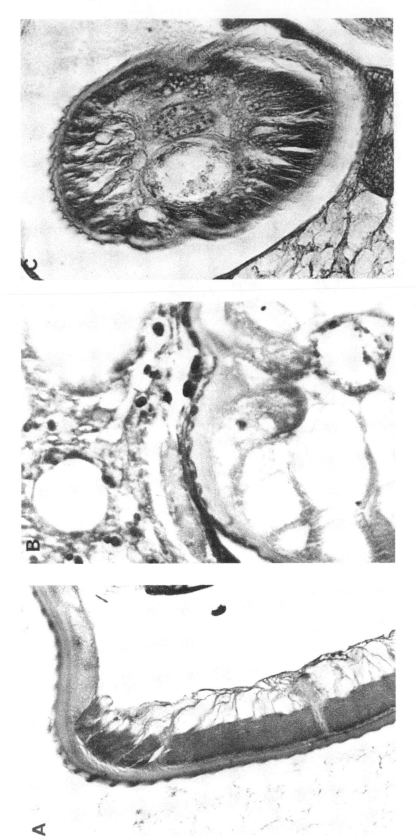

FIGURE 3. Cuticular ridge structure of (A) *D. repens* (magnification × 470), (B) *D. tenuis* (magnification × 675), and (C) *D. ursi* (magnification × 540). (From Gutierrez, Y., *Hum. Pathol.*, 15, 514, 1984. Reprinted with permission from W. B. Saunders Co.)

REFERENCES

1. **Carlisle, C. H. and Atwell, R. B.**, A survey of heartworm in dogs in Australia, *Aust. Vet. J.*, 61, 356, 1984.
2. **Adkins, R. B., Jr. and Dao, A. H.**, Pulmonary dirofilariasis: a diagnostic challenge, *South. Med. J.*, 77, 372, 1984.
3. **Faust, E. C.**, Human infection with species of *Dirofilaria*, *Z. Tropenmed. Parasitol.*, 8, 59, 1957.
4. **Takeuchi, T., Asami, K., Kobayashi, S., Masuda, M., Tanabe, M., Miura, S., Asakawa, M., and Murai, T.**, *Dirofilaria immitis* infection in man: report of a case of the infection in heart and inferior vena cava from Japan, *Am. J. Trop. Med. Hyg.*, 30, 966, 1981.
5. **Nishimura, T., Kondo, K., and Shoho, C.**, Human infection with a subcutaneous *Dirofilaria immitis*, *Biken J.*, 7, 1, 1964.
6. **Moorhouse, D. E.**, *Dirofilaria immitis:* a cause of human intra-ocular infection, *Infection*, 6, 192, 1978.
7. **Tada, I., Sakaguchi, Y., and Eto, K.**, *Dirofilaria* in the abdominal cavity of a man in Japan, *Am. J. Trop. Med. Hyg.*, 28, 988, 1979.
8. **Nelson, R. P. and Thomason, W. B.**, Human dirofilariasis of the bladder, *J. Urol.*, 133, 677, 1985.
9. **Bruijning, C. F. A.**, Human dirofilariasis: a report of the first case of ocular dirofilariasis in the Netherlands and a review of the literature, *Trop. Geogr. Med.*, 33, 295, 1981.
10. **Pampiglione, S., Franco, F., and Trotti, G. C.**, Human subcutaneous dirofilariasis. I. Two new cases in Venice. Identification of the casual agent as *Dirofilaria repens* Raillet and Henry, 1911, *Parassitologia Rome*, 24, 155, 1982.
11. **Robinson, N. B., Chavez, C. M., and Conn, J. H.**, Pulmonary dirofilariasis in man: a case report and review of the literature, *J. Thorac. Cardiovasc. Surg.*, 74, 403, 1977.
12. **Faust, E. C., Thomas, E. P., and Jones, J.**, Discovery of human heartworm infection in New Orleans, *J. Parasitol.*, 27, 115, 1941.
13. **Dashiell, G. F.**, A case of dirofilariasis involving the lung, *Am. J. Trop. Med. Hyg.*, 10, 37, 1961.
14. **Yoshimura, H. and Akao, N.**, Current status of human dirofilariasis in Japan, *Int. J. Zoon.*, 12, 53, 1985.
15. **Ciferri, F.**, Human pulmonary dirofilariasis in the United States: a critical review, *Am. J. Trop. Med. Hyg.*, 31, 302, 1982.
16. **Neafie, R. C., Connor, D. H., and Meyers, W. M.**, Dirofilariasis, in *Pathology of Tropical and Extraordinary Diseases*, Vol. 2, Binford, C. H. and Connor, D. H., Eds., Armed Forces Institute of Pathology, Washington, D.C., 1976, 391.
17. **Tsukayama, C., Manabe, T., and Miura, Y.**, Dirofilarial infection in human lungs, *Acta Pathol. Jpn.*, 32, 157, 1982.
18. **Crawford, G. P. M., Howse, D. J., and Grove, D. I.**, Inhibition of human blood clotting by extracts of *Ascaris suum*, *J. Parasitol.*, 68, 1044, 1982.
19. **Przyjemski, C. and Mattii, R.**, The formation of pulmonary mycetomata, *Cancer Brussels*, 46, 1701, 1980.
20. **Siegelman, S. S., Khouri, N. F., Scott, W. W., Jr., Leo, F. P., and Zerhouni, E. A.**, Computed tomography of the solitary pulmonary nodule, *Semin. Roentgenol.*, 19, 165, 1984.
21. **Welch, J. S. and Dobson, C.**, Antibodies to *Dirofilaria immitis* in Caucasian and aboriginal Australians diagnosed by immunofluorescence and passive Arthus hypersensitivity, *Am. J. Trop. Med. Hyg.*, 23, 1037, 1974.
22. **Sato, M., Koyama, A., Iwai, K., Kawobata, Y., and Kojima, S.**, Human pulmonary dirofilariasis with special reference to the ELISA for the diagnosis and follow-up study, *Z. Parasitenkd.*, 71, 561, 1985.
23. **Hawkins, A. G., Hsiu, J. G., Smith, R. M., Stitik, F. P., Siddiky, M. A., and Edwards, O. E.**, Pulmonary dirofilariasis diagnosed by fine needle aspiration biopsy. A case report, *Acta Cytol.*, 29, 19, 1985.
24. **Kelly, W. T., Firouz-Abadi, A. A., Roszkowski, A., and Zimmerman, P. V.**, Pulmonary dirofilariasis diagnosed by computerised tomography scan controlled percutaneous needle aspiration, *Aust. N. Z. J. Med.*, 15, 656, 1985.
25. **Orihel, T. C. and Beaver, P. C.**, Morphology and relationship of *Dirofilaria tenuis* and *Dirofilaria conjunctivae*, *Am. J. Trop. Med. Hyg.*, 14, 1030, 1965.
26. **Gutierrez, Y.**, Diagnostic features of zoonotic filariae in tissue sections, *Hum. Pathol.*, 15, 514, 1984.
27. **Gutierrez, Y.**, Diagnostic characteristics of *Dirofilaria subdermata* in cross sections, *Can. J. Zool.*, 61, 2097, 1983.
28. **Welch, J. S. and Dobson, C.**, The prevalence of antibodies to *Dirofilaria immitis* in aboriginal and Caucasian Australians, *Trans. R. Soc. Trop. Med. Hyg.*, 68, 466, 1974.
29. **Welch, J. S., Dobson, C., and Freeman, C.**, Distribution and diagnosis of dirofilariasis and toxocariasis in Australia, *Aust. Vet. J.*, 55, 265, 1979.

30. **Dobson, C. and Welch, J. S.,** Dirofilariasis as a cause of eosinophilic meningitis in man diagnosed by immunofluorescence and Arthus hypersensitivity, *Trans. R. Soc. Trop. Med. Hyg.,* 68, 223, 1974.
31. **Desowitz, R. S., Rudoy, R., and Barnwell, J. W.,** Antibodies to canine helminth parasites in asthmatic and nonasthmatic children, *Int. Arch. Allergy Appl. Immunol.,* 65, 361, 1981.
32. **Sim, B. K. L.,** Immunology of filariasis, in *Filariasis,* Bull. No. 19, Mak, J. W., Ed., Institute of Medical Research, Kuala Lumpur, Malaysia, 1983, 65.

APPENDIX

Richard B. Atwell

I. INTRODUCTION

This section is a series of flowcharts and lists based on information presented in the preceding chapters and on specific references given with some figures. There is some duplication due to the interrelated nature of some of the topics. Some of the detail is presented as hypotheses based on logical deductions and it is hoped that some of the ideas will enable researchers to see where future work should be concentrated. Other ideas on some concepts for future prognostic assessment are presented in Chapter 7.

Obstruction
    Heartworm — live and dead
    Thrombosis
    Thromboembolism
    Luminal "plugging"
Obliterative
    Extensive inflammation — mural, luminal, periarterial
    Organizing thrombi and thromboemboli
    Intimal disease, when extensive
    Myointimal proliferation
    Arterial fibrosis and the development of distal lumen fibrotic plugs
    Parenchymal periarterial hemorrhage and secondary fibrosis
    Septal fibrosis
    Pneumonitis (type 3 hypersensitivity?)
Vasomotor
    Loss of vascular elasticity, associated with the secondary fibrosis of inflammation and thrombosis
    Constriction due to release factors from platelets and thrombi, e.g., effect on S2 receptors?
    Secondary constriction due to main pulmonary artery dilatation?[1]
    Reduced effectiveness of histamine on vascular resistance in infected dogs?[2]
    Hyperkinetic effects from hypertension itself, leading to increased muscularization of the pulmonary arterial tree and so further vasomotor potential, probably does not occur in this disease?

FIGURE 1. Hypertensive factors in dirofilariasis. ([1]Laks, M. M,. Juratsch, C. E., Garner, D., Beazell, J., Jengo, J., and Criley, J. M., *Circulation*, 48 (Suppl. 4), 114, 1973. [2]O'Malley, N. A., Venugopalan, C. S., and Crawford, M. P., *Am. J. Vet. Res.*, 46, 1463, 1985.)

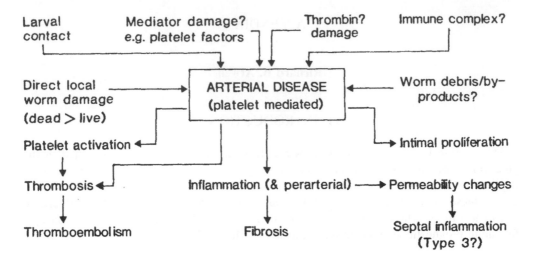

FIGURE 2.   Factors associated with arterial disease.

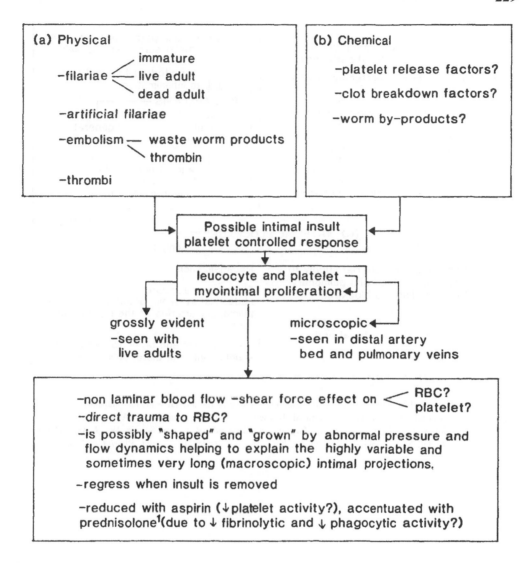

FIGURE 3.   Factors relating to the development and progression of intimal disease. RBC = red blood cell. The effects on plasminogen activation by prednisolone is possibly balanced by the presence of an activator of plasminogen present in filariae. ([1]Boreham, P. F. L., *J. Helminthol.*, 58, 207, 1984.)

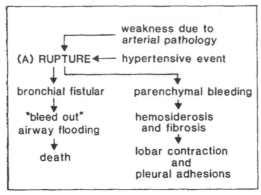
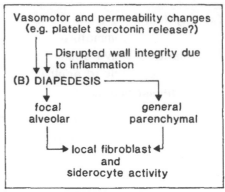
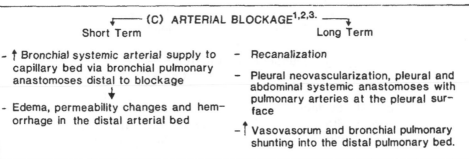

FIGURE 4.   Causes of hemorrhage in dirofilariasis: (A) rupture of the arterial wall, (B) diapedesis through the arterial wall, (C) secondary events to arterial luminal obstruction. ([1]Barie, P. S. and Malik, A. B., *J. Appl. Physiol.*, 53, 543, 1982. [2]Johnson, R. L., Cassidy, S. S., Haynes, M., Reynolds, R. L., and Schulz, W., *J. Appl. Physiol.*, 51, 845, 1981. [3]Jindal, S. K., Lakshminarayan, S., Kirk, W., and Butler, J., *J. Appl. Physiol.*, 57, 424, 1984.)

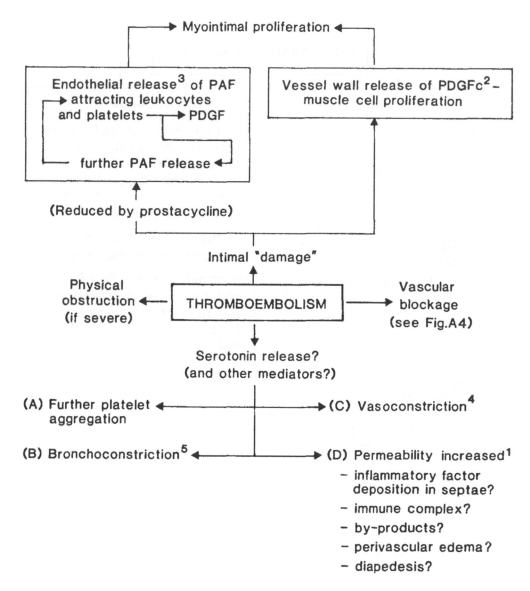

FIGURE 5. Effects of thromboembolism in the dog. Anti-serotonin therapy should block A B C and D; hydralazine, terbutaline, and nitroprusside should reverse C; PAF = platelet-activating factor; PDGF = platelet-derived growth factor; PDGFc = PDGF-like molecules. ([1]Malik, A. B. and von der Zee, H., *Circ. Res.*, 42, 72, 1978. [2]Bowen-Pape, D. F., Ross, R., and Seifert, R. A., *Circulation*, 72, 735, 1985. [3]Benveniste, J. and Chignard, M., *Circulation*, 72, 713, 1985. [4]Haver, V. M. and Namm, D. H., *Blood Ves.*, 21, 53, 1984. [5]Thomas, D., Stein, M., Tanaba, G., Rege, V., and Wessler, S., *Am. J. Physiol.*, 206, 1207, 1964.)

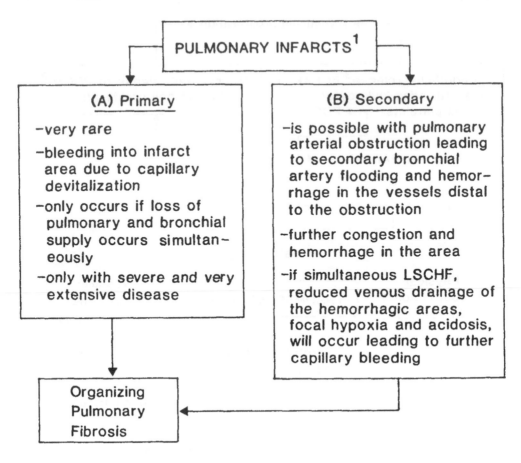

FIGURE 6.   Proposed etiology of pulmonary infarcts. LSCHF = left-sided congestive heart failure. ([1]Dalen, J. E., Haffajee, C. I., Alpert, J. S., Howe, J. P., Ockene, I. S., and Paraskos, J. A., *N. Engl. J. Med.*, 296, 1431, 1977.)

Acute and severe turbulence
    Filarial mass, as in caval syndrome cases
    Tricuspid valve entanglement by filariae
Chronic and low grade turbulence
    Filariae in the blood flow?
    Mild arterial disease — intimal proliferation
    Severe disease — ridge lesions, branch lesions, sacculation, truncation, etc.

FIGURE 7.   Possible causes of turbulence in dirofilariasis.

Seen subclinically in severe cases associated with increased erythrocyte mechanical fragility due to intimal pathology (trauma?) and associated microscopic nonlaminar blood flow (shear effects?)
Seen in caval syndrome and believed to be associated with increased shear forces and with fibrin deposition due to the turbulence associated with the large numbers of worms
Associated with increasing worm burdens based on the extent of renal hemosiderosis
No hemolytic factor in infected canine sera or in worm extracts (see Chapter 11)

FIGURE 8.   Causes of hemolysis in dirofilariasis.

Generalized

Hemolysis in the pulmonary arteries or heartworm excretory products containing iron?

Pulmonary hemosiderosis is not seen distally to completely obstructed pulmonary arteries, so not from a systemic source via the bronchial blood supply

Usually seen in severe cases

Hemoglobin infusion into the pulmonary artery leads to greater microscopic hemosiderosis in the lungs (alveolar) than in the spleen or kidney[6]

Focal

Secondary to alveolar and more generalized bleeding

Siderocytes often seen associated with fibroblasts peripherally in areas of hemorrhage in the lungs[7]

FIGURE 9. Possible causes of pulmonary hemosiderosis. ([1]Buoro, I. B. J. and Atwell, R. B., unpublished data, 1982. [2]Atwell, R. B., Ph.D. thesis, University of Queensland, Brisbane, Australia, 1983.)

Severe L_5 (immature adults) arteritis occurs mainly in distal arteries

Shown to move on angiograms from more distal to more proximal sections of an artery

A factor in the etiology of caval syndrome with large burdens moving toward the right atrial area?

Larger, proximal, arterial diameter reduces effective, worm surface-area contact with the intima, and so there is less resultant intimal damage, compared to greater-intimal contact per worm in smaller distal vessels

Proportional, luminal, obstructive effect per worm is lessened with the larger, cross-sectional area of larger arteries compared to distal arteries, possibly of similar diameter to the worm diameter

Proximal location of filariae probably allows a streamlining effect with worms in the "edge" of the blood flow reducing potential blood turbulence

FIGURE 10. Factors associated with filarial movement to more proximal arteries with maturity.

Dead adult	Granuloma (arterial and periarterial)
	Thrombosis
	Embolism
	Inflammation
	Obstruction (if large debris)
Live adult	Platelet activation, platelet activating factor release?
	Local intimal disease
	By-products? intimal insults? pneumonitis? — distal vasculature
	Turbulence, nonlaminar flow (if large numbers)
	Large numbers of filariae in right atrial area causing turbulence and signs of caval syndrome
	Fibrin net on filariae could also be cause of hemolysis in caval syndrome?
	Valve entanglement (rare)
	Obstruction (only if very large numbers)
	Upset valve function (only if large numbers)
	Ectopic — effect related to locality in host
Immature L_5 adults	Severe inflammation (distal arteries)
	Thrombosis
	Obstruction
	Ectopic — effect related to locality in host
Microfilariae	Microgranuloma with death due to microfilaricides
	Pneumonitis and other organ involvement with microfilarial death in cases of immune-mediated occult disease
	Kidney lesions — trauma, immunological?
	Associated with the DEC reaction
Artificial adult filariae (n = 30)	Local intimal lesions but no pressure or obstructive effects over a 6-week period in uninfected dogs[1]

FIGURE 11. Direct effects of the parasite. ([1]Atwell, R. B. and Buoro, I. B. J., unpublished data, 1982.)

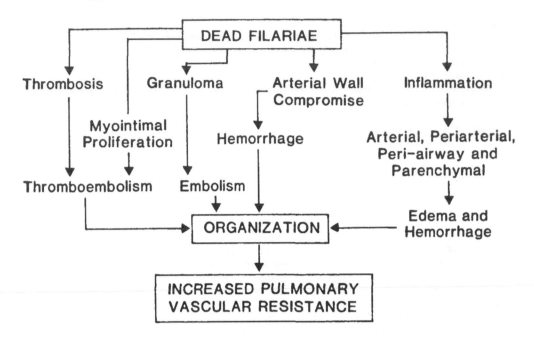

FIGURE 12.   Direct effects of dead filariae.

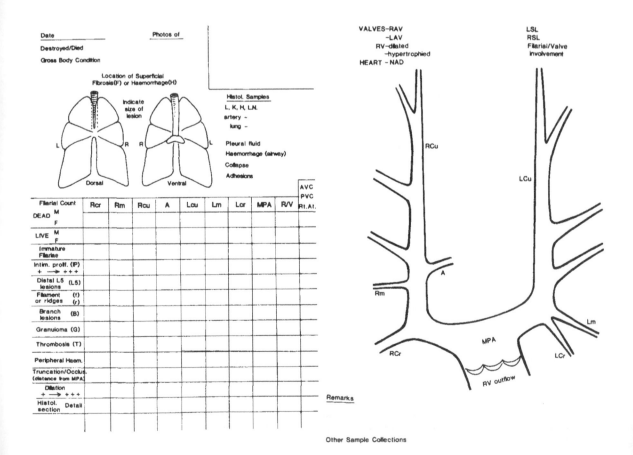

FIGURE 13.   Detailed data collection sheet for use at necropsy.

ACKNOWLEDGMENT

Much of this appendix could not have been prepared without the expert artistic assistance of John McDougall of the Department of Veterinary Medicine, University of Queensland.

INDEX

A

D

M

N

O

P

Printed and bound by CPI Group (UK) Ltd, Croydon, CR0 4YY

22/10/2024

01777632-0007